# ECOLOGICAL MODERNISATION AROUND THE WORLD

## Perspectives and Critical Debates

## Of Related Interest

A GREEN DIMENSION FOR THE EUROPEAN COMMUNITY
Political Issues and Processes
*edited by David Judge*

SUSTAINABLE DEVELOPMENT IN WESTERN EUROPE
Coming to Terms with Agenda 21
*edited by Tim O'Riordan and Heather Voisey*

RIO: UNRAVELLING THE CONSEQUENCES
*edited by Caroline Thomas*

NETWORKS FOR WATER POLICY
A Comparative Perspective
*edited by H. Bressers, L.J. O'Toole, Jr. and J. Richardson*

ECOLOGY AND DEMOCRACY
*edited by Freya Mathews*

DILEMMAS OF TRANSITION
The Environment, Democracy and Economic Reform in Europe
*edited by Susan Baker and Petr Jehlička*

ENVIRONMENTAL MOVEMENTS
Local, National and Global
*edited by Christopher Rootes*

# ECOLOGICAL MODERNISATION AROUND THE WORLD
## Perspectives and Critical Debates

*Edited by*
ARTHUR P.J. MOL and DAVID A. SONNENFELD

Routledge
Taylor & Francis Group
LONDON AND NEW YORK

*First published in 2000 by*
FRANK CASS PUBLISHERS

Published 2013 by Routledge
2 Park Square, Milton Park, Abingdon, Oxon OX14 4RN
605 Third Avenue, New York, NY 10017

*Routledge is an imprint of the Taylor & Francis Group, an informa business*



British Library Cataloguing in Publication Data

Ecological modernisation around the world : perspectives
and critical debates
1. Environmental policy 2. Environmental protection – Political aspects 3. Sustainable development
I. Mol, A. P. J. (Arthur P. J.) II. Sonnenfeld, David A.
363.7

Library of Congress Cataloging in Publication Data

Ecological modernisation around the world: perspectives and critical debates / edited by Arthur P.J. Mol and David A. Sonenfeld.
p. cm.
This group of studies first appeared in a special issue of Environmental politics (issn 0964-4016) 9/1 Spring 2000 published by Frank Cass – T.p. verso.
Includes bibliographical references and index.
1. Environmental policy. 2 Ecology. 3. Economic development – Environmental aspects. I. Mol, A.P.J. II. Sonenfeld, David Allan.

GE170.E26 2000
363.7'05 – dc21 00-022705

ISBN 13: 978-0-714-68113-9 (pbk)
ISBN 13: 978-0-714-65064-7 (hbk)

This group of studies first appeared in a Special Issue of *Environmental Politics* (ISSN 0964-4016) 9/1 Spring 2000

# Contents

# Acknowledgements

This volume could not have been produced without extraordinary co-operation and friendly debate among the authors, nor without the efforts of peer reviewers around the world. Most selections were developed from papers originally presented at meetings of the Research Committee on Environment and Society (RC-24) of the International Sociological Association, Montreal, Canada, July 1998; and the Section on Environment and Technology, American Sociological Association (ASA), San Francisco, USA, August 1998. Riley Dunlap, former president of RC-24, recognised in an early phase the need to have an RC-24 session on ecological modernisation and stimulated the organisation of it. Tom Rudel, Rutgers University, was helpful in supporting the creation of a lively roundtable session at the ASA meetings.

The co-editors of this volume would like to express their appreciation to the managing editors of *Environmental Politics*, especially Neil Carter, University of York, in their endorsement of this project from the beginning. Arthur Mol acknowledges the continuing intellectual contribution to this project of Gert Spaargaren and other colleagues within the Department of Environmental Sociology, Wageningen University; and a number of individual scholars, among whom Fred Buttel and Allan Schnaiberg deserve special mention. David Sonnenfeld's efforts on this project were made possible through the generous financial assistance of a Ciriacy-Wantrup Postdoctoral Fellowship in Natural Resource Economics, University of California, Berkeley; the kind encouragement of Michael Watts, Institute of International Studies, University of California, Berkeley; and support from Eugene Rosa and other colleagues in the Department of Sociology, Washington State University.

# INTRODUCTION

# Ecological Modernisation Around the World: An Introduction

ARTHUR P.J. MOL and DAVID A. SONNENFELD

## Introduction

Many contemporary environmental social scientists and commentators suggest that a major turn occurred in the 1980s with regard to the continuing undermining of sustenance bases of western industrial societies. The Brundtland report [*WCED, 1987*] is often denoted as the codification of that transformation, which was marked by other historical events as well, including the 1992 United Nations Conference on Environment and Development (UNCED). Beyond this common understanding, divergent interpretations have developed on (i) the nature of that transformation, (ii) the actors and actions which have triggered innovations in societies' interactions with external nature, (iii) the extent to which such environmental improvements have reflected changing environmental ideologies and discourses, and (iv) the social and geographical distribution of those changes.

Numerous social scientists have analysed aspects of this turn, looking at, for example, the changing role of the nation-state in safeguarding the environment [cf. *Jänicke, 1993*], and the role of social movements in representing environmental interests *vis-à-vis* economic agents [cf. *Rawcliffe, 1998*]. Few attempts have been made to formulate more general explanations of current transformations of environmental practices, discourses and institutions. One of the more sustained efforts to do so has been a growing number of publications which can be brought together under the label, 'ecological modernisation'. Scholars in different disciplines from around the world have been developing and 'testing' this body of work for nearly two decades.[1]

The present volume aims to contribute to the critical advancement of Ecological Modernisation Theory. It tries to capture and evaluate the state of the art of this approach's analysis of contemporary environmental reforms, while also presenting fresh challenges from outside this school of thought. Although Ecological Modernisation Theory is the central focus of this work, neither we, the editors, nor the individual authors interpret it as the only valid approach. Various articles in this volume give evidence of the limitations of the present state of the art of Ecological Modernisation Theory and the need to confront it with other perspectives to develop it further. Moreover, contributors to this volume vary in their 'support' for this school of thought. Whether engaged or in debate with it, however, the authors find Ecological Modernisation Theory to be one of the more valuable points of reference in contemporary social sciences for analysing society–environment interactions at the turn of the millennium.

In order to provide a conceptual framework in which to situate individual contributions, we first outline the historical emergence of Ecological Modernisation Theory and summarise its key features, before outlining the contents of this volume.

## The Emergence of Ecological Modernisation Theory

Ecological Modernisation Theory was first developed in the early 1980s primarily in a small group of western European countries, notably Germany, the Netherlands and the United Kingdom (UK). Social scientists such as Martin Jänicke, Volker von Prittwitz, Udo Simonis and Klaus Zimmermann (Germany), Gert Spaargaren, Maarten Hajer and Arthur P.J. Mol (the Netherlands) and Albert Weale, Maurie Cohen and Joseph Murphy (UK) made substantial contributions to this scholarship (see the various references in the different contributions to this volume). More recently, empirical studies have also been carried out focusing on among others Finland [e.g., *Jokinen and Koskinen, 1998; Sarinen, forthcoming*], Canada [e.g., *Harris, 1996*], Denmark [e.g., *Andersen, 1994*], Europe [*Neale, 1997*], Lithuania [e.g., *Rinkevicius, 2000* and this volume], Hungary (e.g., Gille, this volume), Kenya [e.g., *Frijns et al., 1997*] and Southeast Asia (e.g., Sonnenfeld; and Frijns *et al.*, this volume). It is the German sociologist Joseph Huber, however, who should arguably be seen as its founder [cf. *1982, 1985, 1991*].

Even in its relatively short existence, Ecological Modernisation Theory has developed with considerable diversity and debate, not only by national background and theoretical foundation,[2] but also chronologically. Reserving a more detailed overview and analysis of this literature for elsewhere, we believe it useful for the present purposes to distinguish at least three stages in the development and maturation of this school of thought.

The first contributions, especially those by Joseph Huber [cf. *1982, 1985*], were characterised by a heavy emphasis on the role of technological innovations in environmental reform, especially in the sphere of industrial production; a critical attitude towards the (bureaucratic) state;[3] a favourable attitude towards the role of market actors and dynamics in environmental reforms; a systems-theoretical and rather evolutionary perspective with a limited notion of human agency and social struggles; and an orientation towards analyses at the level of the nation-state.

The second period, from the late 1980s to the mid-1990s, showed less emphasis on technological innovation as the key motor of ecological modernisation; a more balanced view on the respective roles of states and 'the market' in ecological transformation [cf. *Weale, 1992; Jänicke, 1991, 1993*]; and more attention to institutional and cultural dynamics of ecological modernisation [cf. *Hajer, 1995; Spaargaren and Mol, 1991, 1992; Cohen, 1997*]. During this period, scholarship on ecological modernisation continued to emphasise national and comparative studies of industrial production in the Organisation for Economic Co-operation and Development (OECD) countries.

Since the mid-1990s, the frontier of Ecological Modernisation Theory has broadened theoretically and geographically to include studies on the ecological transformation of consumption; ecological modernisation in non-European countries (newly industrialising countries, less developed countries, the transitional economies in Central and Eastern Europe, but also OECD countries such as the USA and Canada); and global processes. Works in the present volume fit squarely in this third phase of Ecological Modernisation Theory.[4]

Notwithstanding their temporal, national, and theoretical differences, one can gather this scholarship together under the aegis of Ecological Modernisation Theory. Such studies arguably have three broad perspectives in common: (i) moving beyond apocalyptic orientations to see environmental problems as challenges for social, technical and economic reform, rather than as immutable consequences of industrialisation; (ii) emphasising transformation of core social institutions of modernity – be it not beyond recognition – including science and technology, production and consumption, politics and governance, and the 'market', on multiple scales (local, national, and global); and (iii) positioning in the academic field distinct from counter-productivity/deindustrialisation, postmodernist/strong social constructionist, and many neo-Marxist analyses.

We turn next to outlining some of Ecological Modernisation Theory's core themes.

## Core Themes of Ecological Modernisation Theory

From the initial contributions onwards, the aim of Ecological Modernisation Theory has been to analyse how contemporary industrialised societies deal with environmental crises. The core of all studies in the tradition of Ecological Modernisation focuses on (existing and programmed) environmental reforms in social practices, institutional designs and societal and policy discourses to safeguard societies' sustenance bases.

Some authors emphasise that these social transformations in institutions, practices and discourses are paralleled by physical changes in tendencies of environmental disruptions and material flows [e.g., *Jänicke et al., 1992*]. According to these latter scholars, from the mid-1990s onwards, a process of decoupling or delinking of material from economic flows emerged in ecological front-runner[5] nations such as Germany, Japan, the Netherlands, Sweden and Denmark. In a number of cases (countries, industrial sectors, issues), environmental reforms arguably even resulted in an absolute decline of natural resources used and emissions produced, regardless of economic growth in monetary or material terms (amount of products).[6] Considerable debate has emerged about whether these improvements have actually taken place and to what extent any such improvements are structural or incidental.

It is not physical improvements *per se*, however, but rather social and institutional transformations which have been and still are at the core of much current scholarship on ecological modernisation. These transformations can be grouped in five clusters:

(1) *Changing role of science and technology*: Science and technology not only are judged for their role in the emergence of environmental problems but also valued for their actual and potential role in curing and preventing them. Traditional curative and repair options are replaced by preventive socio-technological approaches incorporating environmental considerations from the design stage of technological and organisational innovations. Science and technology are not marginalised despite an apparent growing uncertainty of expert knowledge regarding definitions and causes of, and solutions for, environmental problems.

(2) *Increasing importance of market dynamics and economic agents* (such as producers, customers, consumers, credit institutions, insurance companies, etc.) as carriers of ecological restructuring and reform (in addition to the more conventional categories of state agencies and new social movements that prevail in almost all social theories on the environment.

(3) *Transformations in the role of the nation-state*: More decentralised, flexible and consensual styles of governance emerge, with less top-down, national command-and-control environmental regulation – often referred to as political modernisation [cf. *Jänicke, 1993; Jänicke and Weidner, 1995*]. More opportunities for non-state actors to assume traditional administrative, regulatory, managerial, corporate,[7] and mediating functions of the nation-

states, referred to by some as subpolitical arrangements [cf. *Beck, 1994; Hogenboom et al., 1999*]. Emergent supranational institutions also undermine the nation-state's traditional role in environmental reform.

(4) *Modifications in the position, role and ideology of social movements*: Increasingly, social movements are involved in public and private decision-making institutions regarding environmental reforms, in contrast to having been limited to the periphery or even outside of such processes and institutions in the 1970s and 1980s.[8] Along with this is a partial shift from anti-systemic, demodernisation to reform ideologies. These changes, in turn, have led to debates within social movement organisations regarding tensions of dualistic strategies and ideologies.

(5) *Changing discursive practices and emerging new ideologies*: Complete neglect of the environment and the fundamental counterpositioning of economic and environmental interests are no longer accepted as legitimate positions [cf. *Spaargaren and Mol, 1992; Hajer, 1995*]. Intergenerational solidarity in dealing with the sustenance base has emerged as an undisputed core principle.

Such social transformations feature as central topics of scholarship on ecological modernisation in western industrialised countries, and increasingly elsewhere as well. Two further positions can be identified in these studies. Some scholars [cf. *Weale, 1992; Mol, 1995; Spaargaren, 1997*] use these premises as analytical tools to understand the social dynamics of present-day processes of environmental reform. A second group (including Christoff [*1996*]; Boons [*1997*]; Dryzek [*1997*]) moves beyond this position, claiming that these premises have not only analytical value but also normative merit in outlining desirable and feasible paths for environmental reform – or they contest such premises.

## Contents of this Volume

Essays collected in this volume reflect the core themes outlined above and are a typical product of what we have labelled the third phase of Ecological Modernisation studies. The contributions fall into two general groupings. The first several are primarily oriented towards forwarding ecological modernisation's *theoretical* foundations. They assess historical debates the school-of-thought has been engaged in, extend Ecological Modernisation Theory to address environmental aspects of consumption practices and consumer behaviour, and examine the relationship between national 'knowledge orientations' and ecological modernisation, respectively. The subsequent papers draw from empirical *case studies* to assess the usefulness of this theoretical framework in understanding environmental reform processes in three sets of countries outside north-western Europe. Two contributions focus on western OECD countries (the USA and Finland); two others pay attention to transitional economies in eastern and central Europe (Lithuania and Hungary); the final two concentrate on newly industrialising countries in South-east Asia (Indonesia, Malaysia and Thailand; and Viet Nam, respectively).

In the first selection, Arthur Mol and Gert Spaargaren review both older and newer theoretical debates with which Ecological Modernisation Theory has engaged. During the 1980s, proponents of this approach, including the two authors, demarcated themselves from both 'counter-productivity' (deindustrialisation or small-is-beautiful) and anti-capitalist perspectives. More recently in the 1990s, ecological modernisation theorists clarified differences between their approach of addressing environmental problems and that of social constructionists, postmodernists, and radical (deep) ecologists. Interestingly, some of the common roots and perspectives between Ecological Modernisation Theory and eco-Marxism with respect to social inequalities and ecological restructuring also are explored.

In the following contribution, Gert Spaargaren and Bas van Vliet challenge environmental social scientists to address the environmental implications of consumption practices and consumer behaviour. Rejecting traditional social psychological studies of consumer behaviour as inadequate for the task, they draw from Giddens' structuration theory, Bourdieu's theory of distinction, Warde's and others' sociology of consumption, and Cowan's, Otnes', Shove's and others' work on domestic consumption, to outline an ecological modernisation perspective on the greening of domestic consumption. Spaargaren and van Vliet suggest that processes of ecological modernisation are affecting not only material production, but also increasingly consumption as well, to the level of daily practices of individual households. They call for improved social scientific analyses and policies for understanding and encouraging green consumer behaviour, and outline a conceptual model for doing so.

Maurie Cohen continues his contributions to Ecological Modernisation Theory in the third study, exploring the importance of environmental values and orientations in ecological modernisation, enriching its up till now undertheorised cultural dimensions. He constructs an ideal-typical model of environmental knowledge, combining variations of ecological consciousness and epistemological commitment. Using the Netherlands as a case study, he then suggests the usefulness of national studies of 'environmental knowledge orientations' to predict the likelihood of success of ecological modernisation in different countries.

David Pellow, Allan Schnaiberg and Adam Weinberg question the validity of Ecological Modernisation Theory in the fourth selection, through a case study of the social relations and environmental impacts of an urban waste recycling programme in Chicago, USA. Among others, they challenge one of Ecological Modernisation Theory's core hypotheses: that production processes are increasingly designed and conducted using ecological criteria. They observe that in Chicago, recycling has been placed on an increasingly profit-oriented basis, to the detriment of both employees of that industry and the natural environment. They argue that this case validates Schnaiberg's well-known 'treadmill of production' thesis, in which economic élites increasingly dominate all aspects of society and the environment unless checked by grassroots social movements. The authors call for continuing empirical research from both treadmill of production and ecological modernisation perspectives, including on 'zero-sum moments' of ecological decision-making when 'real and inherent conflicts among stakeholders' must be confronted.

Across the world, Pekka Jokinen examines the impact of Europeanisation on agri-environmental policies and practices in Finland, paying special attention to changes in institutional arrangements, discourses and practices. Although it has been suggested that ecological modernisation is facilitated by processes of globalisation, Jokinen finds mixed results in the Finnish case: discourses have changed, but institutional arrangements have been transformed only marginally. Prior to joining the European Union (EU), the Finnish state devoted considerable resources to supporting progressive agri-environmental practices by Finnish farmers. Under the terms of EU membership this was considered an unfair subsidy, however, and had to be discontinued.

In the different setting of transitional economies, Leo Rinkevicius finds Ecological Modernisation Theory useful in analysing shifts in culture and institutional practices in the Baltic republic of Lithuania. Taking an historical approach, Rinkevicius examines environmentalism in Lithuania under Soviet rule, during national liberation, and in the current transition to becoming a liberal, market-oriented society. Consistent with western European experience, Rinkevicius finds that in Lithuania, too, environmental activism has transformed from its formerly oppositional role to being more integrated in society and institutions. At the same time, such activism in Lithuania retains what he refers to as a 'mixed value-orientation', with both 'eco-managerial' and 'romantic-idealistic' aspects.

Zsuzsa Gille analyses the relevance of contemporary notions of industrial ecology and ecological moderenisation in understanding the development of waste management practices from the Second World War onwards in another transitional European society, Hungary. Provocatively, she argues that Hungary's state socialist leadership established some of the first 'industrial ecological' programmes in the early years of Hungarian socialism, in the 1950s – well before the popularity of such ideas in the West. Waste re-use programmes were established in response to Western blockades of industrial goods exports to Hungary. In later years, Hungary's waste re-use programmes became progressively distorted from their original intent, however, resulting in stockpiling of unused wastes and waste-based products. Today in its postsocialist transition, Hungary faces a very different dynamic, with western European countries and firms striving to make the country a regional 'waste processing zone'. Gille invokes what she argues are progressive aspects of Hungary's earlier waste utilisation programmes, and calls for greater citizen and worker participation in the formulation of contemporary waste management policies and practices.

The final two papers in this volume examine the relevance of Ecological Modernisation Theory to newly industrialising countries in South-east Asia. David Sonnenfeld explores what he posits are both accomplishments and contradictions of ecological modernisation in South-east Asia's pulp and paper sector. Faced with strong local and transnational social movements on one hand, and favourable global market conditions on the other, domestically owned manufacturing firms adopted cleaner process technology in the construction of new facilities. In this case, adoption of cleaner technology, Sonnenfeld suggests, was arguably only partially 'ecologically-modern' in that increased pulp production in South-east Asia was predicated on expanded destruction of natural forests and establishment of fast-growing exotic tree plantations in place of smallholder farming. Greater attention should be paid by technology firms and technology-exporting countries, he argues, to developing ecologically modern approaches for small- and medium-sized enterprises, which provide important sources of employment in developing countries.

Lastly, Jos Frijns, Phung Thuy Phuong, and Arthur Mol examine the applicability of Ecological Modernisation Theory to one of Asia's newest 'tiger' economies, Viet Nam, also in transition from a 'command and control' to a more market-oriented economy. With rapid industrial development and increasing environmental threats, they find that Viet Nam is only in the early stages of developing an environmental regulatory apparatus and policy framework, has a hardly adequate programme for encouraging technological change into more environmentally sound directions, and lacks a strong national environmental movement that can press for ecological reforms. They conclude that Ecological Modernisation Theory has only marginal explanatory power for understanding environmental reform processes in Viet Nam today. Frijns, Phung and Mol do find in Ecological Modernisation Theory, however, a hopeful set of prescriptive principals with which Viet Nam could advance the effectiveness of environmental policy and management structures and practices. But Ecological Modernisation in such a context would differ from the original Eurocentric version.

## Conclusion

Taken as a whole, we believe the contributions to this volume push Ecological Modernisation Theory forward on at least five fronts: its geographical scope, coverage, and applicability; its theoretical stance relative to other environmental social science and policy perspectives; its coverage of dynamics of consumption as well as production; its attention to issues of national and civic culture; and its relevance to transitional and newly industrialising as well as advanced industrial countries.

It is too early to ascertain the full extent of Ecological Modernisation Theory's applicability to different economic, cultural, political-institutional and geographical settings and locations around the world. The studies assembled here report mixed results. They converge, however, in finding that the approach and tools of Ecological Modernisation Theory are useful for social scientific analysis and policy formation, even where all conditions for development of ecologically modern institutions do not yet exist. At the same time, some processes of ecological modernisation are global (even while others are not), and thus this body of theory remains at least partially relevant around the world.

Ecological Modernisation Theory is very much a living and growing school-of-thought. Although reared in the context of political and policy debates in north-western Europe, its intellectual 'stock' becomes more heterodox as its scope and influence expands. Old demarcations, such as with certain forms of neo-Marxism and green politics in the western European context are tested again in new contexts and encounters with similar but different intellectual traditions from other parts of the world, with challenging outcomes. Ecological Modernisation Theory's accommodation with concerns about global and local inequality illustrate this dynamic, as are efforts to 'stretch' it to address or explain dynamics in the vastly different political contexts of transitional societies, both in Europe and Asia.

As a maturing but still young school-of-thought, it is only natural for Ecological Modernisation Theory to expand in exploration of important dimensions of contemporary life inadequately addressed by other environmental social scientific approaches. Here, Ecological Modernisation Theory's roots in classical European social theory are real strengths – in addressing issues of consumption practices and consumer behaviour, taking positions between 'realists' and social-constructivists, understanding the development of cultural institutions and practices, and even taking up such intellectually challenging topics as notions of 'national culture orientations'. In undertaking such efforts, Ecological Modernisation Theory may ultimately not only contribute to environmental social science and policy, but also strengthen linkages between those transdisciplinary pursuits and mainstream social science.

Being still a theory in the making, it should not surprise us that numerous issues remain to be elaborated and addressed by Ecological Modernisation Theory. Key outstanding issues include the nature of requisite political and institutional cultures for successful (and 'reflexive') ecological modernisation; the varying role and importance of environmental and other social movements and non-governmental organisations in ecological modernisation, especially in countries with weaker histories and institutions of popular participation; global and domestic inequalities in capacities to establish and maintain ecologically modern institutions, technologies, and practices; and the dialectics of globalisation in its various outlooks and dimensions, and ecological modernisation.

The final word on Ecological Modernisation Theory has yet to be spoken, due in no small part to the changing nature of environmental reform processes themselves. Much work remains to be done developing, testing and analysing this school of thought, both in general, including its theoretical premises, and more particularly in its applicability to different social systems, political configurations and traditions, and geographical regions around the world. We hope that this volume contributes to such efforts as much as it is the beginning of them.

## NOTES

1. Martin Jänicke claims to have been the inventor of the concept and first introduced the notion of ecological modernisationin the late 1970s during political discussions in the Berlin Community Council, of which he was a member.
2. Theoretical traditions range from systems theory [e.g. *Huber, 1985, 1991*], to institutional analysis [cf. *Mol, 1995*], to discourse analysis [*Hajer, 1995*; also *Weale, 1992*].
3. This perspective was shared by Martin Jänicke in his early writings [cf. *Jänicke, 1986*].
4. But see for instance also some contributions brought together in a volume edited by Spaargaren, Mol and Buttel [*2000*], as well as a special issue of the journal *Geography* (forthcoming).
5. These nations are front runners not so much in absolute terms of minimal environmental additions and withdrawals per country or per capita, but rather in terms of policies that transform existing trends of increasing resource consumption and emissions [cf. *Andersen and Liefferink*, 1997].
6. See, for instance, the studies of Jänicke et al. [e.g. *1992*], the publications of the European Environmental Agency [cf. *1998*], those in the tradition of dematerialisation and Factor 4 (or 10 or more); [cf. *Reijnders, 1998*], those in the tradition of the so-called green or

environmental Kuznets curve (see the special issues of the journals, *Ecological Economics*, 1998; *Environment and Development Economics*, 1996; and *Ecological Applications*, 1996).
7. In the case of the privatisation of parastatal organisations.
8. Although one can question whether it should be analysed as a social movement, the German political party, *Die Grünen* (the Greens), stands as a model for this transformation in ideology and position. The process leading ultimately to a position as coalition partner in the German government from 1998 onwards has indeed been paralleled by major internal debates and struggles.

## REFERENCES

Andersen, M.S. (1994), *Governance by Green Taxes: Making Pollution Prevention Pay*, Manchester: Manchester University Press.

Andersen, M.S. and J.D. Liefferink (eds.) (1997), *European Environmental Policy: The Pioneers*, Manchester: Manchester University Press.

Beck, U. (1994), 'The Reinvention of Politics: Towards a Theory of Reflexive Modernization', in U. Beck, A. Giddens and S. Lash, *Reflexive Modernization: Politics, Traditions and Aesthetics in the Modern Social Order*, Cambridge: Polity Press, pp.1–55.

Boons, F. (1997), 'Organisatieverandering en ecologische modernisering: het voorbeeld van groene produktontwikkeling', paper to the Dutch NSV conference, 29 May 1997.

Christoff, P. (1996), 'Ecological Modernisation, Ecological Modernities', *Environmental Politics*, Vol.5, No.3, pp.476–500.

Cohen, M. (1997), 'Risk Society and Ecological Modernisation: Alternative Visions for Post-Industrial Nations', *Futures*, Vol.29, No.2, pp.105–19.

Dryzek, J. (1997), *The Politics of the Earth: Environmental Discourses*, Oxford: Oxford University Press.

European Environmental Agency (1998), *Europe's Environment: The Second Assessment*, New York/Amsterdam: Elsevier.

Frijns, J., Kirai, P., Malombe, J. and B. van Vliet (1997), *Pollution Control of Small Scale Metal Industries in Nairobi*, Wageningen, Netherlands: Department of Environmental Sociology, Wageningen University.

Hajer, M.A. (1995), *The Politics of Environmental Discourse: Ecological Modernisation and the Policy Process*, Oxford: Clarendon.

Harris, S. (1996), 'The Search for a Landfill Site in an Age of Risk: The Role of Trust, Risk and the Environment', dissertation, McMaster University, Hamilton, Ontario.

Hogenboom, J., Mol, A.P.J. and G. Spaargaren (1999), 'Dealing with Environmental Risks in Reflexive Modernity', in M.J. Cohen (ed.), *Risk in the Modern Age: Social Theory, Science and Environmental Decision-Making*, London: Macmillan, pp.83–106.

Huber, J. (1982), *Die verlorene Unschuld der ÖKologie. Neue Technologien und superindustriellen Entwicklung*, Frankfurt am Main: Fisher Verlag.

Huber, J. (1985), *Die Regenbogengesellschaft. Ökologie und Sozialpolitik*, Frankfurt am Main: Fisher Verlag.

Huber, J. (1991), *Unternehmen Umwelt. Weichenstellungen für eine ökologische Marktwirtschaft*, Frankfurt am Main: Fisher Verlag.

Jänicke, M. (1986), *Staatsversagen. Die Ohnmacht der Politik in der Industriegesellschaft*, München/Zürich: Piper.

Jänicke, M. (1991), 'The Political System's Capacity for Environmental Policy', Berlin: Department of Environmental Politics, Free University Berlin.

Jänicke, M. (1993), Über ökologische und politische Modernisierungen', *Zeitschrift für Umweltpolitik und Umweltrecht* 2, pp. 159–75.

Jänicke, M. and H. Weidner (1995), 'Successful Environmental Policy: An Introduction', in M. Jänicke and H. Weidner (eds.), *Successful Environmental Policy: A Critical Evaluation of 24 Cases*, Berlin: Sigma, pp.10–26.

Jänicke, M., Mönch, Binder, M., *et al.* (1992), *Umweltentlastung durch industriellen Strukturwandel? Eine explorative Studie über 32 Industrieländer*, Berlin: Sigma.

Jokinen, P. and K. Koskinen (1998), 'Unity in Environmental Discourse? The Role of Decision-Makers, Experts and Citizens in Developing Finnish Environmental Policy', *Policy and Politics*, Vol.26, No.1, pp.55–70.

Mol, A.P.J. (1995), *The Refinement of Production: Ecological Modernisation Theory and the Chemical Industry*, Utrecht: Jan van Arkel/International Books.

Neale, A. (1997), 'Organising Environmental Self-Regulation: Liberal Governmentality and the Pursuit of Ecological Modernisation in Europe', *Environmental Politics*, Vol.6, No.4, pp.1–24.

Rawcliffe, P. (1998), *Environmental Pressure Groups in Transition*, Manchester: Manchester University Press.

Reijnders, L. (1998), 'The Factor X Debate: Setting Targets for Eco-Efficiency', *Journal of Industrial Ecology*, Vol.2, No.1, pp.13–22.

Rinkevicius, L. (2000), 'The Ideology of Ecological Modernisation in "Doiuble-Risk" Societies: A Case Study of Lithuanian Environmental Policy', in Spaargaren, Mol and Buttel [*2000*].

Sarinen, R. (forthcoming), 'Governing Capacity of Environmental Policy of Finland: An Assessment of Making the EIA-Law and $CO_2$-Tax', dissertation, Helsinki: University of Technology.

Spaargaren, G. (1997), 'The Ecological Modernisation of Production and Consumption: Essays in Environmental Sociology', dissertation, Wageningen: Department of Environmental Sociology, Wageningen University.

Spaargaren, G. and A.P.J. Mol (1991), 'Ecologie, technologie en sociale verandering. Naar een ecologisch meer rationele vorm van produktie en consumptie', in A.P.J. Mol, G. Spaargaren and A. Klapwijk (eds.), *Technologie en Milieubeheer. Tussen sanering en ecologische modernisering*, The Hague: SDU, pp.185–207.

Spaargaren, G. and A.P.J. Mol (1992), 'Sociology, Environment and Modernity: Ecological Modernisation as a Theory of Social Change', *Society and Natural Resources*, Vol.5, No.4, pp.323–44.

Spaargaren, G., Mol, A.P.J. and F.H. Buttel (eds.) (2000), *Environment and Global Modernity*, London: Sage.

World Commission on Environment and Development (WCED) (1987), *Our Common Future*, Oxford: Oxford University Press.

Weale, A. (1992), *The New Politics of Pollution*, Manchester: Manchester University Press.

# THEORETICAL PERSPECTIVES

# Ecological Modernisation Theory in Debate: A Review

ARTHUR P.J. MOL and GERT SPAARGAREN

Ecological Modernisation Theory has been faced with various challenges from different theoretical perspectives throughout the years. This contribution reviews the various debates ecological modernisation ideas have been engaged in. The article starts with a historical perspective on some of the earlier debates that paralleled Ecological Modernisation from its birth in the early 1980s to its maturation. These initial debates with earlier neo-Marxists and deindustrialisation/counterproductivity theorists were formative for Ecological Modernisation Theory, but are no longer all of similar relevance today. Subsequently we concentrate on more contemporary discussions, which only to some extent reflect similar topics. We will respectively enter into discussions with constructivists and post-modernists on the material foundation of social theory, review and refine the controversies with eco-centrists on radical versus reformist environmental reforms and contribute to neo-Marxist understanding of social inequalities in environmental problems and reform.

## I. Introduction

On several earlier occasions we [*Mol, 1995; Spaargaren, 1997*] – as well as numerous others – have noticed that environmental sociology in particular, and the environmental social sciences more generally, have matured as full-fledged subdisciplines in the last decade. One of the social theories that both profited from and contributed to the maturation of the environmental social sciences is the Ecological Modernisation Theory. Originating from the early 1980s, the Ecological Modernisation Theory has become in a remarkably short time a well-established set of ideas, founded in general social theory and supported by a growing number of case studies.

It should not surprise us that as a new theory, and in becoming one of the more prominent theories within environmental sociology,[1] ecological

This contribution profited much from the comments of David Sonnenfeld, an anonymous referee and the discussions in the Research Committee 'Environment and Society' of the International Sociological Association's World Congress in 1998.

modernisation theory has generated questions and criticisms, from outside and within. If there is anything proponents and opponents of Ecological Modernisation Theory can agree upon, it will be that this theory provides a useful vehicle for organising some of the most pressing contemporary theoretical debates in the environmental social sciences, in a similar way neo-Marxist environmental sociology did in the 1970s and 1980s.

In this review we will contribute to the further maturation of environmental sociology – and the other social sciences – after the turn of the millennium by reviewing the recent debates in which Ecological Modernisation Theory is engaged. Our aim in dealing with some of the more prominent contemporary debates in environmental sociology is (i) to respond more explicitly to the various positions in these debates in environmental sociology, (ii) to clarify further the position of various ecological modernisation theorists in these debates, and (iii) to improve our understanding of the issues at stake.

In reviewing the debates Ecological Modernisation Theory has been engaged in, we will start in section II with a historical perspective on some of the earlier debates that paralleled Ecological Modernisation from its day of birth in the early 1980s to its maturation. These initial debates were formative for the Ecological Modernisation Theory, but are no longer all of similar relevance today. In three subsequent sections we will concentrate on more contemporary discussions, which only to some extent reflect similar topics. We will respectively enter into discussions with constructivists and post-modernists on the material foundation of social theory (section III), review and refine the controversies with eco-centrists on radical versus reformist environmental reforms (section IV) and contribute to neo-Marxist understanding of social inequalities in environmental problems and reform (section V).

## II. Ecological Modernisation's Early Debates

If we want to understand the first debates to which Ecological Modernisation Theory has contributed significantly or in which Ecological Modernisation Theory was a central object, we have to be aware of two interrelated circumstances that prevailed during those debates. First, these initial debates of the (early) 1980s took place against the background of both the state-of-the-art of the environmental debate in the late 1970s and early 1980s and the main or dominant currents in environmental sociology at that time. Second, criticism of Ecological Modernisation Theory was focused on the specific contents and outline of the first phase of Ecological Modernisation Theory (Mol and Sonnenfeld, this volume). And this specific outline should to some extent be understood as a (over)reaction directed at the dominant schools of thought in environmental sociology and the environmental debate in the late 1970s and early 1980s.

Not surprisingly, the major controversial topics in which the Ecological Modernisation Theory was engaged at that time were put forward by the two dominant schools of thought of the 1970s: the counter-productivity or deindustrialisation theorists and the neo-Marxists. In this section we want to argue that on the one hand some of the original issues raised by these scholars can no longer be interpreted as adequate criticism of Ecological Modernisation Theory, although these issues still return frequently. On the other hand, however, we observe that other topics raised during these initial debates return in the 1990s, albeit in different forms and using different conceptualisations, showing both continuity as well as progress in the debates in environmental sociology.

**Deindustrialisation and Technological Fix**

Ecological Modernisation Theory can only be understood by taking into account the debate from which it originates. Debates which were dominated by a theory that can be labelled as demodernisation, deindustrialisation or counter-productivity [*Spaargaren and Mol, 1992; Mol, 1995*]. This latter perspective had a strong position among Western European environmental movements and social scientists in the 1970s. Ecological modernisation challenged the core ideas of the demodernisation perspective.

It was especially during the first phase of Ecological Modernisation Theory in the 1980s that debates concentrated between these two perspectives [cf. *Huber, 1991; Mol and Spaargaren, 1993*]. Ecological Modernisation Theory challenged the environmental movement's traditional idea that a fundamental reorganisation of the core institutions of modern society (the industrialised production system, the capitalist organisation of the economy and the centralised state) was essential in entering a path of long term sustainable development. Building upon more widespread criticism of the modernisation project, counterproductivity theorists such as Otto Ulrich, Rudolf Bahro, Barry Commoner and Hans Achterhuis claimed that it was also environmental and ecological deterioration that could be held as proof of the modernisation project being a dead end.

The adherents of Ecological Modernisation Theory acknowledged the need for some fundamental transformations within the modernisation project to restore some of its structural design faults that had caused severe environmental destruction, but claimed that these transformations do not imply that one has to do away with those institutions of modern society that are involved in the modern organisation of production and consumption. In that sense Ecological Modernisation Theory can be seen in line with more general modernisation theories that were so vehemently criticised by demodernisation theorists, although ecological modernisation theorists managed – especially in their later contributions – to incorporate some of the fierce criticism that challenged Parsonian functionalism and related perspectives [cf. *Von Prittwitz, 1993b; Spaargaren, 1997*].[2]

The central debates on technology are closely related to the controversies between deindustrialisation/demodernisation perspectives and Ecological Modernisation Theory, although neo-Marxists have also contributed to that. Perhaps the most often quoted critique against Ecological Modernisation Theory from its days of origin is related to its technological optimism and its supposed technocratic character. In his analysis of recent contributions to environmental sociology Hannigan [*1995: 184*] claims Ecological Modernisation Theory is 'hobbled by an unflappable sense of technological optimism', a conclusion quite similar to that of the Dutch environmental sociologist Egbert Tellegen [*1991*] and the German sociologist Peter Wehling [*1992*]. Redclift [*1999*] contrasts in a more or less similar way ecological modernisation (as a techno-economic management strategy) with more profound, fundamental and deep cultural transformations.

Maarten Hajer [*1995*] has in some ways incorporated the debate on technocracy within the ecological modernisation project by designing two variants of ecological modernisation: a techno-corporatist ecological modernisation and reflexive ecological modernisation. While in the former ecological reform is purely a techno-administrative affair, the latter points at practices of social learning, cultural politics and new institutional arrangements. A similar attempt is made by Christoff [*1996*], identifying weak (that is, economic-technological) and strong (institutional-democratic) ecological modernisation.

Hajer, Christoff, and others such as Dryzek [*1997*] and Neale [*1997*], in fact closely resemble the distinction made in the early days of Ecological Modernisation Theory by Joseph Huber [1985] between a more technocratic and a more sociocratic development path, albeit that Huber himself was not very consistent in his plea for a more sociocratic version of ecological modernisation [cf. *Spaargaren and Mol, 1992*]. His Schumpeterian model of technology-induced social change gave room for such technology-optimism critique as for instance Peter Wehling [*1992*] has extensively argued for.

More recently, Ecological Modernisation Theory adherents have made numerous efforts to (i) adapt this Schumpeterian model and Huber's original technological optimism, and (ii) show the selective reading of the technocracy criticism regarding later contributions to Ecological Modernisation Theory. In the processes of institutional reform technological transformations have their place, although they are not as central as its critics wants us to believe, and certainly not in the sense that technological change forms the motor of, and determines, these reforms. In addition, the conceptualisation of technology and technological change has widened considerably, from the original add-on technologies that were so severely criticised in the 1970s to 'structural change of socio-technological systems' [cf. *Mol et al., 1991; Jänicke et al., 1992; Neale, 1997; Jokinen and Koskinen, 1998*], making claims about its technocratic character less adequate.

We think that regarding both interrelated topics, debates have changed considerably in character. On the one hand they have moved more to the periphery of the environmental debate, while by the same token they strongly reappear in distinct form using different concepts. The de-industrialisation perspective as an overall theory and alternative has lost most of its attraction in the contemporary environmental debate. Especially since the Brundtland Report that started the third wave of environmental concern, demodernisation perspectives do no longer succeed in challenging the core features of Ecological Modernisation Theory. At the same time, some of the most severe technology criticism of the 1980s has resulted in major changes and refinements of the Ecological Modernisation Theory, making the repetition of similar challenges in the mid-1990s inadequate.

Nevertheless, the continuation of discussions on these kinds of topics can be illustrated by analysing the environmental connotations of the notion of reflexive modernisation. Following the debates on the character of contemporary societies, Ecological Modernisation Theory has more recently been positioned *vis-à-vis* reflexive modernisation ideas, especially in its confrontation with Risk Society Theory [*Mol and Spaargaren, 1993; Von Prittwitz, 1993b; Mol, 1996; Cohen, 1997; Blowers, 1997; Buttel, 2000*].

Initially, the counterposition of Ecological Modernisation Theory in its first phase – as developed by especially Huber and Jänicke – versus Risk Society Theory was emphasised [cf. *Mol and Spaargaren, 1993*]. In fundamentally criticising science and technology, the earlier contributions to Risk Society Theory paralleled deindustrialisation/demodernisation perspectives to a major extent. As Risk Society Theory – via the writings of Ulrich Beck [*1986* and later] – was originally closely associated with Reflexive Modernisation ideas,[3] it should not surprise us that Ecological Modernisation Theory was initially interpreted as being in contradiction with Reflexive Modernisation and as a proponent of the phase of high or simple modernisation that preceded the era of late or reflexive modernisation [cf. *Wehling, 1992*].

More recently, the similarities between Reflexive Modernisation as the umbrella theory, and Ecological Modernisation and Risk Society Theory as its substantial parts, have been highlighted [cf. *Mol, 1996; Hogenboom et al., 1999; Cohen, 1997; Hajer, 1995*]. These similarities refer, among others, to the transformation of the old political institutions of the nation-state in environmental reforms and the emergence of new sub- and supra-national political arrangements, the new role of markets and economic actors in triggering environmental protection, and the increasing uncertainty and insecurity around environmental risks and management strategies following the changing role of science.

Still, some clear distinctions remain between the latter two as commentators outside both schools of thought have recently pointed out [cf. *Hannigan, 1995; Blowers, 1997; Buttel, 2000*].

*The Sustainability of Capitalism*

The possibilities, actuality and desirability of a green capitalism, put first and rather provocatively on the agenda of Ecological Modernisation Theory by Joseph Huber, have resulted in an equally extensive debate. Scholars as diverse as Allan Schnaiberg [*1980*], David Goldblatt [*1996*] and James O'Connor [*1996*] have all, using different concepts, attacked the possibilities of an ecological sound Capitalism. James O'Connor's second contradiction of capitalism, Schnaiberg's treadmill of capitalist production and Goldblatt's criticisms of Giddens' limitation to the industrial dimension of modernity in understanding the environmental crisis, are used to point out the important role capitalism plays in environmental deterioration. Neglecting capitalism and failing to attack the fundaments of the capitalist world order will result in superficial and cosmetic environmental reforms that are unable to resolve the ecological crisis in any fundamental way. Moreover, such measures will rather strengthen the capitalist mode of production as it makes capitalism less in need of a green critique [cf. *Dryzek, 1995*] and it promotes and facilitates the continuation of established socio-economic practices that are to the benefit of those in power [*Blühdorn, 2000*].

Ecological Modernisation Theory deviates from this view, even though its position towards capitalism has changed throughout the various phases of its history (see Mol and Sonnenfeld, this volume). While initially the contribution of capitalism to the 'expansion of the limits'[4] was celebrated by Ecological Modernisation Theory, more recently a nuanced position regarding capitalism is presented. It is not that capitalism is considered to be essential for environmentally sound production and consumption (as neo-liberal scholars want us to believe), nor that capitalism is believed to play no role in environmental deterioration. But rather that (i) capitalism is changing constantly and one of the main triggers is related to environmental concerns, (ii) environmentally sound production and consumption is possible under different 'relations of production' and each mode of production requires its own environmental reform programme, and (iii) all major, fundamental alternatives to the present economic order have proved unfeasible according to various (economic, environmental and social) criteria.[5]

Consequently, mainstream ecological modernisation theorists interpret capitalism neither as an essential precondition for, nor as the key obstruction to, stringent or radical environmental reform. They rather focus on redirecting and transforming 'free market capitalism' in such a way that it less and less obstructs, and increasingly contributes to, the preservation of society's sustenance base in a fundamental/structural way. While it can be argued – as some commentators do – that this debate should be considered as rather abstract and outdated, especially since the 'end of history', it shows at the same time continuing relevance regarding two sets of controversies: (i) it is connected with discussions about presumed shortcomings of Ecological Modernisation Theory in analysing conflicts of interest in environmental reforms, and (ii) more concrete or down-to-earth controversies in contemporary environmental politics and policies are to some extent inspired by the earlier (and to some extent continuing) debate on capitalism. We will turn to both related sets of issues respectively.

Conflictual models of social change that dominate neo-Marxist theories have inspired some authors to emphasise Ecological Modernisation Theory's presumed undertheorised notions of power [*Leroy, 1996*], lack of attention to social contexts and ethical issues [*Blowers, 1997*], neglect of emancipatory concerns [*Blühdorn, 2000*], and absence of human agency [*Smidt, 1996*]. According to these and other scholars, Ecological Modernisation Theory analyses environmental reforms primarily via Schumpeterian, evolutionary models that result almost automatically in the greening of production and consumption, without paying sufficient attention to severe struggles between interests (groups) and to normative, ethical or moral reflections and debates [cf. *Sarkar, 1990; Leroy and Van Tatenhove, 2000; Blowers, 1997: 854*].

In reviewing these debates, we want to draw two conclusions. First, these observations are accurate as far as they relate to the first generation studies in Ecological Modernisation Theory (see Mol and Sonnenfeld, this volume). There is indeed considerable merit in neo-Marxist analyses of environmental conflicts. Ecological Modernisation Theory can profit – and has already to some extent, we would argue – from that in refining its analyses of social change. Secondly, these observations can be considered less adequate as far as they focus on and respond to the more recent general idea in Ecological Modernisation Theory of decreasing *fundamental* conflicts about environmental reform programmes in industrialised countries in the late 1980s and 1990s. This observation by ecological modernisation theorists of environmental interests becoming increasingly 'salonfähig' is, however, not pre-given, but constantly (re)produced by struggles and clashes between diverging interests, changing ideologies, and historical transformations in other social arenas, as ecological modernisation theorists have stressed in theoretical elaborations [cf. *Spaargaren and Mol, 1991, 1992*] and detailed case-studies [cf. *Hajer, 1995; Mol, 1995; Rinkevicius, 2000*]. To the extent that evaluations of Ecological Modernisation Theory qualify these more recent contributions as simple evolutionary and system theoretical projections of the future, we consider them as less adequate.

The fundamental debate regarding capitalism (our second point) also echoes to some extent in more 'down-to-earth' debates for instance regarding the discussions on market versus state involvement in environmental policy (with subjects as diverse as privatisation, the adequacy of distinct policy instruments, state failure versus market failure, deregulation, the modernisation of state governance, etc.) and distributional consequences of environmental problems and reforms. With respect to the first group of topics it can be concluded that they seem no longer to be that controversial, as the growing consensus on the discussions on market based instruments versus command-and-control strategies, the increasing role of non-state actors in environmental policy and the new governance styles that seem to replace the old hierarchic state models, show [cf. *Weale, 1992; Sarinen, forthcoming; Mol, Lauber and Liefferink, forthcoming; Hogenboom et al., 1999; Mol, Spaargaren and Frouws, 1998; Leroy and Van Tatenhove, 2000*]. Also these controversies are decreasingly related to neo-Marxist criticism of capitalism, as the debates on privatisation and deregulation seem to exemplify. We think that neo-Marxist scholars have recently especially proved the relevance of their models in forcefully putting on the agenda the unequal distribution of both environmental problems and the social consequences of environmental policies [cf. *Schnaiberg et al., 1986; Schnaiberg and Gould, 1994; Gould et al., 1996*, also Pellow *et al.*, this volume]. It is in these studies that the so-called conflictual models of social change prove their value. We will deal more extensively with these issues in section V.

*Summary*

Our analysis of some of the initial debates on Ecological Modernisation Theory thus far should not be interpreted as an attempt to devalue or 'silence' criticism of Ecological Modernisation Theory. Our aim has been to make two points. First, we have tried to show how some of the initial debates around Ecological Modernisation Theory have become less relevant or adequate today for several interrelated reasons: (i) Ecological Modernisation Theory profited from these critiques by reforming and refining itself as Mol and Sonnenfeld (this volume) also point out, (ii) the environmental and academic discourse on the environment has changed so that contemporary debates deviate on several points from those in the late 1970s and early 1980s, and (iii) the social circumstances in terms of environmental disruption, actual environmental reforms and the role principal actors play in them, have changed considerably in this period. Where recent commentators persist in reassembling the same points of criticism of Ecological Modernisation Theory, we consider them as 'outdated'. In that sense the debate has made progress.

Secondly, we wanted to stress – and will persist in doing so in the following sections – that no matter how distinct and/or refined contemporary debates involving ecological modernisation seem when compared to their ancestors of the (early) 1980s, we should not overlook the continuities. While on some of the topics indicated above discussion seems to have come to a close, other topics continue to stay – though sometimes in different forms – on the agenda for environmental sociology. And to some extent new theoretical 'alliances' are formed as. For instance, on most of the controversies we will elaborate below (social constructivism and the materialist dimension of environmental problems; deep ecology; social inequalities), Ecological Modernisation Theory parallels neo-Marxist ideas.

## III. The Materiality of Environmental Problems

The first contemporary controversy we want to focus on is the position of the materialist dimension in sociology, and the claim of ecological modernisation theory – as we interpret it – that this materialist dimension should not be reduced to social facts only. To enter this controversy with especially postmodernists and strong constructivists, we need a short historical introduction into to the emergence of environmental sociology via human ecology.[6]

It has become common understanding within environmental sociology that our relationship with nature or the environment cannot be taken for granted any longer but has to be reflexively organised. This can be done in different ways, and within environmental sociology we think at least three major schools of thought can be distinguished in this respect: the human ecology tradition, the ecological modernisation school of thought and postmodern views of the environment. These three environmental sociological perspectives are directly related to the wider sociological debates on the character of modernity. As will become clear in this section, the human ecology tradition can be understood as a reaction to the long neglect by 'mainstream' sociology of the materialist dimension of social practices and institutional developments. Ecological modernisation, together with risk-society theories and some moderate versions of constructivism, are the environmental pendants of reflexive modernisation perspectives. The last tradition, including strong or relativist constructivism, can be labelled green postmodernity perspectives.

*Beyond Human Ecology*

The value of human ecology in all its different forms – and its contribution to the emergence of environmental sociology in the 1970s – is the fact that nature and the environment are no longer simply disregarded or done without. The border, or sometimes the iron wall, between the social and the natural as it was created and sustained by most of the classical sociological thinking was criticised, and a more reflexive mode of relating the social and the natural was argued for. The social should not be treated in isolation from the natural, as modern societies are an inherently 'materialistic' affair. The environment is not the passive realm of risks and opportunities that exists somewhere 'out there', waiting to be used one day and one way for serving mankind. We cannot explain or understand the human project by referring to 'endogenous' or 'internal' social facts only. History is an inherently natural affair, and human ecology in general and the HEP–NEP debate[7] within environmental sociology in particular have contributed to the better understanding of this naturalness of history.

The unsatisfactory element of human ecology – from the classical Chicago School via Peter Dickens' influential book *Society and Nature* [*1992*] up to present forms of so-called 'deep ecology' of which Jagtenberg and McKie [*1997*] are recent representatives – has to do with the tendency to try to restore the interrelationship between the social and the natural world in such a way that they seem to underscore the view that all facts, events, goals, outcomes, patterns etc. *as we know of them*, are socially mediated. There is no such thing as the 'biotic community' when this should mean sub-social or non-social. The naturalness of history is mirrored by the historicity of nature [*Harmsen, 1974*].

Ecological modernisation theory contributes to the redefinition of the borders between modern societies and their social and natural environments. The need for such a redefinition in social theory is fully recognised, and in this respect there is general agreement with the proponents of the HEP-NEP approach and other human ecologists. Ecological modernisation theorists equally argue that the notion of 'environment' should be taken seriously and not left un- or undertheorised by social scientists by first constructing a city-wall as a border between social systems and their 'outside natural environments' and then arguing that 'social facts should be explained by using social facts and factors alone'. What is conceived of as 'social' – for example, that what happens inside the city-wall – cannot be explained without reference to the natural, without taking into account the relationships with the outer-world. In fact, this has become one of the central notions in all contributions to the ecological and reflexive modernisation perspectives.

Within ecological modernisation theory – and much in line with neo-Marxist scholars such as Schnaiberg [*1980*] and Bunker [*1985*] – it is agreed that we must go beyond the social by taking into account naturalness, substance flows, energy flows, materials circulating throughout human societies etc. However, in restoring the analytical priority of the environment we should not throw away the baby with the bath water. The crucial difference between ecological modernisation theory and the human ecologies of different kinds is the contention that we must not replace the former disregard of nature with some form of present-day biologism or ecologism.

The former disregard of nature from the side of most of the classical and post-war sociological theories is linked to the crucial design-fault in some of the major institutional clusters of modern societies [*Giddens, 1990*]. When analysing the industrial mode of production and consumption, the attention of most sociologists used to be focused exclusively on factors such as capital, technology and labour. Environmental factors were regarded as 'external factors' in the sense not only of being 'available for free' but also in terms of being of secondary importance when it comes to explaining the dynamics of industrial production and consumption. When ecological modernisation theorists talk about 'repairing' this design fault of modern industrial production and consumption, they request that environmental factors should not only be taken into account, but also that they are structurally 'anchored' in the reproduction of these institutional clusters of production and consumption.

To illustrate the fact that something more serious is at hand than only 'pricing' things that used to be regarded as 'external costs' – the solution as it is pursued by most of the economists working in the neo-classical tradition – ecological modernisation theorists use the more encompassing vocabulary of 'rationalising production and consumption'. This notion refers to ecological rationalities (such as the closing of substance cycles and extensification of energy-use) that have a meaning 'of their own', implying that they are independent *vis-à-vis* other – for example, economic – rationalities that are involved in the reproduction of production-consumption cycles [cf. *Spaargaren, 1997*]. We see a whole new area of concepts emerging, which try to give this ecological rationality social, economic and political impact: environmental accounting and bookkeeping, annual environmental reports, green GNP, environmental efficiency, environmental productivity, environmental auditing, etc. It is these kinds of concepts that establish a link between ecological modernisation as a general theory of societal change on the one hand and ecological modernisation as a political programme or policy discourse on the other.

The recognition of the need to compare, link and sometimes mate ecological rationalities with other types of rationalities involved in the industrial mode of production, distinguishes the ecological modernisation approach from more 'principled eco-centrist approaches' which ascribe ecological criteria an almost absolute priority above other rationalities (as we elaborate upon below).

*Postmodern Critiques of (Green) Grand Narratives*

Some will conclude from this short outline that we are dealing here with nothing less than a new grand narrative in the making. Isn't the idea of the materiality of social systems, and the accompanying notion of ecological criteria and rationality involved in their reproduction, in principle a trans-historical and trans-cultural concept? Can one reasonably argue that the imperative of the 'sustainability' of social systems is in fact a universal one?

When understood in this way, it makes the fact of postmodern authors being among the most fierce critics of this approach understandable and predictable. Ecological Modernisation Theory is seen as a remnant of the old modernisation theories and an extension of the Enlightenment project, and it have has been especially the knowledge claims that are at the foundation of ecological transformation which have been challenged by postmodern perspectives. Indeed, postmodern critiques of ecological modernisation theory are as fierce as the more traditional critics working from a de-industrialisation perspective used to be. The focus of these postmodernists, however, is no longer on the need for 'dismantling' the institutions of modern societies instead of just 'repairing' them, as the debate on the 'technological fix' character of ecological modernisation would have it. Nevertheless, postmodernist critiques are in some respects even more radical in their consequences than those of counterproductivity theorists, because they question the very fact that sustainability criteria could or should be developed in a feasible way whatsoever.

Blühdorn [*2000*] seems to be a recent, rather radical, exponent of this position by (re)starting the debate on what the ecological problem exactly is, and ending up with the conclusion that environmental problems are no longer there: 'to the extent that we manage to get used to the non-availability of universally valid normative standards, the ecological problems ... simply dissolve'. Environmental change is no longer seen as problematic by large segments of contemporary world society in any universal way. According to these postmodernists this plurality and diversity of environmental problem definitions radically devalues any ecologist critique against modern developments, although this consequence is not yet fully acknowledged by the majority of the members of contemporary, so-called postmodern, societies. Nevertheless, late modern society, according to Blühdorn, cannot escape the transition from a modernist to a 'postmodernist politics of nature' and this fact makes the analytical value of ecological modernisation theory of no use at all for postmodernists.

The main objective of these radical postmodernists seems to be to show that all borders are time- and spacebound 'social constructions' which can be 'played upon' now that we have become aware of this fact in our postmodern times. So also the ways in which the borders between societies and their environments are created and sustained – from the Club of Rome in the early 1970s on to the International Panel on Climate Change -experts[8] of the late 1990s – can and must be criticised in order to 'liberate' us from the grand narratives of which the ecological crisis is only the latest plot. No distinction can be made between more or less 'objective', 'true' or widely held intersubjective understandings of reality. More moderate branches of postmodernism, such as Gare [*1995*], are less radical in their conclusions and rather seem to use postmodernist critiques in arguing for a new grand narrative, in which natural science and scientists do play a role in revealing the environmental crisis and speaking for the environment.

When trying to evaluate the relevance of postmodern perspectives for environmental sociology in general and in relation to ecological modernisation theory in particular, it is important to distinguish between different brands of postmodernism and between the different meanings of the term itself.[9] However, distinguishing different brands of postmodern theory or schools of thought within the postmodern tradition hardly seems to be possible due to the complicating fact that the denial of borders is one of the constituting features of postmodern thinking. Some authors from the reflexive modernisation school-of-thought who are judged influential in postmodern circles, have fiercely rejected the postmodern label. Consequently, it is necessary to be very precise when dealing with certain ideas of authors referred to as postmodern.

### *The Social Construction of Sustainability*

According to postmodern thinking, every grand narrative can and should be deconstructed and shown to be arbitrary to a great extent. Since the need for sustainable development is one of the few problems to be recognised and accepted as a challenge to society all around the world, this seems to be a privileged objective for some postmodern critics.

Within environmental sociology the debate that postmodern authors triggered is reflected in the frequently cited dispute on 'realism' versus 'constructivism'. Several authors have contributed to this debate, thereby referring to postmodern issues and ideas in an implicit or explicit way [cf. *Yearley, 1991; Hannigan, 1995; Dunlap and Catton, 1994*]. Standpoints vary from 'hard' or radical to 'soft' or moderate constructivism. The radical or relativist variant of constructivism seems to have as a particular goal to deconstruct or dismantle the naive beliefs that come along with environmental stories about global change, nuclear waste or soil erosion.[10]

From the observation that the environmental discourse has been changing from the early 1970s to the late 1990s with regard to priorities, definitions and approaches, it is concluded that environmental problems do not have a 'real', 'objective' existence but are instead the result of a process of framing certain social problems by certain social actors in a very specific, sometimes arbitrary way. As these relativist constructivists would have it, sustainability as grand narrative, dominant discourse or 'story line' stands in need for a deconstruction, showing that the story could have been framed otherwise, leading to different kind of conclusions and priorities.

Ecological modernisation theorists are not immune to the kind of epistemological issues touched upon by the relativist constructivists. In his book on ecological modernisation, Hajer [*1995*] seems to end up taking a position which is not too far away from where postmodernists would feel comfortable. In a similar way Peter Wehling [*1992*] evaluates the initial position taken by Huber, Jänicke and other ecological modernists in the 1980s as being insufficiently aware of the limitations of modernisation theory in general and ecological modernisation theory in particular. A more 'reflexive' approach is requested, especially when dealing with the role of science and technology in promoting sustainable production and consumption.

Von Prittwitz [*1993b*], Mol [*1996a*], Cohen [*1997*] and others have addressed the challenge to confront ecological modernisation theory with the debate on late- or reflexive-modernity as it has been developed by Beck, Giddens, Lash and others. Although it is doubtful whether it has ever been the case, under the condition of reflexive modernity the ecological modernisation of production and consumption can no longer be thought of or designed in terms of undisputed facts, values and futures. The ecological risks of reflexive modernity are no longer simply accepted on the authority of (natural) scientists, even more so if they at the same time also claim to have a privileged position in pointing out the best or most promising route towards a sustainable future. Science and technology are indeed disenchanted, and this has some potentially far reaching consequences for the ways in which environmental problems are perceived by lay-actors as well as policy-makers.

The fact of science and technology being no longer undisputed and bereft of that special kind of authority bestowed on them in earlier times should not be confused with epistemological issues that explain the crucial differences that exist between the natural and the social sciences. When environmental problems are discussed, these two major – but, in principle, separate – issues are very often intertwined or dealt with simultaneously. This can be said to be the case when for example the 'social' (for example, 'constructed') character of the climate change narrative – explained in terms of different interest groups, media and environmental movements all contributing to a specific mix of policies – would be presented in a way that tries to prove the more encompassing (postmodern) statement that the environmental crisis is something that is 'invented' by social actors and groups whose interests are served best by making a lot of noise about this or that particular social problem. Blühdorn [*2000*] seem to fall victim to this position in claiming that ecological rationality is nothing more than power politics and big money.

What tends to be denied then is the fact that environmental problems do have a 'real' existence. They belong to the type of problem which needs to be analysed and understood not only as social constructs but also in terms of the language of the natural and biological sciences. If we ignore this fact, we would end up were we started in environmental sociology, namely with the HEP–NEP distinction, with postmodern constructivist environmental sociology as the latest variant of exemptionalist thinking.

## IV. Radical Eco-Centrism versus Environmental Reformism

Some branches of radical eco-centrists and ecologists have questioned Ecological Modernisation Theorists regarding their rather moderate proposals for environmental reform. Some of the proponents and representatives of radical ecological restructuring criticise Ecological Modernisation ideas for not giving the environment pride of place in criticising current social developments and designing future trajectories. On this debate we now want to concentrate, by elaborating on the distinction between ecologism and environmentalism. Consequently, we will focus on the meaning of 'radical' and the various dimensions of radicalism, putting the sharp dichotomy between radicals and reformists into perspective.

According to Andrew Dobson, 'the first and most important point to be made about ecologism is that it is not the same as environmentalism' [*Dobson, 1990: 13*]. The important difference is that ecologism is about being radical while environmentalism definitely is not. Where environmentalism can be seen as a sub-plot in a main story such as liberalism or socialism, ecologism is about fundamental changes in present values or patterns of production and consumption. Ecologism – meaning *real Green* values, *real Green* movements and *real Green* politics – is about 'the desire to restructure the whole of political, social and economic life' [*Dobson, 1990: 3*].

For this reason, from a perspective of (deep) ecologism, the (West) German Greens are no longer Green and neither are Friends of the Earth International nor Greenpeace. They all have taken on the guise of environmentalism, a kind of watered down ideology which is just seeking a cleaner service economy, sustained by cleaner technology and producing cleaner affluence. They are no longer questioning Baconian science, western technology and the 'Promethean project to which the Enlightenment gave birth' [ibid.: 9].

Now Andrew Dobson is not telling us something completely new, nor is he the only one contributing to the debate by firm statements on the present state of affairs in the environmental movement. A decade before, the social ecologist Murray Bookchin wrote an 'open letter' to the ecology movement to express his concern with 'the widespread technocratic mentality and political opportunism that threatens to replace ecology by a new form of social engineering' [*Bookchin, 1980: 79*]. Many founders of the environmental and anti-nuclear movement had by then (USA in the early 1980s) become 'managerial radicals, operating *within* the system in the very name of opposing it'.

Other authors are dealing with the same phenomenon, although making different judgements about green radicalism. Robert Goodin, for example, in his *Green Political Theory* tries to 'rescue the greens from themselves' by showing to them that their unconventional and colourful political style, their principled way of life and their 'signing so unanimously and so firmly on to the principles of deep ecology' [*Goodin, 1992: 43*] are obstacles to the content of green politics becoming accepted by a broader part of the electorate. While agreeing that 'shallow' and 'deep' ecology are really different views, he invites the self-styled greens to be prepared to consider more shallow versions of the deep ecology creed.

But were they to do so, Peter List argues in his book *Radical Environmentalism*, radical environmentalists run the risk of losing just those characteristics that make them different. This is because 'the concept of radical environmentalism derives its meaning partially from its dissimilarity to other forms of environmentalism'. Where 'moderate environmentalism' assumes that the environmental crisis can be resolved by modifying attitudes, changing laws, government policies, corporate behaviour and personal lifestyles, radical environmentalism is different in insisting on the need for *fundamental alterations* in values and structures and for demanding *deep and systematic changes* in philosophy and tactics [*List, 1993: 2*].

It would not be very difficult to find within the environmental literature many more examples of dichotomies aimed at contrasting 'radical ecological worldviews' and 'fundamental structural change' with their pragmatic environmentalist counterparts. The point to be made, however, will be clear also without a prolonged *tour d'horizon*. Within social scientific literature on the environment, the main challenge for sociologists and anthropologists is often thought to arise from that stream of thought which most authors at the same time regard as being the dominant view in present-day environmental discourse: (radical) ecologism. The question that emerges from this overview is how we can understand and explain the positions in the debate on radical ecologism versus moderate environmentalism and how we assess these positions.

Sometimes the position one takes on these matters is treated as a more or less psychological affair. Being in the camp of radical ecologists is about being a pessimist by nature: you do not think the bridge can (or ever will) be built. Optimists, however, just start constructing something, using any bricks and steel available to get the building process going. Pessimism versus optimism should explain why social-democrats were in favour of the MIT-report 'Limits to Growth', while its basic conclusions were rejected by liberals and communist alike. Their perception of the environmental message was determined by their general stance on the 'Enlightenment process' [*Bakker, 1978*].

Others do not even bother to connect catastrophic or cornucopian attitudes with basic western political streams of thought. They just state that: 'optimists include: economists, engineers, physicists, and Europeans' [*Luten, 1980: 130*]. We think, however, that there is more to it than psychology; we should move beyond the position that the assessment of current affairs is just a matter of personal opinion or state of mind. It must be possible to approach these matters from a more sociological point of view, analysing the dynamic and historical relationship between shifting environmental ideologies and ever-changing social realities. In doing so the first step is an analytical refinement of the idea of radical reform, which makes it possible to move beyond the rather crude dichotomies discussed so far.

### *Intermezzo: The Episodic Characterisation of Environmental Reform*

How do we define the exact moment when a process of 'modifying attitudes, changing laws, government policies, corporate behaviour and personal lifestyles' is said to evolve into a state of affairs which can be qualified as the 'restructuring of the whole of political, social and economic life'? [*List, 1993: 2*]. Are we just playing with words or is there something more to be said about the character of the process of environment induced social change. How do we distinguish analytically an ecological revolution from a radical rupture or trend-break, and where do we just speak of a process of incremental environmental change? Obviously a key factor involved here is the time-horizon of the process. What seems a slow, gradual but steady process of change today may turn out to be a wholesale restructuring of industrial society some decades from now. What we need then are criteria which can be used to delineate and categorise the different modes of social change.

One of the key concepts enabling us to analyse social change is the notion of *episode*, introduced by Giddens. 'To characterize an aspect of social life as an episode is to regard it as a number of acts or events having a specifiable beginning and end, thus involving a particular sequence' [*Giddens, 1984: 244*]. The scale of an episode can range from the transition between types of societal totality via modes of change affecting the main institutions of a society to the disruption of daily life which results from going through a divorce. To assess the nature of a specific episode we have to analyse empirically its *origin, type, momentum and trajectory* [*Giddens, 1984: 245*]. Answers must be provided on the kind of structural principles or contradictions (for example, the human–nature relationship) which are at the origin of the episode; the episode has to be typified in terms of its intensity c.q. the degree to which the existing institutions (of what kind?) are reshaped or disrupted; finally things have to be said about the pace and the direction of the changes involved.

When (radical) proposals for environmental reform are put forward without considering the type of questions as formulated above, we think little or nothing can be said about their possible impact in the future development of modern societies. Without going into great detail, we want to clarify in the next two sections the different positions in the debate on radical reforms and specifically assess the mode of social change that is implied in the ecological modernisation perspective *vis-à-vis* its eco-centrist opponents.

### *Environmental versus Social Change*

There seem to be various flavours in radical ecologism but we will focus on those streams that stands for basically two things. First, a critique of the anthropocentric view of the interrelation between humans/society and nature/the environment. Second, a critique of industrial society and its technology for disregarding the (physical) limits to growth/development. As most radical ecologists authors have it, both points of critique – on modern societies' culture and structure respectively – cannot be resolved unless there is something like a revolution, a radical and profound alteration of the basic institutions of modern society. The environmental crisis poses a problem to society which cannot be dealt with using the conventional sociological theories on social change and which cannot adequately be resolved within the present-day institutional make-up of modern societies.

At the roots of the environmental crisis are the culture and structure of western industrial society as they were shaped over two or more centuries. All attempts to remedy the problem without basically questioning the overall structure and culture are bound to fail. If we are precise, radical ecologism is at the same time radical on its environmental goals as radical on the existing social structure and culture. Some authors within the radical ecologism perspective add additional radical claims at the same priority level (on democracy, emancipation, social justice/equality, etc., often neglecting possible conflicts between these priorities), while others give ecological goals pride of place. The basic presumption is usually that radical reforms of society's culture and structure not only contribute to, but are a prerequisite for, these desired goals.

To a major extent Ecological Modernisation Theory shares with radical eco-centrists such as Robyn Eckersley [*1992*], John Dryzek [*1987; 1997*] and the later André Gorz [*1989*] the starting position that environmental claims are subsumed in society's structure and culture and that consequently production and consumption processes should be radically improved regarding their environmental impact. However, it diverges from radical ecologism on two levels. First, Ecological Modernisation perspectives – and these are of course not alone or unique in this – do not give environmental objectives an undisputed priority above other societal objectives. Consequently, environmental reforms should not only be judged on their contribution to preserving the ecosystem, but also on other – sometimes conflicting – social values. And although the current relatively marginalised position of environmental interests (for example, *vis-à-vis* economic interests) allows some priority setting on environmental goals today, this cannot be an indisputable position based on some kind of 'objective' reason. Second, radical proposals for environmental improvement do not automatically entail radical societal change in the sense promoted by eco-centrists. Ecological Modernisation Theory claims that not only the environmental debate, but also actual social practices and institutions involving society-nature interactions, are already transforming to a major extent within the boundaries set by the current institutional order, showing that a tight coupling of environmental improvements and radical social change can at least be questioned. There is no – or better: no longer any – simple one to one relationship between radical environmental goals and radical social transformations, as eco-centrists seem to believe.

On a theoretical level, ecological modernisation theorists have labelled this the growing independence of the ecological sphere and rationality.

Cotgrove [*1992: 110*] and Paehlke [*1989: 190*] have both (independently) analysed the uncoupling of two dichotomies: the left/radical versus conservative politics and ideologies run no longer parallel to the dichotomy on green and anti-green positions. More recently this argument has been echoed by, among others, Giddens [*1994*]. While in the early 1970s being green usually meant propagating radical and left politics and ideologies, this is no longer automatically the case from the mid-1980s onward according to Paehlke and Cotgrove. Ecological Modernisation Theory has extended this analytical observation about the domains of politics and culture/ideology to that of economy, of actual activities of production and consumption. The environment becomes relatively independent (now from the economy), ultimately having as a consequence that a capitalist or rather market-based system of production and consumption does not necessarily contradict significant environmental improvements and reforms in any fundamental way. More production and consumption in economic terms (GNP, purchase power, employment) do not have to imply more environmental devastation (pollution, energy use, loss of biodiversity) [cf. *Mol, 1995; Spaargaren, 1997*]. Within principally the same modern institutional lay-out (a market economy, an industrial system, modern science and technology, a system of welfare states, etc.) we can thus look for – and design – radical environmental reforms. Although the principal institutional lay-out will not change beyond recognition, power relations, pricing, priorities in R&D, investment patterns, and physical planning – to name but a few – will alter significantly following radical environmental reform. In the end, the empirical question will of course remain whether these radical environmental reforms will be sufficient to deal with the – to a large extent socially constructed – criterion of sustainability.

### *Putting Radical Ecology into Practice: 'State of Being' versus 'Code of Conduct'*

The counter-positing of an eco-centric and a techno-centric worldview had its mobilising effects in the early 1970s. In the birth period of modern environmentalism, this newly emerging ideology found itself confronted with a dominant world-view and a mode of production and consumption in which there was no role to play for environmental considerations. In order to establish itself as a counter-ideology, environmentalism/ecologism was more or less forced to focus on a limited set of issues which had the best mobilising potential and which were regarded as the most central elements of the emerging environmental ideology. The question of putting these fundamental principles into practice was either postponed or resolved by adhering to a personal political or communal commitment and a green lifestyle, with the latter aimed more at expressing new environmental values rather than final improvements in society's environmental performance. This diagnosis of the situation during the early seventies and the environmentalists' answers to it looks very familiar to us, but is it still relevant for the state of affairs a quarter-century later?

If we are to believe Andrew Dobson and others, the situation early members of the environmental movement were confronted with, is still largely relevant today. The political practice of contemporary modern society offers few or no points of reference for a deep or radical green political ideology, and the points of entry within the socio-economic sphere are still difficult to discern. In sum, there still are very few possibilities for developing 'dark green' principles into a 'code of conduct' within politics or business. This is why, according to Dobson, it would be better if we accepted the fact that radical ecologism only lends itself to be expressed in a certain state of mind, a kind of contemplative (critical) reflection on reality as it is [*Dobson, 1990: 47–63*]. In short, the attitude that fits ecological radicalism best is a certain *'state of being'* which cannot and should not be translated in a direct way into a *'code of conduct'* .

Ecological Modernisation Theory diverts from such an analysis on two main points. First, radical ecologism underestimates the current environment-induced transformations in social practices and institutional developments in especially industrialised societies. The penetration of environmental considerations into the board rooms of the major political and economic organisations, their nestling on the agendas and their impact on the performance of these organisations and institutions can no longer be analytically neglected, although it can still be criticised as being 'too little, too late'. But the fact that environmental considerations are increasingly institutionalised, and no longer wither away with the first economic depression or crisis, gives radical environmentalists a point of entry to the traditional and dominant institutions and organisations that 'rule the capitalist world-economy'. This observation increasingly resonates in the daily practices of established environmental non-governmental organisations such as Friends of the Earth International, Greenpeace and the World Wildlife Federation.

Second, the separation of a *'state of being'* from a *'code of conduct'* paves the way for a radical but rather noncommittal attitude, that combines 'politically correct' contemplations with environmental destructive productive and consumptive activities. Ecological Modernisation Theory deliberately emphasises the close relationship between analysis and criticism of the current state of affairs on the one side and actual transformations and designs of institutions and social practices on the other. Some critics have argued that this has the risk of becoming too 'narrow-minded', becoming caught too much in the present situation without having the possibility of exploring options beyond the existing and dominant societal paradigms. Although there exists a tension between building on existing structures and patterns of action on the one hand and the final desired *'state of being'* on the other, the opening up of windows for '*realistic* utopian models for the future' [*Giddens, 1990*; emphasis added] seems a better alternative than either the noncommittal attitudes towards radical ecologism of both postmodernists and radical eco-centrists, or the business as usual scenarios of the mainstream neo-liberal political and economic elites.

## V. Social Inequalities and Ecological Restructuring

Whereas in the 1970s environmental activists criticised core institutions from the outside for not taking environmental considerations into account, the 1990s are characterised by parallel processes of penetration and transformation of these core institutions by environmental considerations and interests. This process should of course be understood as neither evolutionary nor deterministic, neither irreversible nor 'smooth', as we argued above. We also pointed out that the social struggles and social inequalities that go together with processes of ecological restructuring have originally found a prominent place in, among others, current neo-Marxist studies; and Ecological Modernisation Theory might profit from that.

In this section, then, we elaborate on the relation between ecological restructuring and social inequalities and struggles within a framework of Ecological Modernisation. In dealing with such inequalities, and positioning Ecological Modernisation Theory more clearly towards them, we can make a distinction between (i) inequalities in the relation between humans and nature; (ii) inequalities as distributional consequences of environmental policies; (iii) inequalities related to environmental risks. Global inequalities, which to some extent cut across the last two types, will be our last focus.

### *Social Inequalities*

The inequalities between humans and nature can be seen as the most abstract form of inequality, and have been a key interest for environmental philosophers. The dominant view among them is that the causes of our environmental crisis are to be found in an anthropocentric culture in which there is no room for an intrinsic value of nature. To stop the exploitation of nature, our culture and our attitudes towards nature have to change dramatically: the 'interests' of nature have to be taken into account.

This rather abstract basic idea has resulted in very practical and essential questions, put forward by what we could identify as the deep ecology movement – ranging from groups on animal liberation through to Earth First!: What is the value of a cow other than its meat and milk? Why do we spend millions of dollars on the well-being of dogs, cats and horses, while millions of other species (cows, pigs, chickens) are largely neglected in terms of well-being? And in line with these emerging claims for animal well-being, is it not about time that we developed rights for/of nature?

We have frequently placed security nets under social classes and groups to protect them against too drastic consequences of our modern society. Just as the growing (complexity of) human interdependencies resulted in basic human rights being protected by these security nets, should the growing complexity and interdependencies in our relation with nature not result in basic rights for non-human entities, secured by institutions?

Unfortunately there seems to be a world of difference between the academic debates among these environmental philosophers and the kind of questions raised by activists and among policy-makers. It would enrich the academic debate if the abstract and fundamental questions of environmental philosophers on the intrinsic value of nature were connected with the concrete context of industrial food production and consumption, and included late modernity's changing power relations within and beyond these sectors that (can) contribute to the reproduction and transformation of unequal human–nature relations. The connection of rather abstract contributions of various environmental philosophers with more substantive analyses on industrial food production and consumption that is so characteristic of Ecological Modernisation Theory, could also amend the latter's often quoted limited definition of nature [cf. *Spaargaren and Mol, 1992; Mol, 1995; Blühdorn, 2000*].

This leads us to the second category of inequalities. It has been especially neo-Marxists that have contributed to our understanding that (i) environmental problems are unequally distributed among groups/classes in modern society, (ii) radical environmental reforms are obstructed by the contemporary capitalist structure of modern society, and (iii) radical environmental reforms in this society often results in unequal consequences or distributional effects. While traditionally the emphasis of neo-Marxists has been on the first two categories, more recently – and following developments in society's environmental performance – a shift has taken place to the last one. Environmental policies and strategies often have dissimilar socio-economic (and sometimes even environmental and health) consequences for distinct economic groups or classes. In a more or less similar way, environmental policies can conflict with other – that is non-economic – social or political priorities aiming to improve the position of various social categories, including women, ethnic minorities, and residents of peripheral regions. It goes without saying that distributional effects in the economic sphere frequently – but not always and not by definition – overlap with those on non-economic issues. These distributional effects of different kinds result in distributional conflicts around environmental policies, as Schnaiberg *et al.* [*1986, 1994*] have started to analyse in detail already more than a decade ago. Distributional conflicts in their turn can frustrate radical environmental reforms and should therefore not been left untouched, even not by the most technocratic proposals for environmental reform.

More environmentally sound houses, cars, food and services are to a major extent still the privilege of the rich (and are used to some extent as a new means of distinction), while the poor are most strongly confronted with raising environmental taxes on water, energy and food. This close link between social classes and environmental reform has for a considerable time paralysed radical proposals for environmental reform due to left-wing opposition (see labour union protests against attacks on polluting industries; social democratic objections to eco-taxes). The neglect of distributional effects by environmental authorities, among others, can also severely affect public support, especially amongst the lower and middle income strata, for these reforms. In a similar way, the emancipation of women has increased private car use considerably, and environmental policies to discourage car use can – and often do – directly interfere with the possibility of young mothers returning in the labour process after giving birth (depending of course on the specific local circumstances and other policies).

While examples of detrimental effects of environmental reforms on the less well-off and disadvantaged minorities are numerous, environmental policies can also support the material and non-material improvements of the poor and disadvantaged social groups, especially in those cases where they have been unequally victimised by environmental risks. The environmental justice movement in the USA[11] has pointed out the process of passing on environmental problems to the poor and/or ethnic minorities for many years. A radical environmental reform of the waste dumps and pollution production complexes in the less well-off areas – as increasingly proposed – would also improve the material and non-material position of these neighbourhoods (for instance, by increasing house prices and reducing health risks).[12]

This second category of inequalities results in conflicts between and interest representation of distinct social groups, as neo-Marxists have shown in numerous empirical studies and theoretical elaborations. Distinct from other, more traditional, social and economic problems, however, environmental conflicts do not follow a predictable path of static opposing parties and interests. Farmers can one day be victims of environmental deterioration (see large infrastructure projects, air pollution), while the next day they turn into polluters and consequently become victims of radical environmental reforms (see pesticides and fertilisers). Or, as can be concluded from social movement research, the environmental movement has no longer a natural enemy, nor a natural ally. The environmental movement has to enter into constantly changing coalitions according to issue, time and place: women's organisations, employers, unions, recreation and leisure organisations, public transport representatives are all one day opponents, while the next day they may be allies in fighting for radical reforms. In that sense, environmental struggles cross traditional (economic and other) interest lines and divisions in society and should be analysed as an independent – that is non-reducible – category, as Ecological Modernisation Theory has suggested.[13] In that sense, those neo-Marxist schemes which claim direct parallels between traditional class struggles and environmental struggles, might prove fruitful in individual empirical cases, but have lost their overall theoretical and analytical value.

This brings us to the third category of inequalities connected to environmental reform. It has been especially Ulrich Beck [*1986*] who contributed to our understanding that modern environmental risks add a new social dimension to the other two inequalities. According to Beck, global environmental risks are democratic, both in the sense that they make no difference for distinct social classes, and in that traditional class differences are no longer adequate to understand the distribution of these risks among the population. Who can escape the greenhouse effect, the mad cow disease, or the pesticides 'circle of poison'? Often new 'class' divisions of environmental inequalities are formed: the vegetarian against the meat eater (the mad cow disease associated with Creutzveldt-Jacobs); the outdoor worker against the indoor employee (skin cancer by UV-b). Buttel [*2000*] and others are of course right in criticising Beck's over statement of the dissolution of classes in the distribution of risks in late modernity. The rich generally still have a better chance of protecting themselves against or escaping from such environmental dangers. Although in some cases '*wegreisen hilft letzlich ebensowenig wie Müsli essen*',[14] for most environmental risks, one can argue that locational patterns, life styles, and economic protection opportunities do make a difference. But Beck is right in observing the tendency that socio-economic categories (classes) and environmental risks no longer run parallel by definition, and in noticing that all members of modern society have – in some way or the other – to 'deal with modern environmental risks' under conditions of increasing uncertainty and the growing inability of the old institutions of science and politics to give final conclusions on how to live and act.

In conclusion, we could state that there is still an unequal distribution of environmental risks, but (i) these divisions follow to some extent new distributional patterns in late-modern society and (ii) these risks affect all in their growing uncertainties about how to cope with them. But – we would add [*Mol and Spaargaren, 1993*] – this is especially true in those cases of a growing complexity of interdependencies on an international or global level.

*Global Inequalities*

The international or global inequalities related to environmental disruption and reform can either be interpreted as a fourth category, or as a combination of the second and third category mentioned above, as the lines of reasoning run parallel to some extent. Similar to the analysis above, insights about the distributional effects of both international/global environmental problems and the reform strategies to combat them, can help us to understand the difficulties and obstacles in reaching global environmental improvements. Although Szasz and Meuser [*1997*] rightly criticise the complete separation of national studies on environmental inequalities and international studies on the unequal access to environmental resources, resulting in missing opportunities for theoretical cross-fertilisation, we should not fall victim to simply 'upscaling' our still basically nation-state oriented conceptual efforts to the global level, as seem so typical of – also more recent – modernisation theories [cf. *Beck, 1996*].

Ecological Modernisation Theory has been developed initially in a limited number of West European countries and its postulates, hypotheses and empirical references still partly mirror this geographical focus. Some authors, for instance, have made a direct link between ecological modernisation and neo-corporatist policy arrangements, suggesting that the latter form an essential precondition in the political modernisation towards self-regulation and participative and consensual policy-making. [cf. *Weale, 1992; Dryzek, 1997; Neale, 1997*].[15] These and other kind of analyses have raised questions regarding the generalisability and value of this theoretical framework for other countries.

Investigations regarding the geographical limitations of Ecological Modernisation Theory initially concentrated on developing countries [cf. *Sarkar, 1990; Mol, 1995;* also Frijns *et al.*, this volume] to be followed by studies on New Industrialising Countries and transitional economies in Central and Eastern Europe. The latter countries were believed to have more in common with the European states which gave the theory its original foundation [cf. *Sonnenfeld, 1996*; also this volume; *Rinkevicius, 2000*; *Mol, 1999b*; Gille, this volume]. The general conclusion to be drawn from these studies is that the (analytical) value of Ecological Modernisation Theory for analysing processes of ecological reform in these non-West European contexts is limited, depending especially on both the specific institutional lay-out of the nation-states in question (in contrast with the West European ones) and the degree of 'environmental institutionalisation' that has already taken place.

Also, at the global level we can identify studies that point to the close relationship between material prosperity on the one hand and environmental disturbances and reforms on the other. Internationally, the poor regions, countries and groups are believed to be among the major sufferers of both environmental threats as well as radical environmental programmes, while they would contribute relatively little to global environmental problems. Examples are to be found on climate change (the Alliance Of Small Island States [AOSIS] being the poor victims), on biodiversity (the tropical rain forest countries, and especially the indigenous people), trade and environment, etc.[16] At the same time we are also witnesses of contrasting studies: the – heavily contested – World Bank study [*World Bank, 1992*] on the Green Kuznets curve arguing that beyond a certain point of development the (relative) contribution to global environmental threats diminishes *vis-à-vis* lesser developed countries, studies analysing the environmental improvements of foreign direct investments in developing countries (see Zarsky [*1999*] for an overview), and studies arguing for and showing the incorporation of environmental considerations in the policies and practices of international organisations and institutions, such as the World Bank, the IMF and the various environmental regimes such as the climate convention [cf. *Haas et al., 1993*], so that environmental reforms go together with economic development in the less wealthy parts of the world. Similar to our evaluation of the national distributional effects of environmental problems and policies, we may also conclude that global environmental inequalities no longer can be seen as, or reduced to, a sub-category of social inequalities.

In an earlier contribution [*Mol, 2000*] we have argued from an ecological modernisation perspective that globalisation processes do have consequences for the environment and the distribution of environmental effects. But while in the 1970s economic globalisation (or global capitalism as some would prefer to call it) was primarily related to increasing the unequal distribution of environmental effects, recent developments imply that no general, overall conclusion can be given on globalisation's net positive or negative environmental effects, or on the distribution of these effects around the globe. The institutionalisation of environmental considerations in some of the major global organisations, institutions and dynamics, the growing use of environmental resources (biodiversity, 'pollution rights', natural resources) by 'peripheral' countries and regions in international (economic) struggles and conflicts, and increasing environmental transparency at the global level result in distinct environmental inequalities *vis-à-vis* those in the 1970s. And this insight about new environmental inequalities which no longer follow established patterns of economic and political inequalities is, in fact, what Ecological Modernisation Theory has to offer the more traditional and well established international studies on environmental conflicts and its distributional effects.

## VI. Epilogue

In reviewing the debates Ecological Modernisation Theory has been engaged in from its date of birth in the early 1980s, we have made a separation between some of the initial controversies, notably with counterproductivity theorists and neo-Marxists, and more recent topics and discussions. We have emphasised how the more recent discussions differed from the initial ones, as both social reality (in terms of institutional developments, social practices and environmental discourses) has changed and the more recent contributions have profited from and are built upon the earlier ones. In that sense, a simple repetition of these initial criticisms against Ecological Modernisation Theory is no longer adequate.

Secondly, we have tried to show what kind of contributions ecological modernisation perspectives can offer regarding three contemporary controversies in environmental sociology: discussions on the material foundation of social theory, debates on radical versus reformist environmental reforms and further understanding of social inequalities regarding environmental problems and reform. In doing so, we have not only tried to clarify the position taken by Ecological Modernisation Theory *vis-à-vis* other schools of thought in environmental sociology (eco-centrists, postmodernists, human ecologists, neo-Marxists, social-constructivists). We also tried to take these debates some steps further by integrating perspectives and explaining the differences that persist between the schools of thought. In doing so we hope to have contributed to the further development of environmental sociology and the environmental social sciences.

Regardless of the position taken in the various debates Ecological Modernisation Theory is engaged in, this contribution shows that the environmental social sciences have developed from a marginal and subsumed area of sociology and other social sciences, as became evident during the HEP-NEP controversies, into a full-fledged subdiscipline. And – in addition to other theoretical perspectives – Ecological Modernisation Theory has made a significant contribution to that development. In that sense, the institutionalisation of the environment in the various social science disciplines mirrors the institutionalisation of the environment in institutions and social practices of modern society.

## NOTES

1. Several observations seem to justify the claim of growing popularity of ecological modernisation theory. The fact that at both the 1998 International Sociological Association and the 1998 American Sociological Association sessions were organised on ecological modernisation theory gives some evidence of its prominent status in environmental sociology. In addition an international conference on Ecological Modernisation Theory was organised in 1998 in Helsinki. In analysing a recent handbook of environmental sociology, Buttel [*1997*] noticed that ecological modernisation theory is slowly becoming one of the few more coherent theories in environmental sociology that is to some extent promoted and defended actively. Recent volumes and special issues of journals that focus on this theory contribute to that [cf. *Spaargaren, Mol and Buttel, 2000; Van der Straaten et al., forthcoming; Geography, forthcoming*].
2. The strongest points of critique against Parsonian functionalism and other modernisation theories are related to the evolutionary models of change and their lack of an adequate theory of action, which do not sound unfamiliar in reviewing the initial contributions to the Ecological Modernisation Theory [cf. *Spaargaren and Mol, 1992*].
3. In his major work *Risikogesellschaft*, Ulrich Beck [*1986*] seemed to intermingle the notions of Risk Society and Reflexive Modernisation, while in his more recent work the notion of Reflexive Modernisation has the connotation of the overall analytical notion, which refers to the more apocalyptic dimensions of the modernisation process.
4. Instead of the 'limits to growth', the expansion of the limits was celebrated by first generation Ecological Modernisation theorists, as technological innovations would constantly move these limits further, making the 'treadmill of production' no longer a fundamental contradiction to sustained environmental quality. More recently, the issue of ecological limits has re-entered environmental sociology in a different form, where strong-contructivists deny the existence of any 'objective' or intersubjective limits, while ecological modernisation theorists and neo-Marxists seem to move more to a nuanced realist position (see section III below).
5. This does of course not mean that capitalist production is sustainable, or that one can no longer analyse the detrimental environmental consequences of a capitalist organisation of production. That can be and is still done. But it does affect the political consequences of and value given to such analyses in terms of their social support and their impact on actual transformations and alternatives.
6. This section draws to some extent on the introductory chapter in Spaargaren, Mol and Buttel [*1999*].
7. The debate between the Human Exemptionalist Paradigm (HEP) and the New Ecological Paradigm (NEP) was triggered especially by Dunlap and Catton in a number of publications [*Catton and Dunlap, 1978; Dunlap and Catton, 1979, 1994*].
8. The IPCC is the umbrella institution under which the scientific community reported on the certainties and uncertainties regarding the greenhouse effect. Growing consensus within the IPCC proved a necessary condition for political decision-making within the Framework Convention of Climate Change.
9. For instance, Jagtenberg and McKie [*1997*] consider themselves postmodernists but on the environment they take an eco-centrist position of the kind that is so vehemently criticised by Blühdorn [*1999*], who also claims to talk from a postmodernist perspective. In a similar way Zygmunt Bauman [*1993*] considers himself a postmodernist, although his definition of environmental problems and his elaborations of desirable solutions resemble deindustrialisation and demodernisation ideas, rather than the postmodernism of Blühdorn and others.
10. Freudenburg [*1999*] made the interesting – and largely valid – observation that constructivists seems to be preoccupied with deconstructing the seriousness of environmental problems, while they completely neglect the deconstruction of those increasing number of groups and ideas that propagate the non-existence of environmental deterioration.
11. Environmental justice ideas have resulted in close collaboration between activists and

scientists. For a recent overview of studies from an environmental justice perspective, see Szasz and Meuser [*1997*].

12. This environmental clean up of less well-off areas might, of course, turn into increasing property values that have as an ultimate result the displacement of the poorer part of the population that could only afford to live there due to low land and house prices. The actual material/economic effect in concrete situations clearly depends on various factors.
13. This parallels the perspective of many so-called New Social Movement theorists, who distinguish environmental and other 'new' movements from the 'traditional' class-oriented labour movements [cf. *Offe, 1986; Dalton, 1990; Jamison et al., 1990*].
14. 'In the end, moving away helps as little as eating muesli' [*Beck, 1986: 97*].
15. In extensive research on the emergence of so-called joint environmental policy-making in European countries with strong corporatist traditions (the Netherlands, Denmark and Austria), we have investigated these close relationships between corporatism and political modernisation [*Mol, Lauber and Liefferink, 2000*]. It proves that there is no simple relationship between (neo-)corporatism and political or ecological modernisation: Austria, being the ideal-type corporatist country, fails some of the typical characteristics of political and ecological modernisation.
16. These relations will be dealt with in more detail in a forthcoming book which analyses the relations between globalisation and environment from an ecological modernisation perspective (Mol, forthcoming).

## REFERENCES


Andersen, M.S. (1994), *Governance by Green Taxes. Making Pollution Prevention Pay*, Manchester: Manchester University Press.

Bakker, E.P. (1978), 'Sociologie en Milieuproblematiek', *Sociologische Encyclopedie* 2, pp.441–6.

Bauman, Z. (1993), *Postmodern Ethics*, Oxford and Cambridge, MA: Basil Blackwell.

Beck, U. (1986), *Risikogesellschaft. Auf dem Weg in eine andere Moderne*, Frankfurt am Main: Suhrkamp.

Blowers, A. (1997), 'Environmental Policy: Ecological Modernisation and the Risk Society?', *Urban Studies*, Vol.34, Nos.5–6, pp.845–71.

Bookchin, M. (1980), *Towards an Ecological Society*, Quebec: Black Rose Books.

Boons, F. (1997), 'Organisatieverandering en ecologische modernisering: het voorbeeld van groene produktontwikkeling', paper to the Dutch NSV conference, 29 May 1997.

Blühdorn, I. (2000), 'Ecological Modernisation and Post-Ecologist Politics', in Spaargaren, Mol and Buttel [*2000*].

Bunker, S. (1985), *Underdeveloping the Amazon: Extraction, Unequal Exchange, and the Failure of the Modern State*, Urbana, IL: University of Illinois Press.

Buttel, F.H. (1997), 'Social Institutions and Environmental Change', in M. Redclift and G. Woodgate (eds.), *The International Handbook of Environmental Sociology*, Cheltenham, UK: Edward Elgar, pp.40–54.

Buttel, F. (2000), 'Classical Theory and Contemporary Environmental Sociology: Some Reflections on the Antecedents and Prospects for Reflexive Modernisation Theories in the Study of Environment and Society', in Spaargaren, Mol and Buttel [*2000*].

Catton, W.R. and R.E. Dunlap (1978), 'Environmental Sociology: A New Paradigm', *The American Sociologist*, Vol.13, pp.41–9.

Christoff, P. (1996), 'Ecological Modernisation, Ecological Modernities', *Environmental Politics*, Vol.5, No.3, pp.476–500.

Cohen M. (1997), 'Risk Society and Ecological Modernisation: Alternative Visions for Post-Industrial Nations', *Futures*, Vol.29, No.2, pp.105–19.

Cotgrove, S. (1992), *Catastrophe or Cornucopia. The Environment, Politics and the Future*, Chichester: John Wiley.

Dalton, R. (1990), *Challenging the Political Order*, Oxford: Oxford University Press.

Dickens, P. (1992) *Society and Nature. Towards a Green Social Theory.* New York: Harvester Wheatscheaf.

Dobson, A. (1990), *Green Political Thought. An Introduction*, London: Unwin Hyman.

Dryzek, J.S. (1987), *Rational Ecology. Environment and Political Economy*, Oxford/New York: Basil Blackwell.

Dryzek, J.S. (1995), 'Environmental Compliance, Justice and Democracy', in R. Eckersley (ed.), *Markets, the State and the Environment: Towards Integration*, London: MacMillan, pp.294–308.

Dryzek, J.S. (1997), *The Politics of the Earth: Environmental Discourses*, Oxford: Oxford University Press.

Dunlap, R.E. and W.R. Catton (1979), 'Environmental Sociology', *Annual Review Sociology*, Vol.5, pp.243–73.

Dunlap, R.E. and W.R. Catton (1994), 'Struggling With Human Exemptionalism: The Rise, Decline and Rivitalization of Environmental Sociology', *The American Sociologist*, Vol.25, pp.5–29.

Eckersley, R. (1992), *Environmentalism and Political Theory. Toward an Ecocentric Approach*, London: University College of London Press.

Freudenburg, W. (1999), 'Social Constructions and Social Constrictions: Toward Analyzing the Social Construction on the "Naturalized" as well as the "Natural"', in Spaargaren, Mol and Buttel [*2000*].

Gare, A.E. (1995), *Postmodernism and the Environmental Crisis*, London: Routledge.

Giddens, A. (1984), *The Constitution of Society*. Cambridge: Polity Press.

Giddens, A. (1990), *The Consequences of Modernity*, Cambridge: Polity Press.

Giddens, A. (1994), *Beyond Left and Right. The Future of Radical Politics*, Cambridge: Polity Press.

Goldblatt, D. (1996), *Social Theory and the Environment*, Cambridge: Polity Press

Goodin, R.E. (1992), *Green Political Theory*, Cambridge: Polity Press

Gorz, A. (1989 [1988]), *Critique of Economic Reason*, London/New York: Verso

Gould, K.A., Schnaiberg, A. and A. S. Weinberg (1996), *Local Environmental Struggles: Citizen Activism in the Treadmill of Production*, Cambridge: Cambridge University Press.

Haas, P.M., Keohane, R.O. and M.A. Levy (1993), *Institutions for the Earth: Sources of Effective Environmental Protection*, Cambridge, MA: MIT Press.

Hajer, M.A. (1995), *The Politics of Environmental Discourse: Ecological Modernisation and the Policy Process*, Oxford: Clarendon.

Hannigan, J.A. (1995), *Environmental Sociology: A Social Constructionist Perspective*, London and New York: Routledge

Harmsen, (1974), *Natuur, geschiedenis, filosofie*, Nijmegen: Sun.

Hogenboom, J., Mol, A.P.J. and G. Spaargaren (1999), 'Dealing with Environmental Risks in Reflexive Modernity', in M. Cohen (ed.), *Risk in the Modern Age*, London: Macmillan, pp.83–107.

Huber, J. (1985), *Die Regenbogengesellschaft. Ökologie und Sozialpolitik*, Frankfurt am Main: Fisher Verlag.

Huber, J. (1991), *Unternehmen Umwelt. Weichenstellungen für eine ökologische Marktwirtschaft*, Frankfurt am Main: Fisher.

Jagtenberg, T. and D. McKie (1997), *Eco-impacts and the Greening of Postmodernity. New Maps for Communication Studies, Cultural Studies, and Sociology*, London: Sage.

Jamison, A., Eyerman, R. and J. Cramer (with J. Laessøe) (1990), *The Making of the New Environmental Consciousness. A Comparative Study of the Environmental Movements in Sweden, Denmark and the Netherlands*, Edinburgh: Edinburgh University Press.

Jänicke, M., Mönch, Binder, H. *et al.* (1992), *Umweltentlastung durch industriellen Strukturwandel? Eine explorative Studie über 32 Industrieländer*, Berlin: Sigma.

Jokinen, P. and K. Koskinen (1998), 'Unity in Environmental Discourse? The Role of Decision-Makers, Experts and Citizens in Developing Finnish Environmental Policy', *Policy and Politics*, Vol.26, No.1, pp.55–70.

Leroy, P. (1996), 'Nieuwe stappen in de Milieusociologie?', *Tijdschrift voor Sociologie*, Vol.15, No.2, pp.67ff.

Leroy, P. and J. van Tatenhove (2000), 'New Policy Arrangements in Environmental Politics: The Relevance of Political and Ecological Modernisation', in Spaargaren, Mol and Buttel (eds.), [*2000*].

List, P.C. (ed.) (1993), *Radical Environmentalism; Philosophy and Tactics*, Belmont: Wadsworth.

Luten, D.B. (1980), 'Ecological Optimism in the Social Sciences: The Question of Limits to Growth', *American Behavioral Scientist*, Vol.24, pp.125–51.

Mol, A.P.J. (1995), *The Refinement of Production. Ecological Modernisation Theory and the Chemical Industry*, Utrecht: Jan van Arkel/International Books.

Mol A.P.J. (1996a), 'Ecological Modernisation and Institutional Reflexivity: Environmental Reform in the Late Modern Age', *Environmental Politics*, Vol.5, No.2, pp.302–23.

Mol, A.P.J. (1999b), Globalisation and Changing Patterns of Industrial Pollution and Control', in S. Herculano (ed.), *Environmental Risks and the Quality of Life*, Rio de Janeiro: UFF.

Mol, A.P.J. (2000), 'Globalization and Environment: Between Apocalypse-Blindness and Ecological Modernisation', in Spaargaren, Mol and Buttel (eds.) [*2000*].

Mol, A.P.J. and G. Spaargaren (1993), 'Environment, Modernity and the Risk Society: The Apocalyptic Horizon of Environmental Reform, *International Sociology*, Vol.8, No.4, pp.431–59.

Mol, A.P.J., Lauber, V. and J.D. Liefferink (eds.) (2000), *The Voluntary Approach to Environmental Policy: Joint Environmental Policy-Making in Europe*, Oxford: Oxford University Press.

Mol, A.P.J., Spaargaren, G. and A. Klapwijk (eds.) (1991), *Technologie en milieubeheer. Tussen sanering en ecologische modernisering*, The Hague: SDU.

Mol, A.P.J., G. Spaargaren and J. Frouws (1998), 'Milieurisico's als nieuw sociaal probleem', in K. Schuyt (ed.), *Het sociaal tekort. Veertien sociale problemen in Nederland,* Amsterdam: de Balie, pp.229–43.

Neale, A. (1997), 'Organising Environmental Self-Regulation: Liberal Governmentality and the Pursuit of Ecological Modernisation in Europe', *Environmental Politics*, Vol.6, No.4, pp.1–24.

O'Connor, J. (1996), 'The Second Contradiction of Capitalism', in T. Benton (ed.), *The Greening of Marxism*, New York: Guilford.

Offe, C. (1986), 'Nieuwe sociale bewegingen als meta-politieke uitdaging', in L.J.G. van der Maerssen *et al.* (eds.), *Tegenspraken, dilemma's en impasses van de verzorgingsstaat*, Amsterdam: Samson, pp.27–92.

Paehlke, R.C. (1989), *Environmentalism and the Future of Progressive Politics*, New Haven, CT and London: Yale University Press.

Redclift, M. (2000), 'Environmental Social Theory for a Globalizing World Economy', in Spaargaren, Mol and Buttel (eds.) [*2000*].

Rinkevicius, L. (2000), 'The Ideology of Ecological Modernisation in "Double-Risk" Societies: A Case Study of Lithuanian Environmental Policy', in Spaargaren, Mol and Buttel (eds.) [*2000*].

Sarinen, R. (forthcoming), 'Governing Capacity of Environmental Policy of Finland: An Assessment of Making the EIA-Law and CO2-Tax', dissertation, Helsinki: University of Technology.

Sarkar, S. (1990), 'Accommodating Industrialism: A Third World View of the West German Ecological Movement', *The Ecologist*, Vol.20, No.4, pp.147–52.

Schnaiberg, A. and K. Gould (1994), *Environment and Society: The Enduring Conflict*, New York: St. Martin's Pres.

Schnaiberg, A. (1980), *The Environment: From Surplus to Scarcity*, Oxford/New York: Oxford University Press.

Schnaiberg, A., Watts, N. and K. Zimmermann (eds.) (1986), *Distributional Conflicts in Environmental-Resource Policy*, Aldershot: Gower.

Smidt, K. (1996), 'Over beschaving en milieu', *Amsterdams Sociologisch Tijdschrift*, Vol.23, No.1, pp.60–81.

Sonnenfeld, D.A. (1996), 'Greening the Tiger? Social Movements' Influence on Adaptation of Environmental Technologies in the Pulp and Paper Industries of Australia, Indonesia and Thailand', dissertation, University of California, Santa Cruz.

Spaargaren, G. (1997), 'The Ecological Modernisation of Production and Consumption: Essays in Environmental Sociology', dissertation, Department of Environmental Sociology, WAU, Wageningen.

Spaargaren, G. and A.P.J. Mol (1991), 'Ecologie, technologie en sociale verandering. Naar een ecologisch meer rationele vorm van productie en consumptie', in Mol, Spaargaren and Klapwijk (eds.) [*1991: 185–207*].

Spaargaren, G. and A.P.J. Mol (1992), 'Sociology, Environment and Modernity: Ecological Modernisation as a Theory of Social Change', *Society and Natural Resources*, Vol.5, No.4, pp.323–44.

Spaargaren, G., A.P.J. Mol and F. Buttel (2000), 'Introduction: Globalization, Modernity and the Environment', in Spaargaren, Mol and Buttel (eds.) [*2000*].

Spaargaren, G., Mol, A.P.J. and F. Buttel (eds.) (2000), *Environment and Global Modernity*, London: Sage.

Szasz, A. and M. Meuser (1997), 'Environmental Inequalities: Literature Review and Proposals for New Directions in Research and Theory', *Current Sociology*, Vol.45, No.3, pp.99–120.

Tellegen, E. (1991), 'Techniek: oorzaak of oplossing van milieuproblemen', in Mol, Spaargaren and Klapwijk (eds.) [*1991: 21–8*].

Van der Straaten *et al.* (forthcoming), *Ecological Modernisation*, London and New York: Routledge.

Von Prittwitz, V. (1993b), 'Reflexive Modernisirung und öfentliches Handeln', in V. von Prittwitz (ed.), *Umweltpolitik als Modernisierungsprozeß. Politikwissenschaftliche Umweltforschung und -lehre in der Bundesrepublik*, Opladen: Leske & Budrich, pp.31–50.

World Bank (1992), *World Development Report 1992: Development and the Environment*, New York: Oxford University Press.

Weale, A. (1992), *The New Politics of Pollution*, Manchester: Manchester University Press

Wehling, P. (1992), *Die Moderne als Sozialmythos. Zur Kritik sozialwissenschaftlicher modernisierungstheorien*, Frankfurt/New York: Campus.

Yearley, S. (1991), *The Green Case: A Sociology of Environmental Issues, Arguments and Politics*, London: Harper Collins.

Zarsky, L. (1999), 'Havens, Halos and Spaghetti: Untangling the Evidence about Foreign Direct Investment and the Environment', paper prepared for the OECD Conference on Foreign Direct Investment and Environment, 28–29 January, The Hague (CCNM/EMEF/EPOC/CIME(98)5).

# Lifestyles, Consumption and the Environment: The Ecological Modernisation of Domestic Consumption

GERT SPAARGAREN and BAS VAN VLIET

Ecological modernisation theory has been developed so far mainly with respect to the production sphere. In order to apply the theory to the sphere of consumption, it needs to be enriched with some of the central concepts from the sociology of consumption. What results is a contextual model of (domestic) consumption which combines an actor-oriented approach with a system-of-provision perspective of consumer behaviour. In conclusion, the relevance of the proposed model for research on the ecological modernisation of domestic consumption is discussed.

## I. Introduction

Environmental sociology has become a well-established sub-discipline within sociology. Over the last decade this discipline has grown into maturity: it has diversified regarding the contributing world regions, the research themes covered and the theoretical or conceptual models being discussed. The debate on globalisation helped to broaden the originally western or eurocentric outlook that came along with its mainly American and European origins in the 1970s. The debate on 'constructivism' versus 'realism' made us aware of some of the complexities surrounding the role of science and technology in the present phase of modernity, referred to by many sociologists as 'reflexive modernity' [*Beck, Giddens, and Lash, 1994*]. The theory of ecological modernisation stimulated theoretical discussion within environmental sociology and contributed to making the field more sensitive to developments in environmental policy-making.

For the future development of environmental sociology, the study of consumption and consumer behaviour constitutes an area of strategic importance. However, environmental sociology has not yet been able to offer promising theoretical perspectives on how to conceive consumer behaviour. The search for determinants of environmentally (un)friendly behaviours – grounded in models from social psychology – came close to a deception. The environmental impacts of what consumers actually do turned out to be very complex. Most studies of environmental consumer behaviour, by focusing on separate products or specific activities, provide little opportunity for consumers to monitor their own behaviour, yet this self-monitoring is essential if consumers are to make their lifestyles more sustainable. So there is a need for new directions to be explored. The emerging field of the sociology of consumption indicates one of the more promising routes in this respect. We would argue that the present day stagnation in the approach to consumer-behaviour could be partly explained by the fact that environmental sociologists have not been able to keep up with the developments that have taken place in the wider disciplines of sociology and anthropology. This is especially true for the study of consumption.

This contribution aims at contributing to the further development of consumer studies within environmental sociology. In section II we develop a formal theoretical approach to consumer behaviour, borrowing extensively from the sociology of consumption. We argue that the study of consumer behaviour should start with the social practices in which actors are involved when pursuing their daily routines. With the help of Anthony Giddens' structuration theory we provide a definition of the formal concept of lifestyle and then go on to discuss the notion of a 'sustainable' lifestyle. The notion of 'system of provision' will serve as key notion when it comes to studying consumption practices from an institutional perspective. In section III the formal model will be refined and extended by applying it to 'domestic consumption'. By taking the household context as a starting point, we outline the general route to (studying) the ecological modernisation of domestic consumption. We conclude in section IV with an outline of a future research program in this area, illustrating the kinds of research questions that are generated by our approach.

## II. Towards a Sociology of Sustainable Consumption[1]

The study of consumption and consumer behaviour has to a large extent been left to economists and (social) psychologists. For a long time, sociologists had regarded consumption and the consumer society as phenomena that mainly deserved criticism. This critical attitude to consumption was shared by many environmental social scientists [*Winward, 1994: 75*]. While the legacy of the Frankfurt School can perhaps partially explain this attitude, this factor alone does not suffice to account for the lack of serious research in this area. A more important explanation can be found in the productivist orientation that dominated sociological theories for so many years. Work, factories, labour unions, the division of labour and the role of technology, these were the phenomena that kept sociologists occupied. When consumption was given any treatment at all, it was regarded mainly as a 'derivative' of production.

The bias on production-related social issues is not restricted to academic circles. Also in the field of environmental politics the productivist approach to consumption has dominated. Policy-frameworks for approaching the 'target-group'[2] of citizen-consumers are derived from the frameworks that have been developed with respect to institutional actors in the sphere of production. Consumers are much more heterogeneous and less professionalised when compared with target-groups in the sphere of production. One cannot conclude a covenant with consumers for the simple reason that consumers do not participate in the neo-corporatist consultation circuits that have become prominent in Dutch – and to some extent European – environmental policy making [*Liefferink and Mol, 1998; Lauber and Hofer, 1997*]. Moreover, consumer-groups are not familiar with the highly specialised, differentiated jargon developed by environmental professionals in government and business circles. It is highly unlikely that a consumer-oriented approach can be generated within the existing frameworks of policy. Therefore, we need a new approach.

A sociological model for guiding consumer-research in environmental sociology and serving as a basis for environmental policy making should allow a precise formulation of concepts of 'environmental behaviour' and 'sustainable lifestyles'. We propose a model rooted in the structuration theory of Anthony Giddens and which can be given a more specific meaning for environmental sociology. The formal concepts of structuration-theory will be connected to some of the substantive themes put forward by the theory of the ecological modernisation of production and consumption. This should prevent some of the problems that come along with using social-psychological models on environmental behaviour as well as economic models on consumer-behaviour. Social-psychological models are strong in stressing the importance of the values and beliefs human agents adhere to. They are, however, weak in connecting individual (motives for) action with the 'wider society'. In other words, they lack a proper scheme for analysing the interplay between 'action' and 'structure' or between 'micro' and 'macro' levels. Economic models are used on various analytical levels, but these models do not pay attention to the 'motives' or 'reasons' of citizen-consumers behind a certain pattern of behaviour. Within the economic theory of 'revealed preferences', everything judged an 'irrational' factor is excluded from conceptual schemes. So the task of the sociological model must be to find a solution in which the main pitfalls of both models are prevented.

Within Giddens's structuration theory the analysis of environmental behaviour focuses principally on the *behavioural* or *social practices* in which human agents participate. Individual behaviour and its underlying reasons, interests and motives, is studied in the context of social practices situated in time and space and shared with others. Beliefs, norms and values regarding (environmentally friendly) action are therefore not assumed to exist in a 'social vacuum' – as they are in the social-psychological model – but in a context. They are analysed as the *rules* which 'belong to' a specific social practice that is shared with others. The (relative) 'power' of the actor to change the course of action is specific for a certain context too, depending on the *resources* that are implied in the reproduction of social practices. Within structuration theory, rules and resources together constitute the structures that are involved in the reproduction of social practices.

A formal theory does not indicate the specific sets of social practices that researchers should focus on. The choice is left to the researcher in question. If the focus is on sustainable lifestyles and consumption patterns in a certain country or region at a certain time, one could include those practices that are recognised by the existing policy planning system as having the most relevant environmental impacts. In section III we will provide a set of practices that is especially relevant for the study of domestic consumption.

FIGURE 1
A CONCEPTUAL MODEL FOR STUDYING CONSUMPTION PRACTICES

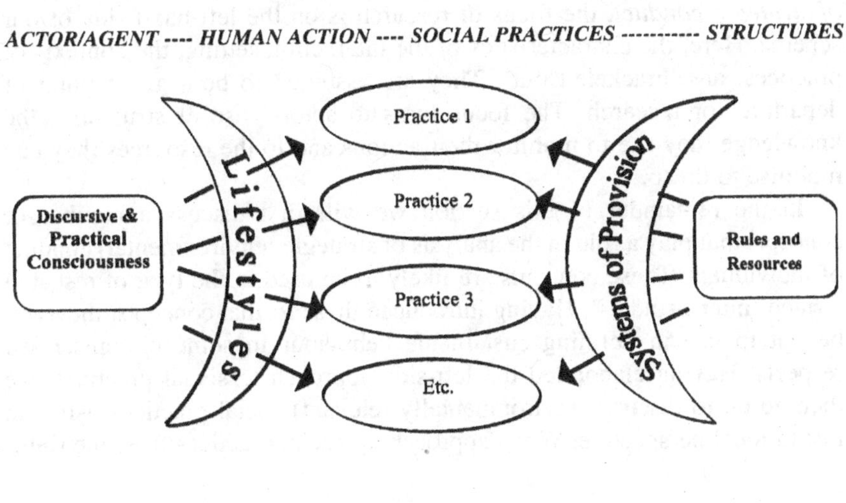

Figure 1 presents a basic 'structurationist' scheme in which (consumption) behaviour is linked to structures that are involved in the reproduction of social practices.

Before going into the relationship between action and structure, we should introduce one of the core notions of structuration theory: the concept of the 'duality of structure'. The idea of the duality of structure refers to the dual character of the rules and resources involved in the (re)production and transformation of social systems. Social systems are sets of social practices, situated in time and space. On the one hand actors are 'forced' in their actions to draw on existing rules and resources. In such cases, structures are 'media' in the sense that they enable a human actor to act. On the other hand these structures are in turn confirmed and reinforced by the actors' very actions. In this sense, structures are both media and outcomes of human action.

Structuration theory has deliberately eliminated the concepts of 'micro-level' and 'macro-level' from the vocabulary it uses to analyse social reality. Throughout the years, too many connotations have become attached to these terms. Micro-processes have become associated with 'subjectivity' and 'freedom of action', whilst macro-processes have been seen as 'objective', as 'structures' that restrain the freedom of action. To avoid the macro–micro terminology, the different lines of approach for research into social practices are referred to as 'institutional analysis' and 'analysis of strategic conduct', respectively. What is at issue here is whether social practices are approached from the 'right' side of our scheme or the 'left' side, as shown in Figure 1. In an *institutional analysis* of social practices the actors' knowledge and skills are 'bracketed out' to focus on the institutions as recurrent reproduced rules and resources. In *the analysis of strategic conduct*, the focus of research is on the left-hand side of our scheme. Here, the characteristics of the interaction setting, the contexts of practices, are 'bracketed out'. They are assumed to be a given point of departure for research. The focus turns to actors' use of structures, the knowledge they use to monitor their actions and to the resources they can mobilise to do so.

In the remainder of this section we will first discuss a number of concepts that play a role in the analysis of strategic (environmental) conduct of individuals. These concepts are likely to be used in the type of research termed 'micro-studies'. Having introduced these formal concepts, they will be put to use in defining sustainable behaviour in some circumscribed respects. Having elaborated the left-side approach to social practices, we then go on to discuss (environmentally relevant) social practices using an institutional perspective. When approaching social practices from the right-side of the scheme in Figure 1, the focus is on the ways in which social practices are embedded in broader socio-technical environments. We will use some of the terminology as developed in the sociology of consumption and the sociology of technology as these concepts may facilitate the institutional analysis of the 'systems of provision' that underpin consumer-behaviour.

*Lifestyles and the Role of Individuals*

The actions performed by individual actors diverge, as suggested in Figure 1, into a large number of distinct social practices, which is not the same as falling apart or disintegrating altogether. The formal concept of 'lifestyle'[3] refers to the specific form of integration brought about by social actors. In their lifestyles, people realise a – partial – integration of the variety of social practices that span their daily lives. Actors 'bind' their distinct (sets of) social practices into a reasonably 'coherent' unity. As a descriptive concept, 'lifestyle' has become synonymous with the classic concept of 'behaviour pattern' [*Mommaas, 1993: 160*]. However, the lifestyle concept does not only refer to the formal process of integration of social practices but also to the 'story' which the actor tells about it. With each life*style* there is a corresponding life *story*, in the sense that by creating a specific unity of practices the actor expresses who he or she is or wants to be. The lifestyle serves to express a person's individual identity, a 'narrative of the self'. Both elements are indicated in Giddens' definition of lifestyle: 'A lifestyle can be defined as a more or less integrated set of practices which an individual embraces, not only because such practices fulfil utilitarian needs, but because they give material form to a particular narrative of self-identity' [*Giddens, 1991: 81*]. Lifestyles refer to the degree of coherence to be found in people's behaviour. The notions of integration and coherence are important because modes of action followed in one context may reasonably differ from those adopted in others. Giddens refers to this phenomenon in terms of different lifestyle segments or lifestyle sectors. 'A lifestyle sector concerns a time-space "slice" of an individual's overall activities, within which a reasonably consistent and ordered set of practices is adopted and enacted' [*Giddens, 1991: 83*]. If a person wants to maintain a certain level of credibility, both for herself and for others, then a certain coherence of the lifestyle and a certain level of integration of actions throughout varying practices will become essential.

*The Concept of Environmental Utilisation Space*

The formal concepts of Giddens' structuration theory need to be connected to the notions of ecological modernisation theory to say something about *sustainable* lifestyles. Until recently, ecological modernisation theory has focused mainly on institutional analysis. The theory is built on the assumption that during the past decades an environmental induced transformation process has emerged. It has led to new 'rules of the game' for the social organisation of production and consumption.

Ecological modernisation, on the one hand, refers to a sociological theory that has chosen the long-term transformation of western society as its object of analysis. Conceived as a 'political programme', on the other hand, it refers to the central perspectives and strategies of environmental policies of the Netherlands and other industrialised countries [*Spaargaren and Mol, 1992; Weale, 1993; Hajer, 1995*]. The nucleus of the theory is the view that new, environment-induced sets of rules and resources constitute *independent criteria* – besides cultural, economic or political criteria – to judge individual and institutional action. In the process of ecological modernisation, the criteria by which an assessment is made of what may count as an ecologically more 'rational' mode of production, gain increasing independence vis-à-vis socio-economic, political or cultural criteria.

An initial analysis of strategic conduct of citizen-consumers – aimed at producing criteria for 'more ecologically rational ways of acting' – can benefit from the lines of thought that have already been developed at the institutional level. For example, the meaning of the concepts 'environmental management' and 'environmental control' in the production sphere may be used in a rather similar way in the sphere of consumption. Environmental management or control by citizen-consumers can be defined as the conscious effort to achieve a reduction in the environmental impacts associated with specific lifestyle characteristics. This effort is based on the acknowledgement of a 'flexible ceiling' covering the total environmental impact of a lifestyle.

The notion of a 'flexible ceiling' or a certain amount of 'environmental utilisation space' requires some further elucidation. The recognition of the fact that there are 'physical limits' to the 'sustenance base' on which the social organisation of production and consumption rests, is combined with the idea of these limits being 'flexible' to a certain extent. The ceiling is set partly by technological and ecological criteria which refer to 'the functional components of eco-systems that play a crucial role in the sustainability of social systems' [*Opschoor and Van der Ploeg, 1990*]. Some of these criteria should be treated as 'core' or 'hard' criteria for sustainability in a circumscribed, eco-technical way, referring ultimately to issues of survival of both ecosystems and social systems. Some other criteria are (more) open to political debate, such as 'quality of life' and 'integrity of nature'. Given this definition, the idea of environmental utilisation space can never be fully operationalised by means of one-dimensional technical-scientific concepts and analytical frameworks. This is true for the application of the idea in the sphere of production. It is equally valid, and perhaps even more so, for application in the analysis of more sustainable lifestyles and consumption patterns.

In policy practice, however, the theoretically nuanced concept of environmental utilisation space is often used in a very limited way. The experiments conducted with the idea of an all-encompassing 'environmental ration card' for citizen-consumers reveal an empiricist reduction of the concept. This is not only unfeasible given our current knowledge, but politically undesirable as well. To our view, the value of the idea of environmental utilisation space lies in its expression of the 'rationale' behind a reduction in the environmental impacts within certain production organisations, consumption patterns or lifestyles. It is a characteristic feature of this rationale that it combines the idea of physical limits with the idea of actively making use of a certain amount of space. Contrary to a concept such as 'limits to growth', the idea of environmental utilisation space evokes the image of available space that can literally and legitimately be 'made use' of.[4] The idea thus refers to the different ways in which environmental resources are being utilised (efficiently or inefficiently, prudently or recklessly) within limits that are set both by (eco)politics and (eco)technology.

The development of more sustainable lifestyles implies that actors (are made to) reconsider all the distinct lifestyle segments or sectors from an environmental management perspective as described above. In doing so they create an 'environmental profile' of the different segments of their lifestyles. The efforts which actors make regarding the greening of their lifestyles may contribute to the level of integration or coherence of their lifestyles. In this process, an optimum distribution or 'rate of exchange' will be pursued by social actors. He or she will look for the optimum exchange rate between the economic, ecological, cultural and social capital – to borrow some concepts introduced by Pierre Bourdieu – that the actor has at his or her disposal at a certain field of action. The assessment of the optimum use of available environmental utilisation space may even result in a process of creative bookkeeping and housekeeping. Actors try to achieve the most beneficial distribution between the different segments of their lifestyles. In an analogy to the sphere of production, 'rational action' is therefore no longer determined by economic criteria alone. The rationally calculating citizen will be just as keen on avoiding environmental risks or on making 'environmental profits' as she is on realising economic benefits or enhancing status.

So much for the approach to social practices from the perspective of the left side of our scheme. To complete our discussion of the scheme, we now turn to analysing the ways in which lifestyle-choices are connected with relevant institutions of society.

*The System-of-Provision Perspective of Consumer-Behaviour*

If lifestyle- and consumption issues are discussed within the sociology of consumption, there are at least two distinct ways in which these concepts can be loaded. The two major branches within the discourse on lifestyles and consumption can be referred to as the horizontal 'distinction perspective' on the one hand and the vertical 'system-of-provision perspective' on the other. From these two perspectives, the vertical perspective in particular can serve well in a sociological analysis of consumption behaviour that stresses the importance of the interconnectedness of production and consumption. Systems of provision should, however, not be analysed from a top-down, productivist orientation but also from a consumer-oriented perspective. We will shortly introduce the 'distinction perspective' before we elaborate upon the 'system-of-provision perspective'.

*The distinction perspective*: Perhaps because of the overriding influence of the work of Pierre Bourdieu, some authors almost exclusively relate lifestyle to the stylisation of daily life. People seek to express their sense of style by a specific choice of (cultural) goods. Lifestyle-choices are a matter of (good or bad) taste, and especially the (lack of) understanding that people have of higher culture or art is regarded as decisive for their performance in the field of distinction. This approach to consumer behaviour is focused on aesthetics, fashion, identity, signs and the dream-world of the modern shopping centres. If concrete products are discussed, the central objective is to establish the ways in which they can bring about a specific form of 'stylised awareness in matters of consumption' [*Lury, 1996*]. Elaborating upon the work of Bourdieu, Featherstone [*1991*] claims that 'the new conception of lifestyle can best be understood in relation to the habitus of the new "petite bourgeoisie" as an expanding class which seeks to preserve and legitimise its own particular dispositions and lifestyle. A class fraction most closely involved in symbolic reproduction' [*Featherstone, 1991: 84*]. These groups set themselves up as the cultural intermediates that promote a general interest in style by 'estheticising' everyday products and commodities. Additionally they popularise commodities from the 'higher culture' to an increasing extent, thus make them available to larger groups of consumers.

The Bourdieu-inspired streams of thought in the sociology of consumption are important for environmental sociologists because they make us aware of the crucial importance of the social or symbolic dimension of consumption. People do not only buy certain products because these products can be used and 'enjoyed' when being 'consumed'. People 'use' goods and services, according to Douglas and Isherwood [*1979*], to relate themselves to other people. To understand why things matter to people, we also have to look beyond the matter itself. The first interest of environmental sociologists is usually on the ecological characteristics of certain products and consumption styles. However, they should also recognise the ways in which these ecological characteristics are shaped by the subjective dimensions of the consumption process.

*The system-of-provision perspective*: The approaches emphasising only the symbolic function of products and services, however, fail to see that changes in consumer culture and consumer behaviour are to an important extent connected with major changes in the organisation of (different sections of) industrial production. The 'horizontal' approach to the dynamics of consumption does not consider the 'history' of the products, their context of origin or their roots in the specific organisation of production-consumption-cycles. Fine and Leopold argue that these contexts of origin should be given a prominent place in what they label as the 'vertical approach' to consumption [*Fine and Leopold, 1993*]. We prefer the label 'system-of-provision perspective' to refer to the vertical dimension because the term provision (and, as we will see in section IV, its corresponding concepts of 'access' and 'use') is more easily connected with a consumer oriented perspective on production-consumption cycles.

This system-of-provision approach 'expects different commodities or groups of commodities to be distinctly structured by the chain or system of provision that unites a particular pattern of production with a particular pattern of consumption'. Empirical research to the changes that took place in product-groups and their corresponding systems of provision, enables us to evade generalist theories which do not recognise the differences between separate clusters of commodities. Existing consumption theories do not pay sufficient attention to differences in the ways commodities are handled in for instance the traffic system, the food system, the energy system, or the housing system. The 'systems-of-provision perspective', makes the 'distinct set of imperatives governing different sets of commodities' visible [*Fine and Leopold, 1993: 4*].[5]

Some systems of provision are very stable and predictable, while others are highly dynamic and unstable. This kind of diversity may be approached by introducing some further theoretical concepts, developed by Alan Warde and others in their contribution to the sociology of consumption. Warde has positioned himself in a debate that started in the 1980s among British sociologists putting forward some of the issues of urban sociology in the sociology of consumption. The organisation of 'collective consumption' and its relation to social inequalities was restated with the 'privatised' versus the 'socialised' mode of consumption with a 'consumption cleavage' as a result. In particular, the contribution of Peter Saunders [*1987*] triggered much discussion, because he counterposed the two modes of provision in a rather specific way. The privatised mode was identified with freedom of choice, high quality of products and services and enhanced autonomy for the consumer. The socialised mode on the other hand came to represent lack of choice, poor quality and bureaucratic (over)regulation.

We are dealing here with a highly politicised discussion with initially little room for nuance. Writing about the market context, Saunders does not spend a single word on the social determinants of consumption behaviour, the market power of producers, or the negative effects of the 'ideology of consumerism'. He simply starts from the premise that 'market choices give meaning to individuals' life through consumption' [*Warde, 1990: 231*]. When, on the other hand, the subject changes to public services, the stress is on dependency, lack of options, inferior quality and obligation. Saunders considers state-provision as 'rest category' in comparison with the market supply of goods and services. To a large extent, the model is derived from the handling of one kind of product, namely the house. Private ownership of a house is supposed to be one of the indicators for the existence of a consumption gap, the dividing line between the new social classes within the consumption culture. Empirical research showed that these new dividing lines would also express themselves in the preference for particular political parties [*Dunleavy, 1980; Saunders and Harris, 1990*].

The superiority of the privatised mode is arguable even in the housing sphere [*Kemeny, 1980*]. One should realise that what is investigated here is indeed an important, yet single consumption good. Private house ownership cannot function as a model for the way the multiform package of goods and services is handled. When commodities other than houses are at issue, the linking of 'ownership' and 'control' and the relation to 'quality' becomes less unambiguous. Another objection to the Saunders model is that it distinguishes only two major forms of consumption. It neglects consumption in a household context as well as the handling of consumption goods in the 'informal' economy of neighbourhood and family networks. Finally, consumption is more than the purchase of goods alone. There are many different ways of supplying products, maintaining products, and disposing of products that are not covered by a bipolar model like that of Saunders.

Alan Warde can be credited for breaking through the dichotomy of 'privatised' versus 'socialised' modes of consumption and making room for analytically more fine-tuned analyses of different modes of provision. Analysis of production-consumption cycles within a systems-of-provision approach allows for mixed forms or modes of provision, accompanied by different modes of access as well as different modes of 'use' or 'enjoyment'. In the next section, we apply Alan Warde's concepts to the dynamics of household consumption.

## III. Analysing Domestic Consumption

Now that we have explored a contextual model of consumption with help from the notion of a duality of structure and the systems-of-provision approach, in this section we use these concepts to elaborate our understanding of the dynamics of domestic consumption. The latter is in many ways broader than consumption alone, as it encompasses many more social practices, including household labour and care. The household encompasses many practices that relate human agents to the use of natural resources, which makes it extremely relevant to study from the point of view of environmental sociology. We will start with discussing the work of Schwartz Cowan that is in many ways in line with a systems-of-provision approach to consumption. By referring to Per Otnes's analysis of household consumption, we will also show that the notion of duality of structure is extremely valuable when it comes to analysing household practices. Before turning to the ecological modernisation of household consumption in the remainder of this article, we end this section with two main issues that seem to be conditional for any change in household practices: those of time-space structures and the standards of comfort, cleanliness and convenience.

### *Industrialisation of the Household*

In her impressive study on the industrialisation of American households, Ruth Schwartz Cowan convincingly illustrated that one cannot treat domestic consumption in isolation from the sphere of industrial production.

> [T]he industrialisation of the home appears to have been composed of millions of individual decisions freely made by householders: the Jones's down the block decided to … buy a washing machine, and the

> Smiths around the corner fired the maid and bought a vacuum cleaner. But the matter is not as simple as that. The Jones's washing machine would not have done them a bit good if the town fathers had not decided to create a municipal water system ... earlier, and if the local gas and electric company had not got around to running wires and pipes into the neighbourhood [*Cowan, 1983: 13–14*].

Cowan made clear that any changes in the way we cook, dress and care are directly connected to changes in the food-system, the clothing-system and the health-care-system. And although some of the major 'women's' tasks were indeed lifted out their domestic context and transmitted to industrial sectors of society, Cowan shows that we are not dealing with a one-way process. Where the gas, electricity and water-systems are involved, some of the 'old' chores (drawing the water) were replaced by new ones (cleaning the bathtub). In the case of the transport-system, Cowan showed that women gradually replaced the transport-services that used to be conducted by the butcher's boy, the baker's man, the doctor and the barber. Because she observed similar processes in other sectors as well, Cowan concluded that the industrialisation process eventually implied 'more work for mother'.

For a number of reasons, the methodology adopted by Cowan can be said to offer a promising perspective also when studying domestic social practices from an environmental point of view. First, she takes a medium to long term perspective. This enables us to think through alternatives to the present socio-technological systems that seem to have a 'fixed' character when viewed only from a short-term perspective. Examples are the centralised water and energy networks that were build up during the last century in most western countries.

Secondly, Cowan lays great stress on the interconnectedness of so-called 'internal' and 'external' processes. She cross-cuts the divisions between the sphere of production and the sphere of consumption. When working with a strict division between these spheres, the meaning of concepts like 'work' or 'industry' tend to be biased because they are thought to be rooted only or primarily in the sphere of production.

Thirdly, Cowan theoretically emphasises and empirically illustrates the very intricate connections between so-called 'technical' and 'social' aspects of domestic social practices.

Finally, Cowan makes us aware of the fact that in the historical development of socio-technological systems, some roads are indeed taken and some others are not. Technological systems and devices sometimes 'fail', not necessarily because they are technically spoken inferior but because they do not fit the diverging interests of companies, trade unions and households at that crucial moment of its emergence.

Cowan's perspective fits very well in the approach developed in the field of the sociology of technology [*Bijker, Hughes and Pinch, 1987*]. With this tradition she shares both some of its innovative and stimulating insights as well as some of its shortcomings. The strong elements are that 'technological systems' are treated as inherently social in character, consisting as they do of both people and artefacts. This approach to technology pays especially close attention to the interrelation between individual elements or entities (artefacts, skills, natural phenomena and also individuals) on the one hand and the technological system as a whole on the other. The shortcomings of the sociology of technology approach seem to apply especially to the system-approach within this tradition. There is a lack of concepts that could describe the role of human agency in the development of technological systems. Considerable attention is given to the culture of the technological system, its 'style' and the way in which the cultural matrix 'guides' technological development. But the role of human agency is not dealt with as something that could explain the dynamics of systems of provision.

The contribution of Cowan is one of the very few examples within this tradition that at least recognises the need to study technological networks also from a consumer-perspective. Cowan uses the word 'consumption junction' to refer to the need to study systems of provision also from a consumer point of view. The consumption junction

> is the place and the time at which the consumer makes choices between competing technologies, and tries to ascertain how the network may have looked when viewed from the inside out, which elements stood out as being more important, more determinative of choices, than the others, and which path seemed wise to pursue and which too dangerous to contemplate [*Cowan, 1987: 263*].

This is the best one can get from the sociology of technology when it comes to conceptualising human agency and its role in technological development.

To take the analysis one step further, we have to deal with questions about the ways in which consumer-choices about different technological systems are connected to consumers' lifestyle-choices and issues of identity. We should recognise the importance of the relationship between 'inside' factors (such as the internal time-space organisation and 'cultural style' of the household) and 'outside' factors (the external systems of provisions like the electric or power networks).

### *The Duality of Structure in Reproducing Domestic Consumption Practices*

As we have argued more extensively in section II, a fruitful way of conceptualising the interplay between agency and structure can be found in the work of Anthony Giddens. The Norwegian sociologist Per Otnes applied some of the basic concepts of Giddens' structuration theory to the practice of household-related consumption. The core concept of the 'duality of structure' is given proper expression in Otnes's phrase that domestic consumption can be summarised as a process of 'being served by, and serving, a number of essentially collective socio-material systems' [*Otnes, 1988: 120*]. Every time we tap some water or switch on the lights, we are making use of the services that are provided by expert-systems. At the very same moment we contribute to their ongoing reproduction. These expert-systems constitute what Otnes calls 'the collective underpinnings of private life' [*Otnes, 1988: 120*]. Without the proper functioning of these expert-systems our daily life would be very difficult to maintain [*Otnes, 1988*; also *Hughes, 1983*]. The collective socio-material systems are a special case of systems of provision. They are special because they imply 'material infrastructures' as an essential element of systems of provision. Once the citizen-consumer has been 'connected' to the water works, the sewage system or the electricity grid, the consumer has become a 'captive' consumer. Captive consumers cannot just shift from one system to another without losing resources (money, knowledge, skills) that have been invested in the present networks. These so called 'sunk costs' prevent citizen-consumers from moving freely between different systems of provision, provided that there would be more than one option available.

Although we 'know' in general about our dependency on collective socio-material systems, we use water and energy in a very routine way. We are not consciously aware of the fact that behind the taps and the power-points there are very advanced and extensive expert-systems. We are only made aware of the character of this relationship when confronted with some event that (temporarily) breaks down our routines. This can be an interruption of the guaranteed supply of water and energy or a removal or home-improvement project which we decide upon ourselves.

In all these cases the taken-for-granted character is replaced by a serious questioning of the existing modes of organising the social practices that constitute our daily actions. These interruptions of daily routines make us (temporarily) alert and very sensitive to consider alternative modes of organising our consumption practices. When going through periods of de-

and re-routinisation, we become aware of the level and the nature of what Giddens calls our 'discursive penetration' of the collective socio-material systems. This awareness can pertain for example to the skills that are needed to serve appropriately the systems, or to the power-relations that are involved in the reproduction of the system. The discursive awareness might also be directed to the environmental impacts that characterise the present make-up of the system.

Studying domestic consumption from a perspective as sketched above has relevance for environmental sociologists for at least one reason. Collective socio-material systems that we serve and by which we are being served, represent to a considerable extent the 'sustenance base' of our daily domestic routines. They are 'material' systems not only in the sense of being technical systems, with technologies, infrastructures, apparatuses, tubes and gadgets. On a more fundamental level they represent the material- and energy-flows which form the physical substratum to our domestic lives. They are involved in the organisation of our intercourse with the environment.

So when people are striving for more sustainable lifestyles and patterns of domestic consumption, the possibilities (not) offered by collective socio-material systems are of strategic importance. When high levels of environmental consciousness meet low levels of 'green innovation' of systems of provision, the result will be a lack of environmental friendly behaviour. On the other hand, domestic agents will only accept more sustainable devices in the field of energy and water under the condition that the devices 'fit' into the overall organisation of their households and lifestyles. There are two main issues concerning household organisation that need some more elaboration: the time–space structure of domestic routines and the accompanying cultural standards. We will discuss both factors separately.

*The Time–Space Structure of Domestic Social Practices*

One may wonder why people accept new technologies, or buy certain products or types of equipment. Douglas and Isherwood [*1979*] suggest as a working hypothesis that they do this to increase their 'personal availability'. This offers a more promising starting point than theories explaining purchasing behaviour from the wish to satisfy needs or the urge to show off, or similar motives. Products are employed to achieve a reorganisation or rationalisation of the household. The products help consumers to set themselves free to carry out other tasks or engage in other activities. To the anthropologist Douglas, the desire to liberate oneself from household tasks is strongly linked to the necessity to be available for activities that increase the status of actors. Status may be obtained from an increase in social capital like visiting friends, organising a big party, and so forth. This specific application of the motive behind the increase of 'personal availability' is less important to our goal than the working hypothesis itself.[6]

An important characteristic of household labour is that it consists in large part of periodically recurring activities or jobs that cannot be put off. Douglas and Isherwood illustrate these 'periodicity constraints' of households by means of two types of activities, without further differentiating these two types analytically. On the one hand there are the household routines better known as 'chores': making beds, cooking meals, cleaning the bathroom, shopping. On the other hand there are activities relating to child care and upbringing: breast-feeding, taking the children to school, caring for an elderly parent living in your home, walking the dog. These are activities that have to be performed according to a fixed pattern in time. As partly complementary activities together they result in a more or less constraining 'pattern of periodicities of household processes'.

Along with the concept of the 'periodicity of household labour', Douglas and Isherwood introduce two further assumptions. First, constraints are held to differ with social class or status groups, according to the principle that 'the higher the status, the less periodicity constraints; the lower the status, the greater periodicity constraints'. Secondly, they are directly linked to the division of roles between men and women. '[T]he most general account of the division of labour between the sexes that fits everywhere would be based on the periodicity of women's work' [*Douglas and Isherwood, 1979; 120*].

High-frequency tasks limit the action radius of people – in this case mostly women – in their efforts to arrive at a greater 'personal availability'. As a consequence, women will try to get burdened with as few of these tasks as possible. Douglas and Isherwood are rather positive in their formulation when they observe that 'anyone with influence and status would be a fool to get encumbered with a high-frequency responsibility' [*1979: 120*]. If possible and affordable, people will try to give these tasks to third parties (baby-sitter, domestic help, window cleaner, etc.) and try to reduce the claim on personal time by 'rationalising' the labour involved, through the application of technology. People make use of commodities and services to free themselves from the most demanding tasks that belong to their specific consumption scale. For this reason, 'scale-facilitating goods' form the key to understanding when we are looking for the relationship between technology and household consumption.

The essential point in the analysis of Douglas and Isherwood is that actors, in their efforts to attain a higher personal availability, are trying to free themselves from 'high-frequency-low-esteem' tasks. To put it more generally: it is about the freeing of domestic agents from fixed time-space slots in everyday life. Agents will not accept technologies that make their domestic routines more 'rigid' in terms of pre-given schedules. They will instead actively strive for those technological and organisational devices that give them greater flexibility in performing domestic tasks. One should remember this when studying more sustainable goods and services offered to domestic consumers. For example the idea of 'pooling' of equipment and resources as an environmentally more beneficial alternative to fully privatised goods and services, may contradict the idea of flexible time-space structures.

*Standards of Comfort, Cleanliness and Convenience*

The perspective derived from the work of Cowan can be regarded as more promising than most of the models used by neo-classical economists because now the focus is no longer on 'isolated choices for isolated products' but instead on 'sets of goods and services' and systems of provision. These sets of goods and services are then connected with specific drives or motives from the side of domestic consumers. In fact this is the kind of approach that Per Otnes has been arguing for [*Otnes, 1988*]. He envisages a social scientific discipline that pays sufficient attention to the spending, the use and the making of products or services rather than their utilisation alone. To give a name to the discipline Otnes proposes 'chreseology of everyday life', which can be interpreted as the doctrine of using, employing and implementing [*Otnes, 1988: 163–4*].

Such sociology of domestic consumption relates specific clusters of products and services to specific modes of provision in the context of specific spatial settings of domestic social practices. Examples of these settings are the kitchen, the bathroom, the hobby-room, or the garden. These are the spatial contexts in which actors make use of cooking and gardening tools that can only (be made to) perform properly when connected to collective socio-material systems. The chreseology of consumption should provide detailed descriptions of the daily routines of domestic consumption. This can enhance our knowledge about the various ways in which people relate to technologies, households relate to expert-systems and the 'private' relates to the 'public'.

Writing in the sociology of technology tradition, Otnes does not investigate in any detail the distinct motives and interests of human agents as to answer the question why they are performing these consumption

FIGURE 2
STANDARDS OF COMFORT CLEANLINESS AND CONVENIENCE MET BY INFRASTRUCTURES, OBJECTS AND CONVENTIONS, USES AND PRACTICES

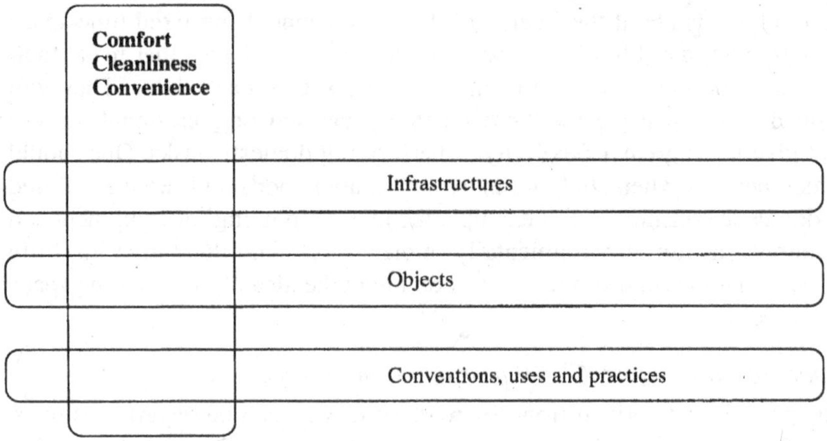


*Source*: Shove [*1997*].

practices. Of course, in their modern, rationalised and stylised kitchen people combine cooking utensils with energy and water to produce a meal. But this obvious fact does not provide us with any insights into the taste for food that comes along with their lifestyles and the cooking-culture which these specific domestic agents represent. Elizabeth Shove, among others, put the cultural aspects of domestic consumption at the centre of analyses [*Shove, 1997; Shove and Southerton, 1998*]. She specified them with standards of 'Comfort, Cleanliness and Convenience' to which people adhere and subscribe. Domestic consumption practices should be analysed not only as a mixture of objects and infrastructures but also as routines that make people meet the standards they think of as 'normal' or 'minimal required' or 'common'. To meet these standards, Shove proposed to categorise the rules and resources as infrastructures (the plumbing system for instance), objects (commodities) and conventions, uses and practices. Figure 2 represents the perspective as proposed by Shove [*1997*].

Shove's analysis is different from the one offered by Douglas and Isherwood. She does not treat the 'rationalisation' of domestic consumption primarily as a process of 'doing away' with as many chores as possible. Shove discussed the buying and selling of freezers, refrigerators, bathtubs, kitchens and heating systems in terms of the level of comfort, cleanliness and convenience that these technologies promise to bring to the household, provided that they are properly connected to the systems of provision that support them. On the one hand, these standards are an expression of the specific lifestyles of people. In that sense they are a highly individual, privatised affair. But the number of showers one takes in a week depends not only on the job one occupies or the amount of sporting activities one is engaged in. It is also to a large extent determined by the standards of cleanliness one is accustomed to. These standards are shared with many others and 'learned' in the course of one's life. It is especially the anthropology of consumption that made us aware of the variation that exists both across time and across different cultures when it comes to standards of comfort, cleanliness and convenience.[7]

So far, we have discussed household consumption with tools derived from structuration theory and the systems-of-provision approach. Additionally, we have argued that any attempt to innovate or change households – from within or from outside – should acknowledge the way these changes fit in the time-space structure of households and the standards of comfort, cleanliness and convenience to which householders adhere. We turn now to the ecological modernisation of household consumption.

## IV. The Ecological Modernisation of Domestic Consumption: Outline of a Research-Program

As pointed out in section II, the essence of ecological modernisation is the 'emancipation' of ecology from culture, politics and economics. Huber [*1982*] situated ecological modernisation as a third phase of industrial development, beginning in 1980, following earlier stages of the 'breakthrough', to 1948, and 'construction' of industrialisation, from 1948 to 1980, respectively. Every new cycle and the social transformations that go along with it, is made possible by a new key-technology to arise. Such new technologies have to be picked up by innovative entrepreneurs who bring about a new wave of industrial innovation. Such an evolutionary theory of social change is of course vulnerable to the same kind of criticism that was formulated with regard to all other modernisation theories.

However, in its essence, the theory had two particularly attractive features. First, it analyses major changes in the organisation of production and consumption in a direct connection with environmental problems. Second, it puts at the centre of attention the institutions that are most important in bringing about the switch over into more sustainable production and consumption cycles: economy and technology [*Spaargaren, 1997: 17–18*]. The ecological modernisation of domestic consumption would stem from the same basic principles: the emancipation of ecology out of the economic and cultural spheres; the reconstruction of domestic practices after a long period of 'industrialisation of the household'; and the introduction of environmental innovations taken up by capable and knowledgeable domestic consumers.

Environmental innovations can only be defined in relative terms, as what is innovative depends on the temporal and spatial context in which they emerge. Innovations might be accomplished in techniques, but also in procedures, financial arrangements, and all other institutions that make the production-consumption-cycle go round. They deserve the adjective environmental if they represent a potential reduction of environmental impact provided that the innovation is applied in the proper context [*Van Vliet, 1998*]. This, however, does not exclude other goals that may be accomplished by the innovation like 'economically more efficient' or 'bringing more comfort'.

It is also important to realise that environmental innovations might come to domestic consumers through all possible modes of provision. Green taxes will mostly be provided through a public mode of provision, while green-labelled food products will mostly be provided through private modes of provision. Some innovations, however, are typically provided through informal modes of provision like the sharing of cars or washing machines with neighbours or neighbourhood-composting schemes. The ecological modernisation of domestic consumption, in other words, is more than only a derivative of ecological modernisation in the sphere of production. In fact, many environmental innovations that have been introduced by private companies and public utilities (such as organic food products, green electricity schemes or grey water systems) during the last decade are said to be primarily initiated by the wish of the consumer.

We will conclude our argument by discussing a more detailed version of the model introduced in section II, followed by an indication of the route to be followed in future research concerning the ecological modernisation of domestic consumption.

Figure 3 presents our model again, with examples in the centre of the figure of the kind of social practices relevant to the study of domestic consumption. The 'formal' model of social practices within the duality of structure is given 'empirical' content by introducing environmental innovations that are developed in the production-consumption cycles in modern industrial societies. These environmental innovations in the form of techniques, procedures or social arrangements might be introduced in the cycle of production and consumption by different modes of production and provision and may be accessed and used also in various ways.

FIGURE 3
OUTLINE OF A RESEARCH MODEL FOR STUDYING THE ECOLOGICAL MODERNISATION OF DOMESTIC CONSUMPTION

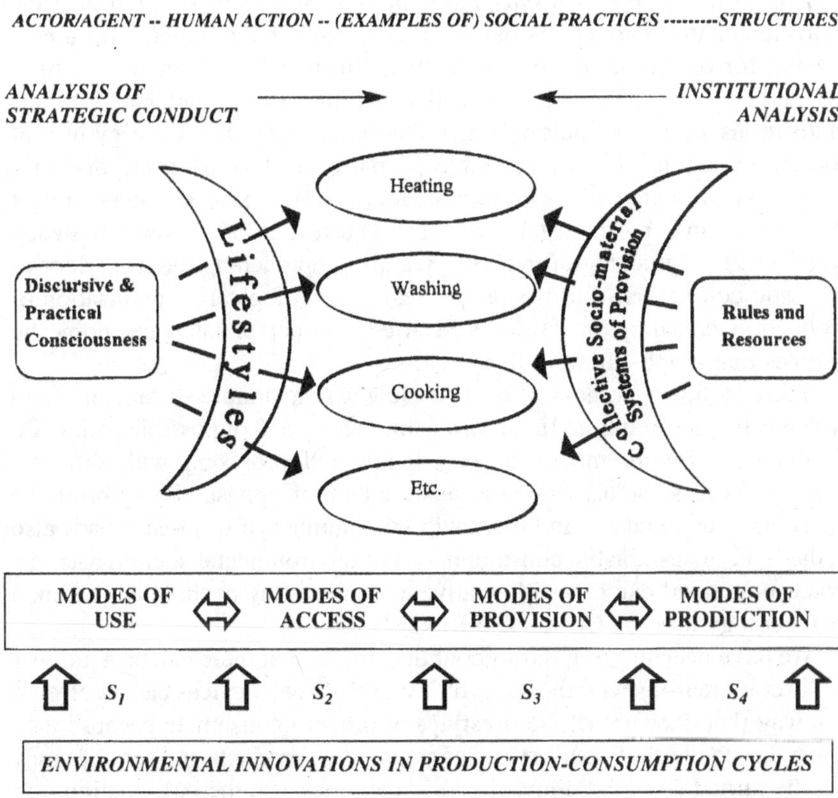


On the left side of the model, human actors – aiming at a reduction of the environmental impacts of their lifestyles – are dependent on the environmental innovations made available to them through the systems of provision. On the right side of the model, companies, public utilities and governmental agencies involved in the development of more sustainable goods and services are dependent on human actors. They have to recognise environmental innovations as relevant 'tools' that fit their lifestyles and their internal domestic organisation as well as their specific standards of comfort, cleanliness and convenience.

We will argue that over the last 15 years or so the number of environmental innovations that became available for domestic consumers has increased significantly, at least in a number of European countries [*Raman et al., 1998*]. Not only have the numbers increased, but also the character of the innovations has been changed for the better. Consumers looking for opportunities to 'green' their lifestyles and domestic routines are getting 'served' much better than in the 1970s and early 1980s. Innovations which can help to lower the direct and indirect energy use[8] of households are no longer accessible primarily to a small group of 'early innovators' who are willing to sacrifice lots of time, space and money to get things done in a more sustainable way. The results of a slow but steady process of the 'greening of industry' are also approaching the front door of domestic consumers of all social classes. The ecological modernisation of production–consumption cycles will affect, sooner or later, the domestic routines that shape our daily lives.

There is, however, not one best possible way to a more sustainable form of domestic consumption. In fact we witness a myriad of possible pathways, combining different modes of production and provision with different modes of access and use. We now seem to be in the phase of 'exploring the roads that can be taken' and there will be a number of dead-end roads also in the field of sustainable consumption. For environmental sociologists, the 'roads not taken' can be as illustrative and interesting as the roads taken, a methodology used by Cowan and Bijker before.

We have been arguing throughout this article that there can be a number of different reasons why things 'won't work'. Green devices can be offered in a way that does not fit the lifestyles of modern consumers because they effect the standard of comfort in a negative way. Products and services may also be turned down by domestic consumers because the introduction, use and maintenance of the devices do not fit the time-space routines of modern social life. Sometimes (green) products are already a failure at the production line because they are designed and produced from a narrow engineering perspective. Or well-designed products which in principle would fit modern lifestyles fail because of their mode of provision, when consumers are not acquainted with the channels used or when they have lost their trust in the providers. All these theoretical possible reasons behind the 'innovation-failures' can and should be explored in concrete empirical research focused on the ecological modernisation of domestic consumption.

Such empirical research might have as its focus the environmental innovations themselves, one or more systems of provision where innovations emerge or one or more household practices that are expected to be changed after innovation. For example, the application of solar energy could be studied by starting with exploring the history of its development, its potential impact on existing energy providing systems they substitute and the role of social actors such as manufacturers, energy utilities, and local governments in speeding up or slowing down the diffusion of solar energy devices. This should provide insight into how and why household consumers might get involved in solar energy and which modes of provision and production there are to adhere to. Focussing on systems of provision, one might study the chances for ecological restructuring of the energy system and the role of domestic consumers, who through their shared social practices partially define the course of development within these systems of provision. If the ecological restructuring of energy sectors is accompanied with the diffusion of solar energy, the study might emphasise the most appropriate modes of access and uses there are for domestic consumers to get involved in solar energy schemes.

We would propose the use of the notion of 'slots' in doing such empirical research. A slot refers to the extent to which innovations would fit in household contexts. In Figure 3, they are indicated as S1 to S4. They represent a lifestyle-slot, indicating the margins in between which renewal of lifestyle segments would affect other lifestyle segments; a slot for domestic time–space structures, which indicate the extent to which changes in this balance are likely to be accepted; a slot for the standards of comfort, cleanliness and convenience and lastly a slot for the modes of provision and production. Under the premise that one does not use them in a too mechanistic way, the notion of slots can be preferred to the classical social-psychological explanation of the 'failure to adopt' from the side of the consumer, as it is not only the consumer's (lack of) 'environmental consciousness' that explains stagnation in the modernisation process.

Finally, most studies in the sphere of the greening of consumption focus on the stagnation and the lack of innovation, while in general studies on the greening of production (also) address the (partial) successes in the modernisation process. By introducing the notion of slots we also run the risk of being focused too much on the failures. It should be noted however, that slots refer not only to situations of non-fit but to situations of 'fit' as well. When, for example, the lifestyle-slots of green devices are discussed, the focus can be on the way in which these device can contribute to the 'narrative of the self' of the domestic consumers. In a recently finalised research project called DOMUS (Domestic Consumption and Utility Services),[9] we explored which country- and sector-specific routes towards the greening of domestic consumption are being pursued and what kinds of consumer-involvement at different levels have been developed so far. While we would argue that a process of ecological modernisation of domestic consumption is emerging at least in some European countries, one cannot yet make firm predictions on what these future roads look like. As Cowan showed us, this can be assessed properly only 50 years from now. Meanwhile, environmental sociologists should keep on doing empirical research in this area, perhaps revising Cowan's statement:

> The ecological modernisation of the home appears to have been composed of millions of individual decisions freely made by environmentally conscious households: the Jones's down the block are considering a solar energy roof and the Smiths around the corner did away with their car and their dish-washing machine. But the matter is not as simple as that. The Jones' and Smiths' individual choices would not have done them a bit of good if the town leaders had not decided on sustainable building programs and call-a-car projects several years earlier, and if the electricity and water companies had not got around to connecting local solar systems to the central grid and running grey- and rainwater pipes into the neighbourhood.

## NOTES

1. A more extended version of the argument presented in this section can be found in Spaargaren [*1997: Chs. 5 and 6*].
2. The concept of 'target group' has been developed within Dutch environmental policy planning to refer to groups which are 'targeted' in a specific way on behalf of policy-makers because they have a distinct environmental and social profile. Examples of target groups include farmers/agricultural sector; energy sectors; different sectors of industry; small- and medium-scale enterprises and also consumers.
3. Mommaas has indicated that the term lifestyle or style of living is as old as sociology itself and was given a central place in the works of Veblen, Weber and Simmel, in particular [*Mommaas, 1993: 159–81*]. Giddens only began to use the lifestyle concept in his later works, as part of his structuration theory on the one hand and as a 'hinge' between the formal theory and his discussion of (late) modern society on the other. We will use the concept of lifestyle as it is used by Giddens in the context of his formal theory.
4. Another considerable advantage of the term 'environment utilisation space' is the fact that, contrary to the terms used in the limits-to-growth debate, it refers to the interest of the environment as an 'independent' concern, which cannot be reduced to economic categories.
5. Fine and Leopold correctly criticise the generalist character of many consumption theories. However, the authors do not succeed in proposing a solution to this problem at a theoretical level, because they take refuge in empirical research that does not go beyond illustrating that differences between the food and the clothing sector do exist. They fail to draw conclusions from these empirically established differences that would also apply to other systems of provision. As a consequence, there is not much difference between their study and 'normal' sector studies.

6. In our view, the wish or necessity for a more intensive participation of women in the labour process is an equally important motive. In addition, the increase of 'obligations' or activities in the fields of leisure, recreation and tourism may be mentioned as a determining factor.
7. Hal Wilhite [*1997*] for example was able to show how cultural standards concerning 'lighting the home' differ considerably in Japan and Norway or the United States, resulting in different patterns of domestic energy-consumption.
8. The distinction between direct and indirect energy use of households is used in the HOMES-project [*Noorman and Schoot Uiterkamp, 1998*] to distinguish between on the one hand the energy which is 'directly' used to heat and light the house and the energy which is stored in the products and packages on the other.
9. DOMUS is an EU-funded (DG XII) research project on citizenship involvement in the ecological modernisation of public utility sectors in the European Union. Participating countries are the UK, Sweden and the Netherlands.

## REFERENCES


Beck, U., Giddens, A. and S. Lash (1994), *Reflexive Modernization: Politics, Tradition and Aesthetics in the modern Social Order*, Cambridge: Polity Press.

Bijker, W.E., Hughes, T.P. and T. Pinch (1987), *The Social Construction of Technological Systems, New Directions in the Sociology and History of Technology*, Cambridge: MIT Press.

Cowan, R.S. (1983), *More Work for Mother: The Ironies of Household Technology from the Open Hearth to the Microwave*, New York: Basic Books.

Cowan, R.S. (1987), 'The Consumption Function: A Proposal for Research Strategies in the Sociology of Technology', in Bijker, Hughes and Pinch [*1987: 261–80*].

Douglas, M. and B. Isherwood (1979), *The World of Goods: Towards an Anthropology of Consumption*, London: Allen Lane.

Dunleavy, P. (1980), *Urban Political Analysis: The Politics of Collective Consumption*, Basingstone: Macmillan.

Featherstone, M. (1991), *Consumer Culture and Postmodernism*, London: Sage.

Fine, B. and E. Leopold (1993), *The World of Consumption*, London: Routledge.

Giddens, A. (1991), *Modernity and Self-Identity. Self and Society in the Late Modern Age*, Cambridge: Polity Press.

Hajer, M.A. (1995), *The Politics of Environmental Discourse: Ecological Modernisation and the Regulation of Acid Rain*, Oxford: Oxford University Press.

Huber, J. (1982), *Die Verlorene Unschuld der Ökologie. Neuen Technologien und superindustrielle Entwicklung*, Frankfurt/Main: Fisher.

Hughes, T.P. (1983), *Networks of Power: Electrification in Western Society, 1880–1930.* Baltimore, MD: Johns Hopkins University Press.

Kemeny (1980), 'Home Ownership and Privatization', *International Journal of Urban and Regional Research*, Vol.4, pp.372–88

Lauber, V. and K. Hofer (1997), 'Business and Government Motives for Negotiating Voluntary Agreements: A Comparison of the Experiences in Austria, Denmark and the Netherlands', paper for the workshop on the 'Effectiveness of Policy Instruments for Improving EU Environmental Policy Implementation', Bern, 27 Feb. – 4 March 1997.

Liefferink, J.D. and A.P.J. Mol (1998), 'Voluntary Agreements as a Form of Deregulation? The Dutch Experience', in U. Collier (ed.), *Deregulation in the European Union. Environmental Perspectives*, London: Routledge, pp.181–97.

Lury, C. (1996), *Consumer Culture*, Cambridge: Polity Press.

Mommaas (1993), *Moderniteit, Vrijetijd en Stad: Sporen van maatschappelijke transformatie en continuiteit*, Utrecht: Van Arkel, pp.159–81.

Noorman, K.J. and T. Schoot Uiterkamp (eds.) (1998), *Green Households? Domestic Consumers, Environment and Sustainability*, London: Earthscan

Oldenziel, R. and C. Bouw (1998), *Schoon Genoeg, Huisvrouwen en Huishoudtechnologie in Nederland 1898–1998*, Nijmegen: SUN.

Opschoor, J.B. and S.W.F. van der Ploeg (1990), 'Duurzaamheid en kwaliteit: Hoofddoelstellingen van Milieubeleid', in *Het Milieu: denkbeelden voor de 21e eeuw.* Zeist: Kerckebosch, pp.81–128

Otnes, P. (1988), *The Sociology of Consumption*, Atlantic Highlands, NJ: Humanities Press International.

Raman, S., Chappells, H., Klintmann, M. and B. van Vliet (1998), *Inventory of Environmental Innovations in Domestic Utilities: The Netherlands, Britain & Sweden* (Domus report 2), Wageningen, WAU.

Saunders, P. (1987), *Social Theory and the Urban Question*, Second Edition, London: Unwin Hyman.

Saunders, P. and C. Harris (1990), 'Privatisation and the Consumer', *Sociology*, Vol.24, No.1,

Shove, E. (1997), 'Notes on Comfort, Cleanliness and Convenience', paper for the ESF workshop on 'Consumption, Everyday Life and Sustainability', Lancaster, 5–8 April 1997.

Shove , E. and D. Southerton (1998), 'Frozen in Time: Convenience and the Environment', paper for the second ESF workshop on 'Consumption, Everyday Life and Sustainability', Lancaster, 27–29 March 1998.

Spaargaren, G. (1997), *The Ecological Modernization of Production and Consumption* (Essays in Environmental Sociology), Wageningen: WAU

Spaargaren, G. and T. Mol (1992), 'Sociology, Environment and Modernity. Towards a Theory of Ecological Modernization', *Society and Natural Resources*, Vol.5, No.4, pp.323–44.

Van Vliet, B.J.M. (1998), *Analysing Environmental Change in the Relations between Utility Sectors and Domestic Consumers* (DOMUS report 1), Wageningen: WAU.

Warde, A. (1990), 'Production, Consumption and Social Change: Reservations Regarding Peter Saunders' Sociology of Consumption', *International Journal of Urban and Regional Research*, Vol.14, No.2, pp.228–48.

Weale, A. (1993), 'Ecological Modernisation and the Integration of European Environmental Policy', in J.D. Liefferink, P. Lowe and A.P.J. Mol (eds.), *European Integration and Environmental Policy*, London: Belhaven.

Wilhite, H. (1997), 'Cultural Aspects of Consumption', paper for the ESF workshop on 'Consumption, Everyday Life and Sustainability', Lancaster, 5–8 April 1997.

Winward, J. (1994), 'The Organized Consumer and Consumer Information Co-operatives', in R. Keat *et al.* (eds.), *The Authority of the Consumer*, London: Routledge, pp 75–90.

# Ecological Modernisation, Environmental Knowledge and National Character: A Preliminary Analysis of the Netherlands

MAURIE J. COHEN

Comparative research regarding the varying capabilities of the advanced countries to pursue strategies consistent with ecological modernisation has typically focused on institutional and economic capacity. Though these features certainly shape preparedness for this policy programme in important ways, they do not provide sufficient insight into the ability of individual nations to meet the rigorous requirements of ecological modernisation. More specifically, this policy programme is dependent upon a firm commitment to science and a preference to address environmental problems in technological terms. However, environmental decision-making, particularly among members of the lay public, is predicated upon numerous epistemologies and countries exhibit considerable variation in their propensity to align their policies with strict rational reasoning. This study introduces a typology that identifies four archetypal environmental knowledge orientations: rational ecologism, Prometheanism, Arcadianism, and ecocidal mysticism. The concept of national character offers one approach for operationalising this typology for empirical application and a case study of the Netherlands suggests that the Dutch style of interpreting indeterminate environmental data is broadly consistent with the principles of ecological modernisation.

## Introduction

Throughout the past century the environmental movement in most advanced nations has had a Janus face. On one hand, Romantic ruminations, and the critique of industrial society that they frequently embody, have powerfully shaped contemporary ecological sensibilities. Although emphases surely vary, inherent in most of these reflections is a disdain for science, technology, and prevailing forms of societal organisation. According to this broad composite of viewpoints, we should value the natural environment for

The author gratefully acknowledges valuable commentaries on previous drafts of this study by Joseph Murphy, Arthur Mol, David Sonnenfeld, and this journal's two referees.

its sublime, impenetrable qualities, as well as its ability to enliven the human spirit and to inspire the imagination. Adherents harken back to purportedly simpler times before economic instrumentalism and rational knowledge fragmented social cohesion and alienated human beings from nature. This world-view frequently treats the environment as a backdrop against which to theorise about the moral condition of humanity. Eco-warriors protesting road construction in Britain, Earth First! wilderness preservationists in the United States, and staunch opponents of genetically-modified foodstuffs in Germany give political voice to these demodernisation perspectives.

There also exists a second, more pragmatic mode of thought that informs prevailing environmental understanding. Rather than reject technical knowledge as dehumanising, this alternative approach embraces science and technology as central to the amelioration of ecological concerns. We can locate the origins of this view in scientific forestry as it developed in Germany and France during the middle of the nineteenth century. This cast of mind continues to be extraordinarily influential in present-day environmental policy-making and it finds expression in managerialist strategies for promoting sustainable development, as well as in more commonplace activities concerning the regulation of water quality and the protection of endangered lands.

The emotive tradition was initially important for raising public environmental awareness in many countries and politically mobilising people around environmental issues, particularly during the 1960s and 1970s. As governments have grappled since then with the difficult issues associated with the formulation and implementation of environmental legislation, the utilitarian approach has become increasingly dominant.[1] In recent years, the application of science and technology to environmental problems has become more pronounced as influential actors across a wide portion of the political spectrum strive to reconcile conflicting objectives between responsibility for the environment and continual economic expansion. It is out of these efforts that current interest in such topics as industrial ecology and strategic environmental management springs [*Ayres and Ayres, 1996*; *Schmidheiny, 1992*]. Scholars interested in the social theoretical implications of these developments have adopted the term 'ecological modernisation' as a general descriptor of such trends [*Weale, 1992; Mol, 1995; Spaargaren, 1997; Hajer, 1995*].

Some authors envision ecological modernisation essentially as an ineluctable process of social transformation [*Jänicke, 1985; Simonis, 1988*]. This deterministic approach is premised upon a three-phase course of historical evolution beginning with a period of largely agriculturally-based pre-modernity, proceeding through to a stage of modernity organised around industrial production, and finally culminating with ecological modernity

[*Cohen, 1997*]. The terminal epoch in this progression consists of an intensive phase of super-industrialisation predicated upon closed-loop manufacturing systems and advanced environmental technologies that make it possible to repair the so-called design flaws of the transitional modern period.

Another group of ecological modernisation theorists maintains that this developmental process is far from inevitable. Moreover, because different advanced countries bring to this challenging process a diverse range of capabilities, progress across this range of nations will be patchy and uneven [*Dryzek, 1997*].[2] Previous efforts to examine national preparedness for ecological modernisation have generally focused on the institutional and economic factors most relevant to this transformation. For example, Martin Jänicke and his colleagues have developed the concept of 'environmental capacity' principally on the basis of these features [*Jänicke and Weidner, 1997*]. Researchers have to date devoted significantly less attention to the cultural and knowledge-based potentialities of different countries for ecological modernisation.[3]

Because of its dependence on a willingness to respond to environmental problems in technical terms, a strong public resolve for science and technology is a *sine qua non* for ecological modernisation [*Cohen, 1998*]. However, such knowledge coexists in society with many other forms of understanding. Non-scientifically trained lay people in contemporary society – a category that includes most policy-makers and managers – employ a diverse array of epistemologies to interpret ecological information and science is only a single component of the complex framework that orients environmental decision-making. As Huber and Pedersen [*1997*] note in a recent contribution to the anthropological literature, 'People experience their environment in a variety of different contexts and rely pragmatically on different sets of ideas which need not be consistent.' Accordingly, any effort to assess the potential of particular policy programmes must be mindful of how society offsets rational claims pertaining to the environment against other assertions.

To open some new ground in the comparative analysis of ecological modernisation the following discussion introduces a typology that identifies four archetypal environmental knowledge orientations and explains their implications for the pursuit of initiatives consistent with this policy programme. As part of an effort to make more explicit the connection between environmental knowledge and ecological modernisation the discussion then proceeds to a preliminary study of national character in the Netherlands. This case is germane because the Dutch appear to possess some notable qualities that are pertinent for the rational interpretation of environmental information and give the country special potential to move in this direction. The concluding section explores the utility of using the concept of national character to pursue issues related to ecological modernisation and environmental policy-making more generally.

## Environmental Knowledge and Science

Anthropologists have employed the notion of environmental knowledge most extensively to refer to the epistemologies that indigenous peoples use to interpret biophysical information pertaining to landforms, natural resources, agricultural practises, medicinal products, and so forth.[4] The motivation behind this work has been to distinguish traditional environmental knowledge from modern science, or conversely, to argue for the artificiality of the distinction between them [e.g., *Agrawal, 1995; Murdoch and Clark, 1994*]. These are surely worthy initiatives, but the objective here is to focus predominantly on the uptake of science in advanced industrial nations.

It has become routine practice to assume that just because a country has attained a relatively high level of socio-economic development, respect for science is extensive and the march toward rationality has pushed to the margins all alternative forms of knowledge – or in any case will manage to do so in due course. Veneration of science is common not only among natural and physical scientists, but such celebration is found with near-equal prominence among policymakers and orthodox social scientists. All of these social groups evince a strong inclination to embrace the so-called 'deficit' model of the public understanding of science [*Irwin and Wynne, 1996*]. Used to explain the frequent divergence between ordinary individuals and specialists in evaluating environmental risks, this model gives rise to a paternalistic concern that the lay public is insufficiently informed on scientific matters. The policy responses that flow from this assessment normally consist of calls for better communication of scientific information. In some quarters, a more combative critique suggesting that most people are fundamentally irrational is not unusual. Proponents of this more zealous view argue for the exclusion of scientifically uninformed members of the public from participation in efforts to formulate or implement environmental policies.

There has been no shortage of effort to repudiate the deficit model and the oversimplified explanations of the divide between lay and expert appraisals of environmental issues that emanate from it [e.g., *Wynne, 1996*]. To relate these perspectives to the current discussion we must first recognise that the role of science in informing environmental knowledge is much more complicated than an inadequate public understanding of scientific facts. Science, because of the legitimacy it commands, has become an effective source of social control in modern societies and both governments and other sources of authority (for example, industrial managers) use it to give credence to controversial decisions. This scientisation of policy has a rebounding effect in that science invariably becomes politicised [*Eden, 1998;* also *Beck, 1992*]. As such, science is not simply a rigorous methodology for interrogating nature, but it also becomes an ideology that people can appropriate (or reject) to define their personal identities. For instance, by the way they use science in public discourse people send social signals to one another in much the same way they would by wearing a certain article of clothing.

The use of science as an ideology to shape one's personal identity does not necessarily have to be purely symbolic and at certain historical junctures there have been very strong material incentives for choosing to align oneself with this epistemology. For example, in Britain, during the early and mid-nineteenth century, aspirant members of the country's growing middle class affiliated themselves closely with science, often at considerable personal expense and inconvenience, because such knowledge was a necessary prerequisite for social and economic advancement. At the same time, the comfortably ensconced gentry normally expressed incontrovertible contempt for professionalised science and its practitioners because they viewed this form of knowledge as too vocational and a threat to their elite sense of probity [*Cohen, 1999*].

Contemporary society is surely more complex than this reading of life in Victorian Britain and there are multifarious material and cultural incentives working at cross-purposes to induce people to accept or reject claims flowing from scientific sources. The central issue, however, that transcends particular geographic and temporal contexts is that unleashed from the laboratory science becomes a tool for the fulfilment of any number of social and political objectives. When members of the lay public endorse or negate scientific assertions, particularly concerning the creation of environmental knowledge, they are not necessarily engaging in the substance of scientific reasoning. Rather, people are aligning themselves with science on the basis of how it resonates with their ethical sensibilities and values.

## Evaluating Environmental Knowledge Orientations

We can begin an evaluation of environmental knowledge orientations by noting that two axial dimensions condition individual propensities for the interpretation of ecological information. On one hand, people do not assimilate environmental data in a dispassionate manner. Instead, values, emotions, and an implicit understanding of what philosophers over the ages have referred to as the 'good life' guides the process. I term this orienting dimension *ecological consciousness*. On the other hand, ecological information is unavoidably ambiguous and the ways in which people absorb it is a function of the specific interpretive filter they use to differentiate the consequential from the inconsequential. We can refer to the particular organisational schema used to accomplish this task as an *epistemological commitment*. The following discussion examines each of these concepts in greater detail.

*Ecological Consciousness*

Let us initially turn our attention to consideration of the first axial dimension for assessing an environmental knowledge orientation. It is common to equate ecological consciousness with membership in environmental organisations and the display of behavioural tendencies consistent with some measure of responsibility for future generations. Public opinion surveys generally assess this disposition through questions designed to evaluate the extent to which respondents privilege environmental protection over other political objectives. For example, it has become common to encounter instruments that put forth questions of the sort 'Are you an environmentalist?' and to interpret high scores in response to such queries as conveying a pronounced ecological consciousness. No doubt these 'studies' are of quite limited value.

Some researchers have devised more sophisticated surveying methods that seek to place the issue in a context in which respondents must, at least conceptually, make trade-offs [*Kempton, Boster and Hartley, 1995; Dunlap, Gallup and Gallup, 1993*]. One of the most comprehensive comparative investigations of this type is the 1990–91 World Values Survey that examined ecological consciousness in more than three dozen countries. Ronald Inglehart [*1995*] has analysed the relevant portions of this survey and formulated an Environmental Protection Index based upon replies to four survey statements:

(1) I would be willing to give part of my income if I were sure that the money would be used to prevent environmental pollution;

(2) I would agree to an increase in taxes if the extra money is used to prevent environmental pollution;

(3) The government should reduce environmental pollution, but it should not cost me any money;

(4) Protecting the environment and fighting pollution is less urgent than often suggested.

Inglehart classified respondents who 'agreed' or 'strongly agreed' with the first and second items and 'disagreed' or 'strongly disagreed' with the third and fourth items as expressing 'high' support for environmental protection. The Scandinavian countries and the Netherlands ranked highest on this measure of ecological consciousness among the sample of surveyed nations.

While discrete survey items such as those used in the World Values Survey may lend themselves to the specific needs of measuring public opinion, it is questionable whether these instruments, in the absence of any sensitising framework, capture the essence of ecological consciousness in any true sense. More pointedly, three of the four items that Inglehart uses to construct his index are expressed in narrowly economistic terms that equate ecological consciousness with respondents' preparedness to make abstract financial compromises. A more meaningful assessment of ecological consciousness must go beyond purely economic criteria to embrace values and ethics. As Aldo Leopold [*1977 (1949)*] wrote more than a half-century ago, true ecological consciousness requires an understanding of 'intellectual emphasis, loyalties, affections, and convictions'. This more emphatic conception of humans' relationship to their natural surroundings incorporates a desire to preserve the integrity of ecological systems and to give precedence to lifestyle practices consistent with the aims of responsible environmental stewardship. Approached in such expansive terms, we can conceive of ecological consciousness as ranging along a scale from 'weak' to 'strong'.

It is customary to look solely at this axial dimension when measuring public environmental attitudes. However, the framing of people's environmental resolve in such a manner is incomplete because it focuses only on the value-based aspects of their commitment and fails to consider the interpretive filters through which they assimilate ecological information. In other words, people do not simply premise their relationship to the environment on a moral code, but simultaneously condition it by a preference for a particular epistemology.

*Epistemological Commitment*

As previously noted, over the course of the past several decades science has come to occupy an especially privileged position in all advanced countries and it is common to laud – and indeed emulate – nations that appear to be scientifically adept. Such endorsement derives from the fact that expertise in science is tantamount to being economically competitive, sophisticated, and modern. Despite the widespread appeal of this outlook, there is a minority view that attracts attention from time to time. One critique of the ascendancy of science asserts that the dominance of this epistemology does not derive from a superior capacity for ascertaining truth, but rather stems from the collusive aspirations of major industrial corporations to disenfranchise the lay public from democratic governance. The marginalisation of non-scientific knowledge, proponents of this standpoint contend, is especially prevalent in the domain of environmental policymaking because it is here that we find being articulated the most dangerous threats to the mechanisms powering economic growth. As a result, the neutralisation of lay perspectives, through the dissemination of recondite scientific expertise, is deemed to be of considerable importance.

Though we would be negligent to dismiss this interpretation summarily, it is useful to recall the key role of scientists such as Rachel Carson and Paul Ehrlich in first placing the environment on the policy agenda during the 1960s and the influence of the environmental movement in stressing the need to evaluate ecological dilemmas scientifically. None the less, this particular debate, despite its obvious allure, is not the issue I want to take up here. My interest instead is in pursuing the more provocative contention that there is no *prima facie* reason to ground environmental knowledge in science. We could suitably use theology, or any one of a large number of alternative epistemologies, to organise and make sense of ecological information. Let us not forget that within the broad sweep of history, science has occupied the favoured position for an exceedingly short period of time.

To simplify this task of differentiating among a virtually limitless number of knowledge systems we can divide the various frameworks available for interpreting environmental data into two broad categories. On one hand, we have rational-scientific epistemologies that are rooted in the intellectual revolution of the seventeenth century and that rely on the accumulation of purportedly objective facts. These forms of knowledge involve the use of exacting methodologies for collecting data and subjecting them to careful scrutiny. The ultimate goal of practitioners working under the auspices of this knowledge system is to develop axioms that are universally applicable across time and space.

On the other hand, there exists a more disparate category of knowledge that encompasses numinous[5] and aesthetic epistemologies. With origins ranging across extensive geographic and historical distances, this diverse grouping includes religious knowledge traditions, folk and indigenous wisdom, local and tacit knowledge, and knowledge movements in art, music, literature, and philosophy. Though emphases surely vary, all knowledge systems within this classification generally share a tendency to privilege individual intuition, emotion, and the power of the human spirit. Some of these epistemologies are – like science – premised on the systematic assembly and quantitative assessment of empirical evidence, while others are more quixotic and fanciful.

There is a clear tension between rational-scientific and numinous-aesthetic epistemologies and this uneasy relationship has attracted no shortage of commentary [e.g., *Snow, 1993 (1959)*]. In fact, the two classifications to a large degree derive internal coherence from their antipodal relationship with one another and are most readily defined in terms of this diametric dependence. For instance, Max Weber [*1946*] described in a famous oration science's disenchanting effects and the tendency of rational knowledge to vacate the purposefulness from human lives. Motivated by a thirst for meaning, so this critique contends, people turn toward numinous and aesthetic forms of knowledge.

*Four Environmental Knowledge Orientations*

When we contrast our two axial dimensions, ecological consciousness and epistemological commitment, we obtain the fourfold typology depicted in Figure 1. Each of the typology's quadrants corresponds to an archetypal

FIGURE 1

TYPOLOGY OF ENVIRONMENTAL KNOWLEDGE ORIENTATIONS

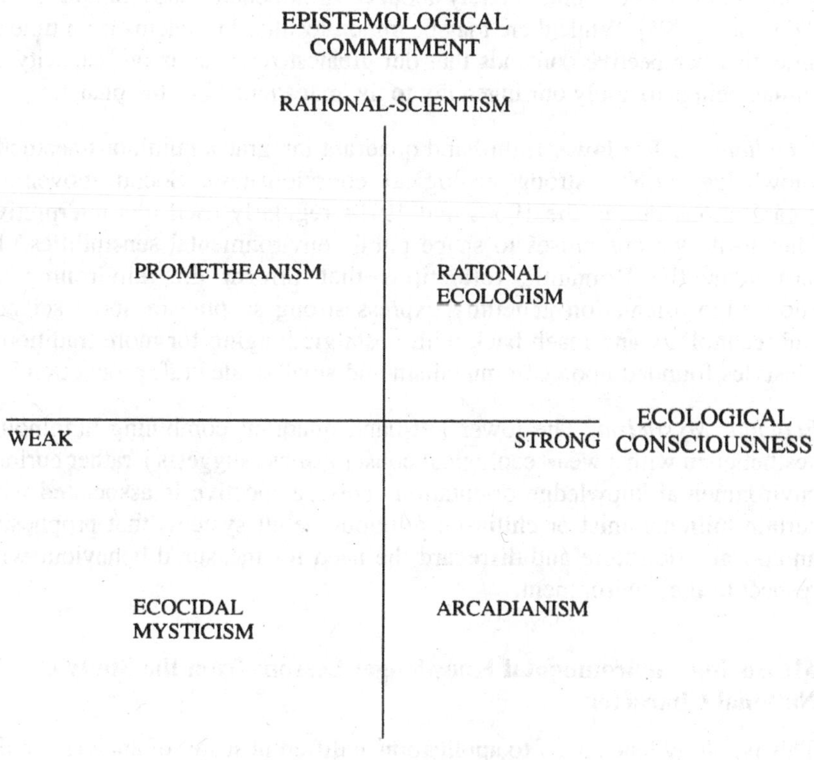

environmental knowledge orientation that provides a distinct organisational schema for interpreting ecological information.

*Rational ecologism*: The upper right-hand quadrant, combining high resolve for both science and the environment, represents an environmental knowledge orientation that attempts to harmonise these two commitments. Assimilated through this interpretive filter, ecological information is tempered by an implicit understanding that rational knowledge will provide the intellectual resources from which it will prove possible to address contemporary ecological concerns. Simultaneously optimistic, technologically focused, and environmentally aware, this orientation has a tendency to view progress in salutary terms.

*Prometheanism*: The orientation depicted in the upper left-hand quadrant processes environmental data through an interpretive filter that has a strong disposition toward science and a weak ecological consciousness. This is the lens that fervent technological enthusiasts such as Julian Simon and Aaron Wildavsky use to discount anxiety about environmental issues [*Simon, 1986; Wildavsky, 1995*]. While there may be minor, localised problems from time to time, this perspective contends that our greatest resource is our capacity as human beings to apply our ingenuity to the management of the planet.

*Arcadianism*: The lower right-hand quadrant integrates numinous-aesthetic knowledge with a strong ecological consciousness. Social movement organisations during the 1960s and 1970s regularly used this interpretive filter in its various guises to shape public environmental sensibilities.[6] In particular, the Romantic reflections that inform this environmental knowledge orientation generally express strong scepticism about science and technology and reach back with nostalgic longing for more traditional lifestyles founded upon communalism and small-scale craft production.[7]

*Ecocidal Mysticism*: The lower left-hand quadrant combining numinous-aestheticism with a weak ecological consciousness suggests a rather curious environmental knowledge orientation. This perspective is associated with certain millenarianist or chiliastic religious belief systems that prophesise an apocalyptic future and disregard the need for measured behaviour with respect to the environment.

## Measuring Environmental Knowledge: Lessons from the Study of National Character

This typology lends itself to application at different scales of analysis. In the first instance, we could use it to classify individuals by developing appropriate metrics with which to gauge their epistemological commitment and ecological consciousness. For example, to measure the extent to which people align themselves either with rational-scientific or numinous-aesthetic knowledge we could draw on an extensive body of literature regarding the communication of scientific information. Over the course of the past forty years, researchers have used social surveys to evaluate the public understanding of science along three dimensions: interest, attitudes, and knowledge [e.g.,. *Bauer, Durant, and Evans, 1994*]. Because of the widespread belief that economic competitiveness depends on an enthusiastic endorsement of rational knowledge, national governments and multinational organisations regularly facilitate the administration of these instruments to monitor the public status of science. Additionally, researchers have conducted considerable work along the lines of the World Values Survey discussed above to construct a measure of ecological consciousness and we could appropriate such techniques to assess this axial dimension at a micro-level of analysis.

Far more challenging is the use of this typology to describe the way in which individual environmental knowledge orientations aggregate across a population and give rise to a collective disposition for assimilating ecological information.[8] The aim here is to speak about a specific country as being prone towards, say, Prometheanism or Arcadianism, and to suggest that its public has an inclination for assimilating environmental data in ways that are roughly consistent with one of these archetypal characterisations. This does not mean to suggest that all members of a society will have a similar propensity; within a nation people will of course align themselves with environmental knowledge in diverse ways. These variations stem from mundane individual differences, as well as from the extent to which people are variously disposed toward a single coherent archetypal characterisation rather than toward a combination of two or more environmental knowledge orientations. It is also likely that people will adjust their perspective according to the circumstances of a particular situation.[9] Moreover, due to disparities in political power, efforts to assess the environmental knowledge orientation of specific nations are much more complicated than simply massing individual dispositions within a given population.

Despite these methodological complications we can advance the theoretical conjecture that the non-trivial aggregation of individual environmental knowledge orientations across a country gives a distinctive stamp to how a nation construes ecological information. Although this propensity will no doubt be difficult to discern, it subtly shapes the interpretation of environmental data in crucial ways.

The question however remains whether we can capture the two axial dimensions of the environmental knowledge typology without violating the canons that govern the conduct of orthodox social science research. Understanding of a nation's ecological consciousness must be grounded in an appreciation of how people use local cultural resources to infuse meaning into life and assessment of a country's epistemological commitment requires careful observation, untold numbers of casual conversations, and development of an affinity that only comes from living in a place for an extended period. Though few mainstream social scientists would deny the existence of cross-national variations in how people assimilate ecological information, the strength of their professional resolve for universalism makes them suspicious of the procedures necessary to expose these dispositions. The conventional view contends that methodologies based on the expertise of a single researcher limit opportunities for replication and lead to the production of idiosyncratic scholarship.[10] It is precisely for these reasons that academe (except for anthropology and a segment of sociology) has side-stepped the study of environmental knowledge.[11]

This sort of discussion begins to encroach quite closely upon the realm of inquiry generally known as the study of national character. Situated at the confluence of several social science disciplines, we can trace contemporary interest in this area back to the work of a small number of eminent anthropologists and psychiatrists who developed national personality profiles of the various combatants during the Second World War. Heavily grounded in the application of psychoanalytic theory, these scholars used largely anecdotal evidence to compile descriptive reports for government agencies. The Office of War Information in the United States commissioned the most significant pieces of this work and drew on the talents of leading academic figures such as Margaret Mead and Ruth Benedict [*Hoffman, 1992*]. This research attempted to explain, for instance, Japanese tidiness or Russian tempestuousness in terms of the respective child-rearing practises commonly employed in the two societies. While the extent to which these views might have aided the war effort is debatable, by the 1950s they began to come under severe criticism.

It was not until the following decade, and publication of a celebrated book by Almond and Verba [*1963*], that the comparative study of national character was placed on sounder methodological footing. These authors' research, influenced by a searching desire to understand why democracy failed in certain countries during the years leading up to the Second World War, indicated that nations have different propensities for sustaining representative government. This work highlighted the importance of civic obligation, public trust in government institutions, and interpersonal associativeness as critical to the maintenance of stable democracies. Although Almond and Verba's insights attracted considerable attention among political scientists and others at the time, its emphasis on culture was out of step with general intellectual trends. Since the mid-1960s, the social sciences have evinced a distinct preference for political and economic explanations of social change, a development that we can trace to the increasing popularity of theories of rational action and the expanding use of computer-aided techniques for manipulating quantitative data [e.g., *Renn, Jaeger, Rosa and Webler, 1999*]. The study of national character also appears to have been an early victim of political correctness. While it was permissible to discuss the values that contributed to, say, the Swedish welfare state, it was unacceptable to highlight – as Banfield [*1958*] did to notorious effect – the clannishness and insularity of southern Italians.

During this interval, sociologists and psychologists also took up the study of national character, toiling under the rubric of 'social structure and personality' [*House, 1981;* also *Schooler, 1996*]. Alex Inkeles has been at the forefront of much of this work and he and his colleagues have tried to move the field away from speculation based on impressionistic interpretations to more rigorous statistical analysis. The primary vehicle for achieving this reform has been the formulation of the concept of 'modal personality' to describe the distribution of individual attitudes within a national population [e.g, *Inkeles, 1997*]. However, prevailing theoretical and empirical orthodoxy has forced these researchers to spend several decades swimming against the scholarly tide.

Throughout this period there has been no shortage of critics of the study of national character who have contended that the concept is deterministic and leads to homogenisation (for a review, see Schooler [*1996*]). Moreover, because authors have tended to emphasise the relative immutability of culture, research into shared traits is typically equated with conservatism and a predilection for the *status quo*. Some writers working in this area, due to their tendency to perpetuate journalistic impressions and hackneyed stereotypes, have inadvertently compounded the potency of this critique. None the less, despite its chequered history, there are indications that national character is now becoming more central to contemporary social science concerns.

The end of the cold war and the political reconfigurations taking place in its aftermath, as well as the increasing influence of transnational bodies such as the European Union, have contributed to the current upsurge of interest in the study of national character. This new wave of research often couches its subject matter in modified terminology – for example, 'national identity' and 'political culture' have largely replaced 'national character' – to give the enterprise a new veneer and to create distance with the tainted tradition of its predecessors [*Fukuyama, 1995; Huntington, 1996; Lipset, 1996; Putnam, 1993*]. To a large extent, this recent turn is motivated by a recognition that conventional analyses fail to capture important features of the human experience and that political institutions, educational and language traditions, and economic systems do indeed create certain shared dispositions. Importantly, this new generation of work is less triumphalist and attempts to overcome the more reasonable criticisms alleging that such endeavours are necessarily reductionistic, blind to class and regional differences, and prone to the expression of staid caricatures. Notwithstanding this encouraging trend, the preceding summary makes clear that any effort to employ the concept of national character is bound to be a treacherous affair.

There is further reason to tread gingerly. In an era when processes of globalisation seem to be so prominent it is perhaps convenient to presume that the singular features of nations are becoming tenuous. Some commentators have gone so far as to suggest that the diffusion of electronic communication and the growth of international exchange breaks down parochialism and imparts identities consistent with 'global citizenship' [*Giddens, 1998*]. While national boundaries are certainly becoming more porous, we should question whether this process is necessarily leading to greater homogeneity. Trade in information and cultural symbols is surely not a new development as countries have long borrowed freely from abroad, adjusting various influences to suit local circumstances [e.g., *Pells, 1997*]. Still, the pace and breadth of current forms of exchange seem to be original, suggesting that there may be some credence to the claims of globalisation proponents.

## Environmental Knowledge in National Content: The Case of the Netherlands

### *Background and Rationale*

The comments above suggest that a thoroughgoing study of environmental knowledge orientations as attributes of national character is bound to be, even for a single country, a labour- and time-intensive exercise. Standard methodologies and data sources are fraught with inadequacies and it is necessary to devise innovative techniques for conducting the necessary evaluations. Given the provisional nature of our fourfold typology of different dispositions for interpreting ecological information, it is prudent to proceed, at least initially, by testing the robustness of this approach using a more expeditious approach. Accordingly, by making use of secondary literature pertaining to the traits of one uniquely situated country, namely the Netherlands, the following section provides a preliminary application of the typology.

A sizeable body of published work has accumulated on Dutch national character in recent decades and we can usefully exploit it to gain some cursory insights into the country's environmental knowledge orientation.

Analysis of the Netherlands is also likely to expose the potential of this approach for gauging the capacity of different countries for ecological modernisation as most comparative policy research identifies the Dutch as having some notable capabilities.[12] As such, it is instructive to examine whether the study of national character might enable us to glean any additional understanding of the Netherlands' ability to implement initiatives consistent with this policy programme for pursuing environmental improvements. Furthermore, there are credible reasons to suspect that the concept of national character is more robust for small, relatively homogeneous countries and the Dutch case presents us with a favourable context in these respects.

As a preface to what follows, it is useful to note that the Netherlands has since the mid-1980s emerged as an international leader in the environmental field [*OECD, 1995*]. Though official rhetoric has at times exceeded *bona fide* accomplishments, the country has pioneered the use of several creative approaches to environmental policy-making (for example, negotiated covenants, voluntary agreements, and target-group consultation) to chart a relatively progressive reform path. To its credit, the Dutch government has also adapted its regulatory framework to overcome conventional single media interventions and to pursue more integrated responses for formulating environmental policy. Moreover, the Netherlands has been at the forefront of efforts to devise measurable indices of sustainability to gauge progress moving toward the challenging targets set forth in its sequence of National Environmental Policy Plans [*Carley and Christie, 1992, Ch.13; Bennett, 1991*]. At the community level the Dutch are pursuing a number of interesting initiatives. For example, neighbourhood groups are now organising households into so-called eco-teams to implement local programs to reduce material consumption and waste generation [*Aarts, Goudsblom, Schmidt and Spier, 1995*]. Analyses of the sources driving environmental policy-making in the Netherlands have typically focused on the country's strong basis for effective planning and tradition of consensual politics [*Bressers and Plettenburg, 1997; OECD, 1995*]. The following discussion seeks to reach beyond these institutional features to explain the country's environmental performance in terms of certain traits that shape how the Dutch public assimilates ecological information.

We can embark on our examination by making the unremarkable observation that the Dutch landscape is subject to extensive management, a situation that stems from the country's small size, high population density, and intensive agricultural practices. The Netherlands also has a heralded reputation for controlling and manipulating nature as part of its epic struggle to reclaim land from the sea. In fact, approximately one-half of the country's land mass is below sea level. It was apparently common at one time to hear the Dutch brag that 'in other countries God made the land but in Holland the Dutch made it for themselves' (quoted in Ginkel [*1993: 54*]). By all accounts, this tradition of assiduously using science and technology to manage water and land has shaped a utilitarian sensibility toward the environment. Furthermore, the Dutch mentality is not especially compromised by more idyllic or escapist modes of thought, despite the influence of a nineteenth-century locally-inspired school of Romantic painting.[13] Religious passions in the Netherlands never inspired the same forms of emotive expression as occurred, for example, in Germany and contemporary Dutch ecological thought displays a palpable pragmatism and anthropocentrism.

The study of Dutch national character was an especially fertile area of inquiry during the 1930s and 1940s and several prominent social scientists active in the Netherlands during this period, including Sebald Rudolf Steinmetz and Johan Huizinga, contributed to this body of literature.[14] Most of this work described the Dutch as phlegmatic, reticent, moralistic, tolerant, and thrifty. Other qualities that provided frequent points of referral in these efforts to delineate the typical Netherlandish temperament included strong senses of independence and perseverance informed by a love of freedom.[15] Commentators also regularly emphasised the Dutch penchant for tidiness and domesticity.[16]

To identify the historical forces that have shaped the Dutch personality students of national character have frequently focused on the necessity of peasants to band together in their fight to hold back the water. The sociologist Bierens de Haan explained that

> the infinite stretches of water and land … must have influenced the national character and made it more open to personal piety … [A] people belongs to and is shaped by the lay of its land, with water playing an important role in the process (quoted in van Heerikhuizen [*1982: 111*]).

Scholars seeking to locate the origins of the consensual qualities that pervade contemporary Dutch politics, frequently invoke the battle against the water. However, closer examination reveals considerable ambivalence on this point. For instance, the Dutch sociologist J.P. Kruijt observed that

> [t]he early organization of the drainage districts does certainly indicate that the struggle against the water did make people join forces, but they did so in numerous extremely small groups – and instead of solidarity among these groups there was petty squabbling, sometimes even *within* the group (italics in original) (quoted in van Heerikhuizen [*1982: 106*]).

We can usefully exploit other observations by Kruijt who focused his analyses of Dutch national character on the formative role of religion and the country's social history. He noted that the dispositions normally associated with Calvinism in the Netherlands, namely a pious orthodoxy combined with respect for liberal humanism, had existed long before they received a formal theological expression. Conditions were conducive to the development of such traits because of the relatively weak hold of feudalism in the lands that today comprise the Netherlands [*van Heerikhuizen, 1982: 111–12;* also *Schama, 1987: 40*]. Ruth Benedict, the American anthropologist who played a pivotal role in the early development of the modern study of national character, also conducted research on the Dutch during the Second World War. Benedict wrote briefing reports for military officials preparing to invade the Netherlands to oust the occupying German forces. To avoid conflict between American soldiers and Dutch civilians the planners of the operation wanted to give the troops a sense of the Dutch people [*Ginkel, 1992*]. Because the exigencies of war limited Benedict's ability to conduct actual fieldwork, she compiled her analyses 'at a distance' through the use of a variety of inventive means that included interviews with émigrés, reviews of archival documents and underground newspapers, and translations of folk songs. Her reports described the Dutch as, among other things, moralising, individualistic, freedom-loving, tolerant, melancholy, thrifty and tidy. Although Benedict's portrayal departed in some respects from those of her Dutch contemporaries (for instance, she appears to have confounded the country's religious divisions with class segregation), her assessment closely conformed with the accounts prepared by native scholars.[17]

More recent evaluations of Dutch national character are offered by Alex Inkeles [*1997: 371–5*] who, as noted above, has had a long-standing interest in this general area. On the basis of quantitative social surveys conducted during the early 1980s, Inkeles reports that the Dutch public maintains a strong sense of burden, restlessness, and constraint from life's forces. A striking statistic was that during the 1980s more than 70 per cent of the Dutch public indicated that they felt 'anxious at home', making the Netherlands at the time of the survey the most disquieted country in the European Union.[18] More optimistically, and in seeming contradiction to these dispositions, the Dutch ranked second in a sample of European nations in life satisfaction.

Despite the obvious debates that such representations are likely to engender, the issue remains as to how we might make use of these depictions to explore how the Dutch construe ecological information and create environmental knowledge. At this point we can draw in turn on some social survey data pertaining to epistemological commitment and environmental consciousness, although we should continue to keep in mind the caveats outlined earlier.

*Epistemological Commitment*

It is difficult to formulate direct survey questions to assess epistemological commitment because the concept is not likely to be especially meaningful to respondents. Most practicable is to unpack the concept into components that more readily lend themselves to the constraints of survey research. We can disaggregate epistemological commitment into three constituent elements: (1) attitude regarding the beneficence of science and technology; (2) outlook on the future of human prospects; and (3) willingness to defer to scientific expertise. We can expect respondents' resolve for either rational-scientism or numinous-aestheticism to condition their views on each of these more readily measurable sub-themes (refer to Table 1).[19]

First, concerning the beneficence of science and technology, rational-scientism sees these sources of ingenuity in uncontestably positive terms as they provide the wellsprings to ensure the continual enhancement of the human condition. In contrast, numinous-aestheticism typically views science and technology in a more pessimistic light and strives to reinfuse enchantment into human experience.

TABLE 1

COMPONENTS OF EPISTEMOLOGICAL COMMITMENT

| Epistemological Perspective | Rational-Scientific | Numinous-Aesthetic |
|---|---|---|
| Attitude regarding the beneficence of science and technology | Positive | Negative |
| Outlook on the future of human prospects | Optimistic | Pessimistic |
| Willingness to defer to scientific expertise | Willing | Unwilling |

Second, on the extent to which humans might need to limit their expectations the rational-scientific perspective acknowledges the potential for minor disruptions, but claims our inventiveness will enable us to surmount all challenges. By comparison, numinous-aestheticism contends that human beings are presently stretched to their limits and it is necessary to respect certain thresholds to prevent our hubristic tendencies from pushing us into chaos.

Finally, the two perspectives differ in their stances toward the purveyors of scientific and technological knowledge. Rational-scientism holds the special expertise of scientists and technologists in high regard and demonstrates an uncompromised willingness to defer to the judgements of these specialists. Alternatively, numinous-aestheticism emphasises the importance of local context and the acquisition of knowledge through individual experience. This perspective necessarily questions the validity of proffered expertise that is not anchored within particular geographic and temporal parameters.

We can assess the Dutch disposition on each of these components of epistemological commitment by turning to the *Eurobarometer*, a public opinion survey periodically administered in member countries of the European Union, and comparing the results for the Netherlands with a small sample of neighbouring countries.[20] To gauge public attitudes concerning the beneficence of science and technology we focus on a survey item that asked whether science and technology would create more opportunities for future generations. The results displayed in Table 2 indicate that among the countries in our sample the Dutch are quite confident about the efficacy of science and technology.

TABLE 2

ATTITUDES REGARDING THE BENEFICIENCE OF SCIENCE AND TECHNOLOGY – PERCENTAGE OF RESPONDENTS THAT SAY THAT THANKS TO SCIENCE AND TECHNOLOGY, THERE WILL BE MORE OPPORTUNITIES FOR FUTURE GENERATIONSS

| *Percentage that says there will be more opportunities for future generations* | France | Belgium | Netherlands | Germany | Denmark | Britain |
|---|---|---|---|---|---|---|
| Strongly agree | 16.4 | 18.1 | 34.3 | 22.5 | 32.6 | 20.0 |
| Agree to some extent | 44.3 | 37.3 | 41.4 | 44.8 | 36.6 | 45.2 |
| Neither agree nor disagree | 21.6 | 26.9 | 13.5 | 24.2 | 12.4 | 14.8 |
| Disagree to some extent | 11.1 | 10.8 | 6.9 | 6.7 | 12.4 | 13.8 |
| Strongly disagree | 6.5 | 6.9 | 3.9 | 1.9 | 6.1 | 6.2 |

*Source*: Eurobarometer 38.1 (Nov. 1992).

The second component of epistemological commitment concerning public outlooks toward the future of human prospects is ascertainable by looking at the extent to which respondents in the Netherlands feel technological progress can mediate the conflict between increasingly affluent lifestyles and environmental deterioration. As shown in Table 3, the Dutch (along with the British) are relatively certain that technological progress will make possible still greater levels of consumption.

A third component of epistemological commitment, willingness to defer to scientific expertise, can be assessed by examining whether respondents

TABLE 3

EXISTENCE OF LIMITS TO GROWTH – PERCENTAGE OF RESPONDENTS THAT SAYS TECHNOLOGICAL PROGESS WILL MAKE POSSIBLE HIGHER LEVELS OF CONSUMPTION AND, AT THE SAME TIME, AN UNPOLLUTED ENVIRONMENT

| *Percentage that says technological progress will make possible more consumption* | France | Belgium | Netherlands | Germany | Denmark | Britain |
|---|---|---|---|---|---|---|
| Strongly agree | 7.6 | 10.2 | 10.4 | 4.9 | 10.2 | 5.6 |
| Agree to some extent | 27.6 | 21.7 | 31.1 | 24.4 | 24.2 | 36.1 |
| Neither agree nor disagree | 25.1 | 29.4 | 17.8 | 28.8 | 17.7 | 22.2 |
| Disagree to some extent | 24.8 | 23.5 | 21.5 | 23.5 | 22.0 | 25.2 |
| Strongly disagree | 14.9 | 15.2 | 19.2 | 18.4 | 25.9 | 10.9 |

*Source*: Eurobarometer 38.1 (Nov. 1992).

TABLE 4

WILLINGNESS TO DEFER TO SCIENTIFIC EXPERTISE – PERCENTAGE OF RESPONDENTS THAT SAYS BECAUSE OF THEIR KNOWLEDGE, SCIENTIFIC RESEARCHERS HAVE A POWER THAT MAKES THEM DANGEROUS

| *Percentage of respondents that says scientific researchers are dangerous* | France | Belgium | Netherlands | Germany | Denmark | Britain |
|---|---|---|---|---|---|---|
| Strongly agree | 27.9 | 24.1 | 32.5 | 32.3 | 31.5 | 19.9 |
| Agree to some extent | 36.5 | 31.2 | 37.2 | 39.8 | 35.3 | 42.3 |
| Neither agree nor disagree | 19.8 | 23.3 | 11.5 | 14.7 | 13.2 | 10.9 |
| Disagree to some extent | 10.6 | 12.3 | 11.3 | 11.2 | 12.6 | 17.1 |
| Strongly disagree | 5.2 | 9.1 | 7.5 | 2.1 | 7.4 | 9.7 |

*Source*: Eurobarometer 38.1 (Nov. 1992).

TABLE 5

RESPECT FOR SCIENTIFIC EXPERTS – PERCENTAGE OF RESPONDENTS THAT CONSIDER SCIENTIFIC RESEARCHERS TO BE THE PROFESSIONALS FOR WHOM THEY HAVE THE MOST RESPECT

| *Percentage of respondents that says they have the most (or second most) respect for scientific researchers* | France | Belgium | Netherlands | Germany | Denmark | Britain |
|---|---|---|---|---|---|---|
| Respect most | 36.4 | 24.0 | 15.1 | 21.4 | 19.5 | 12.8 |
| Respect second most | 31.8 | 27.3 | 22.6 | 19.0 | 14.6 | 31.5 |

*Source*: Eurobarometer 38.1 (Nov. 1992).

view scientists as socially responsible individuals. Interestingly, there appears to be widespread consternation for science across the entire sample, but this unease was especially pronounced in the Netherlands (as well as in Germany and Denmark) (refer to Table 4). We can further pursue this issue by considering the different levels of public respect that exist in each country for scientists in comparison to other professionals (judges, medical doctors, lawyers, businesspersons, journalists, bankers, engineers, and architects). The Dutch (again along with the Germans and Danes) express relatively low esteem for scientific experts (refer to Table 5).

### *Ecological Consciousness*

When we turn our attention toward the second axial dimension of the environmental knowledge orientation typology, namely ecological consciousness, we find that on virtually every survey conducted in recent years to measure environmental mindfulness the Netherlands ranks as one of the most cognisant countries in the world. The World Values Survey placed the Dutch near the top of its national ranking of public support for environmental protection (exceeded only by Sweden and Denmark) (refer to Table 6) [*Inglehart, 1995*]. Other assessments have also assigned the Netherlands to the topmost tier of countries expressing support for rigorous environmental standards [e.g., *Mertig and Dunlap, 1995*]. According to such evaluations there appears to be little question about the relatively strong support the Dutch public assigns to this objective.[21]

TABLE 6
PUBLIC SUPPORT FOR ENVIRONMENTAL PROTECTION – PERCENTAGE OF PUBLIC SCORING 'HIGH' ON ENVIRONMENTAL PROTECTION INDEX

| Country | Percentage(1) |
|---|---|
| Denmark | 65 |
| The Netherlands | 64 |
| Britain | 42 |
| Germany (W) | 41 |
| Belgium | 33 |
| France | 30 |

*Source*: World Values Survey as cited in Inglehart [*1995*].

### *Discussion*

We can now interpret these social survey findings against the background provided by the studies of national character. In terms of epistemological commitment, the Dutch display a comparatively robust confidence in the ability of science and technology to ameliorate ecological concerns and to enhance prevailing standards of living without compromising ecological integrity. Additionally, public respect for science in the Netherlands is among the most resolute in the world and the country's commitment to this knowledge system remains relatively unchallenged by other claims-making authorities [*Cohen, 1998*]. Interestingly, by all indications Dutch surety about science is not undiscriminating and, as suggested by the survey items focusing on trepidation for scientific practitioners, the public tempers its enthusiasm for scientific knowledge with a reflexivity of the limits of rational expertise.

It can be argued that contemporary environmental sensibilities in the Netherlands do not spring from a recently manifest constellation of values.[22] Rather they have their origins in long-standing cultural conventions that emphasise order and tidiness. This compulsion for high sanitary standards combines with a strong sense of righteousness to give unique structure to prevailing Dutch ecological consciousness.[23] This synthesis of attributes – thoughtful scientific resolve, cleanliness, and moralistic and anxious tendencies – suggests the Dutch environmental knowledge orientation conforms quite closely to rational ecologism. Such a disposition is strongly consistent with ecological modernisation.

We should not presume that other advanced nations, simply because they are situated at a stage of development similar to the Netherlands, also have an environmental knowledge orientation consonant with rational ecologism. As we have seen, distinctive social and biophysical features have given a characteristic form to the interpretive filter that the Dutch public uses to construe ecological information. Different factors would certainly prove relevant for other countries we might choose to consider. For instance, the particular way that the German public interprets environmental data is the outcome of a distinct array of historical influences and is more directly conditioned by Romantic thought [*Riordan, 1997; Cohen, 1999a*]. Though more careful consideration would obviously be necessary, these tendencies probably push Germany's environmental knowledge orientation more closely toward Arcadianism. To take another example, the United States combines a relatively strong scientific worldview with a modest ecological consciousness (as compared to northern Europe) and this suggests that the country has an environmental knowledge orientation roughly concordant with Prometheanism [*Dunlap and Mertig, 1992*]. Again, more detailed research would be required to substantiate these intuitive observations.

## Conclusion

Prometheanism and Arcadianism have served as the warp and weft of modernity. Public faith in science, technology, and continual progress has regularly been tempered by the critical scrutiny of more circumspect traditions. Throughout the past century there have been repeated attempts to merge these two environmental knowledge orientations and one need only scratch the surface of the historical record to find illustrations of this tendency. Let us look briefly at a few prominent representatives who took up this cause. Karl Gayer, an indefatigable proponent of 'natural forestry' at the turn of the twentieth century, spearheaded one of the first efforts to pursue this objective. To harmonise economic and ecological principles, Gayer argued that diverse forest growth was beneficial both because it more closely approximated pristine nature and it protected against volatile market conditions [*Radkau, 1997*]. At approximately the same time, George Perkins Marsh, an American diplomat and dilettante geographer, decried the inefficiency of prevailing methods of exploiting natural resources and called for the application of greater rationality. He described the unbridled economic prerogatives of timber companies as an unnecessary drain on profitability and sought to infuse into their practises the halcyon sensibilities of Henry David Thoreau. Gifford Pinchot, the father of America's national forest service, later become the champion of many of these ideas in the United States [*Lowenthal, 1958*]. While the effects of Pinchot's reformist zeal were most pronounced in land management, his brand of environmental instrumentalism invariably spilled over into other policy areas, namely water pollution control and public health. Ecological modernisation, the new voice of rational ecologism, represents yet another attempt to merge our conflicting Promethean and Arcadian tendencies [also *Dryzek, 1997; Hajer, 1996*].

As an environmental policy programme, ecological modernisation is contingent upon optimism in science and technology, confidence in human resourcefulness, and strong respect for expert appraisal. Nations that have an environmental knowledge orientation that privileges these notions in the interpretion of ecological information will have a heightened capacity for pursuing initiatives that are consistent with this pattern of reform. It bears repeating, resolve for rational ecologism is not in and of itself a sufficient basis for ecological modernisation, but in combination with the appropriate political and economic capacity it is likely to catalyse the requisite social adjustments.

Though elusive and difficult to pin down using orthodox research techniques, a country's distinct environmental knowledge orientation is largely contingent upon two axial dimensions, namely ecological consciousness and epistemological commitment. As we saw in the Dutch case study, it is readily possible to identify elements of national character that contribute to these two attitudinal dispositions. The fact that commentators over an extended period of time have consistently invoked the same descriptive terms to describe the Dutch should give us some confidence in the plausibility and the relative indominatability of a Nederlandish temperament. The intent is not to suggest the existence of a universal Dutch outlook, but rather to make the far less ambitious claim for the persistence of some fairly unique attitudes among the inhabitants of the area previously known as the United Provinces. Furthermore, this discernible mentality gives a distinctive timbre to public affairs, most particularly as they concern science, technology, and the environment. As momentum continues to build internationally around an 'eco-modernist' agenda (at least among the world's most advanced countries), the qualities that comprise the Dutch character, foremostly a strong resolve for rationalism and moderation, enhance the country's capability to successfully chart a path toward a future roughly consistent with the principles of ecological modernisation.

In looking toward how these insights pertaining to the Netherlands might help to shape a research agenda regarding ecological modernisation it is instructive to observe that scholars have generally pursued the study of environmental capacity within a rather narrow public policy purview instead of seeing it as a topic of wider social science import. Though some scholars working in this area have pursued a social theoretical line of argumentation, there has been little effort to appreciate ecological modernisation as a more general form of modernisation. As Max Weber, Karl Polanyi and Robert Merton have famously observed, modernisation – whether agricultural, scientific, industrial, or ecological – entails fundamental changes of mentality. This recognition obviously reinforces the centrality of national character in the process of social evolution. To take industrial modernisation as an example, people began during the early nineteenth century to see the world in vastly different terms. More specifically, there was a growing dependence on clock time and punctuality as dictated by expanding manufacturing operations. Nations and regions with predispositions that facilitated these changes invariably had a competitive advantage in the race to industrialise [*Landes, 1993*]. If advanced countries are indeed modernising now along ecological lines, it would be useful to identify the features of this process that have analogues in prior time periods and to examine how they have been reproduced over time [*Jamison and Baark, 1999*]. We could achieve these objectives by linking the study of ecological modernisation to these larger themes in the social sciences.

In conclusion, we can reflect on the relevance of this discussion for the theory of ecological modernisation. It has become common in recent years to differentiate between 'weak' and 'strong' versions of ecological modernisation [*Dryzek, 1997;* also *Harvey, 1996: 377–83*]. The weak model is largely grounded in the redesign of manufacturing systems to limit the adverse affects of industrial activities and there are indications that some countries are making significant progress along these lines [*Simonis, 1994; Moomaw and Tullis, 1994; Cohen, 1999b*]. Strong ecological modernisation, however, suggests a more earnest programme, one that reaches beyond industrial ecologists' comparatively mundane forms of adjustment. Strong ecological modernisation entails a reflexive process of social learning, one that recognises the value inherent in scientific ways of interpreting the world, but simultaneously maintains a critical scepticism about the virtues of abstruse forms of expertise. This later approach attempts to integrate a plurality of knowledges and opens up greater opportunities for public engagement [*Funtowicz and Ravetz, 1992;* also *Christoff, 1996; Hajer, 1995*]. Strong ecological modernisation is best viewed as an emergent process of social and institutional reconstitution rather than the pursuit of a prefigured realignment. It is in evaluating and comparing the potentialities of different countries to embark on such a journey that efforts to highlight the environmental dimensions of national character are likely to be most informative.

## NOTES

1. Indeed, it can be argued that since the eighteenth century the scientific understanding of the environment – especially as constituted in terms of 'nature' – has grown increasingly more pervasive. The current discussion is confined to the domain of environmental policymaking rather than this much more formidable context of the prevailing Western worldview.
2. Although scholars have come to apply the theory of ecological modernisation to countries at various stages of socio-economic advancement, its original formulation was limited to the most highly advanced nations. This paper adopts this narrower purview and contends that ecological modernisation as a prescriptive vision for transcending modernity should only apply to the world's most developed countries. This definition of ecological modernisation does not mean to suggest that certain of the theory's strategic elements such as ecological taxation, environmental accounting, and so forth are not relevant for developing nations. Rather the contention is that as a policy programme for the large-scale transformation of industrial societies, ecological modernisation is only applicable to advanced nations. For examples highlighting the application of the theory of ecological modernisation to developing nations refer to Sonnenfeld (*this volume*) and Frijns, Phung, and Mol (this volume).
3. For efforts to highlight the importance of social and cultural contexts in the pursuit of sustainable development more generally see Burgess, Harrison and Filius [*1998*]. Jamision, Eyerman, Cramer and Læssøe [*1990*] discuss the role of environmental knowledge, or cognitive praxis, for purposes of social mobilisation.
4. Some anthropologists differentiate between 'environmental knowledge' and 'ecological knowledge' using the former term to refer to the epistemological frameworks employed by traditional environmental knowledge and the later to modern Western forms of environmental expression. This distinction is less salient for current purposes.
5. *Webster's Third New International Dictionary* defines 'numinous' as 'sacred, magical, holy, inspiring reverence, spiritual, mysterious, and incomprehensible'. The term derives from the spirit Numen believed by animists to inhabit natural objects or phenomena.
6. More recently, environmentalism as a social movement has experienced a process of professionalisation that has turned most major organisations away from humanistic critiques

of industrial society in favour of more scientistic, legalistic, and economistic approaches. In the American context, groups such as the National Resources Defense Council and the Environmental Defense Fund are prototypes of this shift. I am indebted to one of the referee's for reminding me of this distinction.

7. It bears notice that Martin Lewis [*1992*] employs the terms 'Prometheanism' and 'Arcadianism' to distinguish divergent strands of American environmental sensibility, though the specific definitions he attaches to these two expressions differ from those used here.
8. Kempton, Boster, and Hartley [*1995*] have demonstrated that even in a large, diverse country such as the United States environmental sensibilities display a relatively high degree of national coherence.
9. Relevant to this point is the analysis of Harrison and Burgess [*1994*]. See also Sagoff [*1988*] who argues that our perspective for relating to the environment is mediated by whether we are acting as consumers or political actors.
10. None the less, several social scientists have begun to draw attention to the inadequacy of measuring public dispositions toward science and the environment using structured surveys. Through the use of more contextualised techniques such as in-depth discussion groups, ethnographic research, and so forth it is, as Harrison, Burgess and Filius [*1996: 220*] explain, 'possible to study the kind of arguments that people mobilise to rationalise their environmental practises, and to do so in a way that replicates the social contexts of everyday life'. Refer also to Irwin [*1995*], Wynne [*1995*], and Couch, Kroll-Smith and Kindler [*2000*].
11. Though this is not the place for a detailed discussion, it is appropriate to note that several humanistic disciplines, most prominently perhaps the philosophy of science and literary criticism, have in recent years begun to generate important insights pertaining to the formulation of environmental knowledge.
12. Ecological modernisation in the Netherlands has received considerable attention among environmental policy researchers in recent years. International interest in the Dutch case was first perked by publication of the country's ambitious National Environmental Policy Plan in 1989 and initial analysis suggested that the Dutch did indeed appear to posses some unique institutional capacity. In more recent years, the conclusions from studies pertaining to the Dutch experience have become more indeterminate. Native Dutch researchers [*Hajer, 1995; Liefferink, 1997*] have been keen to dampen enthusiasm suggesting that the country is pursuing especially novel environmental strategies.
13. Romanticism more generally in the Netherlands was a relatively muted movement and except for Willem Bilderdijk failed to produce a figure of European stature. For further discussion, see Rupke [*1988*].
14. Bart van Heerikhuizen [*1982*] advances the argument that it was the threat of German National Socialism (and the eventual occupation of the Netherlands during the Second World War) that sparked interest in the subject of national character among Dutch scholars during this period. He contends that at such times it becomes necessary to articulate a rationale for a distinguishable collective identity. I base much of the following discussion on this work.
15. The Dutch term *bedaard* is frequently used to describe the characteristic Dutch personality. The historian Robert Fruin, writing in 1871, suggested that this word 'has no real equivalent, in its full significance, in any other language. It merges into other terms: circumspect in deliberations, slow to act, cool when things go well, patient when things go poorly, persevering in the face of opposition, cool and collected in the face of unhappiness, not boisterous in the face of enjoyment.' Quoted in van Heerikhuizen [*1982: 105*].
16. It is perhaps worth noting that the Dutch word *schoon* means both clean and beautiful, a linguistic connection that suggests the possible existence of an etymological basis for contemporary ecological consciousness in the Netherlands. See van Heerikhuizen [*1982: 108*].
17. According to Ginkel [*1992: 60*], Benedict did not use the same sources as her Dutch contemporaries and appears to have had no direct knowledge of the work of Steinmetz, Kruijt, Huizinga, and other Dutch scholars.
18. The survey from which this datum is derived was conducted in 1981, suggesting that the anxiety reported by Dutch respondents may have been attributable to financial concerns

caused by the economic recession that gripped the country at the time rather to unease stemming from environmental apprehensions. See Table 11.4 in Inkeles [*1997: 374*].

19. These three features do not exhaust the possible characteristics into which we could disaggregate epistemological commitment and the specific selection of these elements is attributable to the availability of data. Despite the 'convenience' of this approach, the three facets arguably capture the essential qualities of this dimension of environmental knowledge.
20. Survey sponsors administered this particular *Eurobarometer* survey in November 1992 to approximately 13,000 people over the age of 15 in more than a dozen countries.
21. It is notable that Dunlap, Gallup and Gallup [*1993*] found the strength of Dutch environmental sensibilities to be considerably weaker than the other instruments reported above, though the reasons for this deviation are not readily apparent.
22. This is the essential thesis of Ronald Inglehart [*1990*] who argues that advanced nations such as the Netherlands have experienced a 'post-materialist' value shift as a result of post-Second World War prosperity.
23. Henk Te Velde [*1996*] argues that with the loss of political status as a great international power, the Netherlands found its *raison d'être* in moral pretension.

## REFERENCES


Aarts, Wilma, Goudsblom, Johan, Schmidt, Kees and Fred Spier (1995), *Toward a Morality of Moderation: Report for the Dutch National Research Programme on Global Air Pollution and Climate Change*, Amsterdam: Amsterdam School for Social Science Research.

Agrawal, Arun (1995), 'Dismantling the Divide Between Indigenous and Scientific Knowledge', *Development and Change*, Vol.26, No.3, pp.413–39.

Almond, Gabriel A. and Stanley Verba (1963), *The Civic Culture: Political Attitudes and Democracy in Five Nations*, Princeton, NJ: Princeton University Press.

Ayres, Robert and Leslie Ayres (1996), *Industrial Ecology: Towards Closing the Materials Cycle*, Cheltenham: Edward Elgar.

Banfield, Edward (1958), *The Moral Basis of a Backward Society*, Glencoe, IL: Free Press.

Bauer, Martin, Durant, John and Geoffrey Evans (1994), 'European Public Perceptions of Science', *International Journal of Public Opinion Research*, Vol.6, No.2, pp.163–86.

Beck, Ulrich (1992), *Risk Society: Toward a New Modernity*, London: Sage.

Bennett, Graham (1991), 'The History of the Dutch National Environmental Policy Plan', *Environment*, Vol.33, No.7, pp.6–9, 31–3.

Bressers, Hans Th. A. and Loret A. Plettenburg (1997), 'The Netherlands', in Martin Jänicke and Helmut Weidner (eds.), *National Environmental Policies: A Comparative Study of Capacity-Building*, Berlin: Springer, pp.109–31.

Burgess, Jacquelin, Harrison, Carolyn M. and Petra Filius (1996), 'Rationalizing Environmental Responsibilities: A Comparison of Lay Publics in the UK and the Netherlands', *Global Environmental Change*, Vol.6, No.1, pp.215–34.

Carley, Michael and Ian Christie (1992), *Managing Sustainable Development*, London: Earthscan.

Christoff, Peter (1996), 'Ecological Modernisation: Ecological Modernities', *Environmental Politics*, Vol.5, No.3, pp.476–500.

Cohen, Maurie J. (1999), 'Science and Society in Historical Perspective: Implications for Social Theories of Risk', *Environmental Values*, Vol.8, in press.

Cohen, Maurie J. (2000), 'Sustainable Development and Ecological Modernisation: National Capacity for Rigorous Environmental Reform', in Denis Requier-Desjardins, Clive Spash and Jan Van der Straaten (eds.) *Social Dimensions of Environmental Decision Processes*, Dordrecht: Kluwer.

Cohen, Maurie J. (1998) 'Science and the Environment: Assessing Cultural Capacity for Ecological Modernization', *Public Understanding of Science*, Vol.7, No.2, pp.149–67.

Cohen, Maurie J. (1997), 'Risk Society and Ecological Modernisation: Alternative Visions for Post-Industrial Nations', *Futures*, Vol.29, No.2, pp.105–19.

Couch, Stephen R., Kroll-Smith, Steve and Jeffrey Kindler (2000), 'Discovering and Inventing Extreme Environments: Sociological Knowledge and Publics at Risk', in Maurie J. Cohen (ed.), *Risk in the Modern Age: Social Theory, Science, and Environmental Decision-Making*, Basingstoke: Macmillan, pp.173–95.

Dryzek, John S. (1997), *The Politics of the Earth: Environmental Discourses*, Oxford: Oxford University Press.

Dunlap, Riley and Angela Mertig (eds.) (1992), *American Environmentalism: The U.S. Environmental Movement, 1970–1990*, Philadelphia, PA: Taylor & Francis.

Dunlap, Riley E., Gallup, George H. Jr. and Alec M. Gallup (1993), *Health of the Planet*, Princeton, NJ: Gallup International Institute.

Eden, Sally (1998), 'Environmental Issues: Knowledge, Uncertainty, and the Environment', *Progress in Human Geography*, Vol.22, No.3, pp.425–32.

Fukyama, Francis (1995), *Trust: The Social Virtues and the Creation of Prosperity*, New York: Free Press.

Funtowicz, Silvio O. and Jerome R. Ravetz (1992), 'Three Types of Risk Assessment and the Emergence of Post-Normal Science', in Sheldon Krimsky and Dominic Golding (eds.), *Social Theories of Risk*, Westport, CT: Praeger, pp.251–73.

Giddens, Anthony (1998), *The Third Way*, Cambridge: Polity Press.

Ginkel, Rob van (1992), 'Typically Dutch: Ruth Benedict on the National Character of Netherlanders', *Netherlands Journal of Social Sciences*, Vol.28, No.1, pp.50–71.

Hajer, Maarten A. (1995), *The Politics of Environmental Discourse: Ecological Modernization and the Policy Process*, Oxford: Clarendon Press.

Hajer, Maarten A. (1996), 'Ecological Modernisation as Cultural Politics', in Scott Lash, Bronislaw Szerszynski and Brian Wynne (eds.), *Risk, Environment, and Modernity: Towards a New Ecology*, London: Sage, pp.246–26.

Harrison, Carolyn M. and Jacquelin Burgess (1994), 'Social Constructions of Nature: A Case Study of Conflicts Over the Development of Rainham Marshes', *Transactions of the Institute of British Geographers*, Vol.19, No.3, pp.291–310.

Harrison, Carolyn M., Jacquelin Burgess and Petra Filius (1998), 'Environmental Communication and the Cultural Politics of Environmental Citizenship', *Environment and Planning, A*, Vol.39, No.8, pp.1445–60.

Harvey, David (1996), *Justice, Nature, and the Geography of Difference*, Oxford: Blackwell.

Hoffman, Louise E. (1992), 'American Psychologists and Wartime Research on Germany, 1941–1945, *American Psychologist*, Vol.47, No.2, pp.264–73.

House, John S. (1981), 'Social Structure and Personality', in Morris Rosenberg and Ralph H. Turner (eds.), *Social Psychology: Sociological Perspectives*, New York: Basic Books, pp.525–61.

Huber, Toni and Poul Pedersen (1997), 'Meteorological Knowledge and Environmental Ideas in Traditional and Modern Societies: The Case of Tibet', *Journal of the Royal Anthropological Institute*, Vol.3, No.3, pp.577–98.

Huntington, Samuel P. (1996), *The Clash of Civilizations and the Making of a New World Order*, New York: Simon & Schuster.

Inglehart, Ronald (1990), *Culture Shift in Advanced Industrial Society*, Princeton, NJ: Princeton University Press.

Inglehart, Ronald (1995), 'Public Support for Environmental Protection: Objective Problems and Subjective Values in 43 Societies', *PS: Political Science and Politics*, Vol.28, No.1, pp.57–72.

Inkeles, Alex (1997), *National Character: A Psychosocial Perspective*, New Brunswick, NJ: Transaction Books.

Irwin, Alan (1995), *Citizen Science: A Study of People, Expertise, and Sustainable Development*, London: Routledge.

Irwin, Alan and Brian Wynne (eds.) (1996), *Misunderstanding Science? The Public Reconstruction of Science and Technology*, Cambridge: Cambridge University Press.

Jamison, Andrew and Erik Baark (1999), 'National Shades of Green: Comparing the Swedish and Danish Styles of Ecological Modernisation', *Environmental Values*.

Jamision, Andrew, Eyerman, Ron, Cramer, Jacqueline and Jeppe Læssøe (1990), *The Making of the New Environmental Consciousness: A Comparative Study of the Environmental Movements in Sweden, Denmark, and the Netherlands*, Edinburgh: Edinburgh University Press.

Jänicke, Martin (1985), *Preventive Environmental Policy as Ecological Modernization and Structural Policy*, Berlin: Berlin Science Center.

Jänicke Martin and Helmut Weidner (eds.) (1997), *National Environmental Policies: A Comparative Study of Capacity-Building*, Berlin: Springer.

Kempton, Willett, Boster, James S. and Jennifer A. Hartley (1995), *Environmental Values in American Culture*, Cambridge, MA: MIT Press.

Landes, David (1993), *The Unbound Prometheus: Technological Change and Industrial Development in Western Europe from 1750 to the Present*, Cambridge: Cambridge University Press.

Leopold, Aldo (1977 [1949]), *A Sand County Almanac*, Oxford: Oxford University Press.

Lewis, Martin W. (1992), *Green Delusions: An Environmentalist Critique of Radical Environmentalism*, Durham, NC: Duke University Press.

Liefferink, Duncan (1997), 'The Netherlands: A Net Exporter of Environmental Policy Concepts', in Mikael Skou Andersen and Duncan Liefferink (eds.), *European Environmental Policy: The Pioneers*, Manchester: Manchester University Press, pp.210–50.

Lipset, Seymour M. (1996), *American Exceptionalism: A Double-Edged Sword*, New York: W. W. Norton.

Lowenthal, David (1958), *George Perkins Marsh: Versatile Vermonter*, New York: Columbia University Press.

Mertig, Angela G. and Riley E. Dunlap (1995), 'Public Approval of Environmental Protection and Other New Social Movement Goals in Western Europe and the United States', *International Journal of Public Opinion Research*, Vol.7, No.2, pp.145–56.

Mol, Arthur P.J. (1995), *The Refinement of Production: Ecological Modernization Theory and the Chemical Industry*, The Hague: CIP Data Koninklijke Bibliotheek.

Moomaw, William and Mark Tillis (1994), 'Charting Development Paths: A Multicountry Comparison of Carbon Dioxide Emissions', in Robert Socolow *et al.*, *Industrial Ecology and Global Change*, Cambridge: Cambridge University Press, pp.157–72.

Murdoch, Jonathan and Judy Clark (1994), 'Sustainable Knowledge', *Geoforum*, Vol.25, No.2, pp.115–32.

Organisation for Economic Cooperation and Development (1995), *Environmental Performance Review: The Netherlands*, Paris: OECD.

Pells, Richard H. (1997), *Not Like Us: How Europeans Have Loved, Hated, and Transformed American Culture Since World War II*, New York: Basic Books.

Putnam, Robert (1993), *Making Democracy Work: Civic Traditions in Modern Italy*, Princeton, NJ: Princeton University Press.

Radkau, Joachim (1997), 'The Wordy Worship of Nature and the Tacit Feeling for Nature in the History of German Forestry', in Mikuláš Teich, Roy Porter, and Bo Gustafsson (eds.), *Nature and Society in Historical Context*, Cambridge: Cambridge University Press, pp.228–39.

Renn, Ortwin, Jaeger, Carlos and Rosa, Eugene A., and Thomas Webler (1999), 'The Rational Action Paradigm in Risk Theories: Analysis and Critique', in Maurie J. Cohen (ed.), *Risk in the Modern Age: Social Theory, Science, and Environmental Decision-Making*, Basingstoke: Macmillan, pp.35–61.

Riordan, Colin (1997), *Green Thought in German Culture: Historical and Contemporary Perspectives*, Cardiff: University of Wales Press.

Rupke, Nicholas A. (1988), 'Romanticism in the Netherlands', in Roy Porter and Mikuláš Teich (eds.), *Romanticism in National Context*, Cambridge: Cambridge University Press, pp.191–216.

Sagoff, Mark (1988), *The Economy of the Earth*, Cambridge: Cambridge University Press.

Schama, Simon (1987), *The Embarrassment of Riches: An Interpretation of Dutch Culture in the Golden Age*, London: Collins.

Schmidheiny, Stephen (1992), *Changing Course: A Global Business Perspective on Development and the Environment*, Cambridge, MA: MIT Press.

Schooler, Carmi (1996), 'Cultural and Social-Structural Explanations of Cross-National Psychological Differences', *Annual Review of Sociology*, Vol.22, pp.323–49.
Simon, Julian L. (1986), *The Ultimate Resource 2*, Princeton, NJ: Princeton University Press.
Simonis, Udo E. (1994), 'Industrial Restructuring in Industrial Countries', in Robert U. Ayers and Udo E. Simonis (eds.), *Industrial Metabolism: Restructing for Sustainable Development*, New York: United Nations University Press, pp.31–54.
Simonis, Udo E. (1988), *Beyond Growth: Elements of Sustainable Development*, Berlin: Edition Sigma.
Snow, C.P. (1993 [1959]), *The Two Cultures*, Cambridge: Cambridge University Press.
Spaargaren, Gert (1997), *The Ecological Modernization of Production and Consumption*, Wageningen: Wageningen Agricultural University.
Te Velde, Henk (1996), 'The Debate on Dutch National Identity', *Dutch Crossing*, Vol.20, No.2, pp.87–100.
van Heerikhuizen, Bart (1982), 'What is Typically Dutch? Sociologists in the 1930s and 1940s on the Dutch National Character', *Netherlands Journal of Sociology*, Vol.18, No.2, pp.103–25.
Weale, Albert (1992), *The New Politics of Pollution*, Manchester: Manchester University Press.
Weber, Max (1946), 'Science as a Vocation', in H.H. Gerth and C. Wright Mills (eds.), *From Max Weber: Essays in Sociology*, Oxford: Oxford University Press, pp.129–56.
Wildavsky, Aaron (1995), *But Is It True? A Citizen's Guide to Environmental Health and Safety Issues*, Cambridge, MA: Harvard University Press.
Wynne, Brian (1995), 'Public Understanding of Science', in Sheila Jasanoff, Gerald E. Markle, James C. Petersen and Trevor Pinch (eds.), *Handbook of Science and Technology Studies*, London: Sage, pp.361–88.
Wynne, Brian (1996), 'May the Sheep Graze Safely: A Reflexive View of the Expert-Lay Knowledge Divide', in Scott Lash, Bronislaw Szerszynski, and Brian Wynne (eds.), *Risk, Environment, and Modernity: Towards a New Ecology*, London: Sage.

# CASE STUDIES FROM AROUND THE WORLD

*Advanced Industrial Countries*

# Putting the Ecological Modernisation Thesis to the Test: The Promises and Performances of Urban Recycling

DAVID N. PELLOW, ALLAN SCHNAIBERG and ADAM S. WEINBERG

## I. Introduction: Testing the Ecological Modernisation Thesis

In this contribution, we challenge one of the core hypotheses of ecological modernisation theory: that the design, performance and evaluation of processes of production are increasingly based on ecological criteria, in addition to economic criteria [*Mol, 1995, 1996; Spaargaren, 1997; Spaargaren and Mol, 1992*]. Ecological modernisation theory has quickly captured a substantial following among environmental sociologists in Europe [*Mol, 1995, 1996; Spaargaren, 1997; Spaargaren and Mol, 1992*] and increasingly in the United States [*Cohen 1998;* Sonnenfeld, this volume].[1]

Proponents of the ecological modernisation theory have been critical of other streams of environmental social science for focusing narrowly on the capitalist character of production, thereby ignoring much of the more robust and ecologically driven industrial aspects of production. In contrast, ecological modernisation proponents have argued from a meso-level institutional analysis. This approach gauges the extent to which the ecological sphere has become an independent sphere in corporate decision-making. Mol, for example, proposes that there are three analytic spheres, each of which is really a network characterised as a social system 'in which actors engage in more or less permanent, institutionalized interactions' [*1995: 63*]. The three spheres are:

(1) policy networks that concentrate on industry–government relations using a political-administrative perspective;

This contribution is based on a paper initially prepared for roundtable discussion at the American Sociological Association annual meetings, San Francisco, CA, August 1998. The comments of two anonymous reviewers are greatly appreciated. The authors also strongly acknowledge and appreciate the continued intellectual companionship of Arthur Mol and David Sonnenfeld; the errors remaining are theirs.

(2) economic networks that concentrate on economic interactions via economic rules and resources between economic agents in and around an industrial sector; and
(3) societal networks that concentrate on the relations between the economic sector and civil society organisations.

Ecological modernisation theorists have sought to demonstrate that, within each network, there are significant environmentally-induced institutional transformations [*Mol, 1995*].

Ecological modernisation theorists argue that industrial society is in a new era that began in the 1980s [*Spaargaren, 1997*]. As in previous eras, this one has been marked by new technologies, innovative entrepreneurs, and farsighted financiers who are bringing about a 'new wave of industrial innovation' [*Spaargaren, 1997: 17*]. This period, referred to as 'reconstruction', is marked by the emergence of an ecological sphere that exists independently from the other three spheres above. In perhaps the most concise statement, Arthur P.J. Mol argues:

> [T]he ecological modernisation theory focuses on the growing independence, 'emancipation' or empowerment of the ecological perspective or sphere from the basic three analytical spheres or perspectives in modern society: political, economic, and socio-ideological or societal [*1995: 64*].

To the extent that this is true, then we can state that the environment has been emancipated from these other spheres, and is beginning to constitute its own sphere – hence, *ecological modernisation*. Thus ecological modernisation is both a theory of industrial change and a 'normative theory' or 'political program' [*Mol, 1995; Spaargaren and Mol, 1992*]. As a theory of industrial change, ecological modernisation suggests that we have entered a new industrial revolution, one of radical restructuring of basic production processes along ecological lines [*ibid.*]. As a political programme, ecological modernisation advocates resolving environmental problems through 'harmonizing ecology and economy' [*Simonis, 1989*], through 'superindustrialisation' rather than deindustrialisation [*Spaargaren and Mol, 1992*].

In this study, we use the example of post-consumer waste recycling in the US to challenge these theoretical-political assumptions. Recycling is an interesting example, because one would expect it to be an ideal case of ecological modernisation. Remanufacturing post-consumer waste into new products should represent a shift by waste firms towards an ecologically-sustainable form of waste processing. We argue that, in fact, recycling did originally emerge in the United States during the 1960s as a social movement with a strong ecological basis. However, we trace the movement's goals as they quickly became transformed into an industrial practice with narrow market interests. Contrary to ecological modernisation theorising, recycling started as a *low*-technology practice with an emancipated ecological sphere. As new technologies have brought firms and financiers into the market, however, the ecological sphere has been recaptured and subsumed under the economic sphere. We offer three critiques of ecological modernisation, arising from our case studies:

First, there is no compelling evidence that the environment has been emancipated from the economic in decision making criteria. In fact, recycling demonstrates the robust character of capitalist production in at least two respects:

(1) the ability of market criteria to dominate the agenda, even in the face of strong public support for ecological protection, and
(2) the inability of ecological interests to penetrate organisational logics even when market opportunities exist.

Second, the modernisation of recycling appears to lead only minimally to a very narrow set of ecological gains. There is some reduction of natural resource withdrawals as virgin materials are replaced with recycled materials. These gains, however, come at the expense of more ecologically-sound forms of waste disposal. Recycling tends to drive waste reuse programmes out of the marketplace.[2] We refer to this as a quantitative gain at the expense of qualitative outcomes.[3] Furthermore, this form of 'modernisation' has brought workers into closer contact with environmental hazards. Workers sorting recyclables have to function in increasingly unsafe conditions, marked by close contact with biohazards and other toxic substances. Thus, there is an increase in environmental risks for workers along with some reduction in natural resource withdrawals.

Third, ecological modernisation is focused narrowly on ecological issues to the neglect of other, equally important components of the social process such as social equity concerns. We argue that any social theory that purports to account for social process has to incorporate issues of social equity and political-economic power. Ecological modernisation was earlier criticised for inadequately addressing equity issues in the developing world. We argue that ecological theorising has the potential to bypass social equity issues in the developed world, as well. This is particularly important for our case study because recycling production processes often have negative social equity outcomes.

In the sections that follow, we examine recycling as a potential model of ecological modernisation. We begin with an overview of the recent history of the modernising recycling sector. We follow with an in-depth examination of recycling in Chicago, to document what social processes modernisation actually creates. While case-based research always leaves itself open to the critique of generalisability [*Ragin, 1987*], we primarily use this case to dimensionalise and problematise a central claim of ecological modernisation theory. We selected the modernising recycling industry in Chicago, because it represents that sort of programme that is becoming commonplace in major urban areas. The case emerges from a larger research effort [*Schnaiberg, Weinberg and Pellow, 1998; Weinberg, Schnaiberg and Pellow, 1998*], where we have systematically examined a variety of recycling programmes that are emerging in the United States. Ultimately, we will leave it up to the reader and future empirical research to draw conclusions about generalisabilty to larger cases. We will come back to this issue in the concluding section of the study.

## II. Ecological Modernisation in the US: The Case of Recycling

### *The Modernising Recycling Industry*

Recycling has a long history in the United States. It can be dated back well into the last century, when recent immigrants collected and mended rags and other waste items, often for resale to low-income families that could not afford new items [*Melosi, 1981*]. Modern recycling first emerged in the late 1960s. The original programmes emerged from social movements active during that time. Small-scale social movement groups set up drop-off centres, where the homeless, immigrants, and low-income populations could retrieve items from trash cans in a process that became known as 'dumpster diving.' The materials were then taken to recycling centres, where they were 'dropped-off' in return for money. The social movement group then resold the materials to scrap metal dealers and other small local remanufacturing firms. The community-based drop-off centres allowed marginal social populations to squeeze out an existence, in a context where employment options were limited.

In the late 1980s, modern recycling arrived on the US urban policy agenda with considerable fanfare. It emerged from a confluence of factors. Recycling emerged at exactly the moment when waste disposal firms needed to negotiate a truce with ecological organizations, local communities, and the state [*Schnaiberg, 1997; Weinberg, Schnaiberg and Pellow, 1998*]. The early 1980s seemed to be a supportive climate for waste firms. The Reagan administration swept into power with an expressly anti-environmental and pro-business orientation. The new administration openly flouted environmental agencies, allowed the violation of existing laws, and lobbied to undercut and/or prevent the passage of new legislation [*Landy et al., 1990; Szasz, 1994*]. Waste firms were able to rely increasingly on various forms of 'sanitary landfills'. They were also operating local or regional sites for toxic waste disposal, through incineration and/or 'safe' landfilling [*Portnoy, 1991*]. By the mid-1980s the political climate was more tenuous. National opinion polls showed high levels of public anxiety about pollution and human health. Public anxiety was sustained by a succession of technological disasters, including the nuclear meltdowns at the Three Mile Island nuclear reactor in the US, and the Chernobyl reactor in the Ukraine [*Kitschelt, 1986*].

Environmental groups used these fears to mobilize their constituencies. By the mid-1980s, it had become harder to site new landfills [*Portnoy, 1991*]. Both constituencies, however, were also aware of and concerned about toxic chemicals. This was a period of time when a number of well-publicised cases highlighted the dangers of toxic waste contamination around the nation and the world. These included Love Canal in New York and Times Beach in Missouri. Internationally, citizens read about the dispersion of radioactive wastes from Chernobyl's nuclear power plant, as well as the toxic disaster at the Union Carbide plant in Bhopal, India. There was also a host of more localised incidents [*Brown and Mikkelsen, 1997 (1990)*] in the same period, often discovered in the process of implementing the Resource Conservation and Recovery Act of 1976. This increasingly mobilised people to reject *any* waste facilities in their backyards.

This newly focused attention on waste facilities made siting new landfills or incinerators nearly impossible [*Szasz, 1994*]. The anti-toxics and environmental movements successfully prevented the siting of hundreds of landfills across the United States during the 1980s, prompting many observers to declare a 'landfill crisis'. While most media outlets interpreted this crisis as stemming from a lack of landfill space, its origin was actually rooted in the increase of political resistance against siting rural and urban dumps.

In contrast, there had been a 'business as usual' transnational transfer of many wastes from industrial to underdeveloped countries. Nationally, there was also a redistribution of hazardous wastes from more affluent northern US cities to less affluent Native American reservations, as well as to more rural and southern areas [*Moyers, 1992*]. Within metropolitan areas, hazardous waste sites were increasingly centred in poorer and often minority communities and neighbourhoods. This in turn raised new social and political concerns about environmental injustice in the disposal of wastes [*Portnoy, 1991*]. Environmental injustice claims offered yet another obstacle to siting and expanding new or existing landfills to include both municipal and industrial wastes.

The infamous Mobro 4000 and its 'Garbage Barge' journey further exacerbated this controversy. In 1987 a garbage barge filled with municipal waste from New York City sailed down the East Coast of the United States. It continued to the Bahamas, Belize, and Mexico, being denied entry at each port. After 6,000 miles of sailing, the ship returned to New York City, where its wastes originated, finally destined to be buried on Long Island itself.

This 'landfill crisis' created a new set of urban bedfellows. The state needed to find ways to alleviate pressure to site new landfills. Firms had to deal with shifting consumer perceptions about waste disposal. And environmental groups were anxious to harness renewed public enthusiasm into a proactive environmental practice. Recycling emerged as a solution that would serve everyone's agenda. The central promise was a new 'urban alchemy'. Private firms would take garbage and turn it into a resource. Rather than *paying* to place waste into landfills, municipalities would instead now *make* money selling garbage to firms; they would in turn re-sell the materials to other remanufacturing firms, who would make new products.

In the process, firms would save municipalities money and create jobs. They would also preserve natural resources. Ecological withdrawals would be minimised as recycled materials would be substituted for virgin materials. Ecological additions would be reduced as fewer materials would be placed into landfills. Cities across the country worked with industry to create recycling programmes. At this point, the political stage actually shifted away from pre-existing environmental groups. Recycling served a variety of political agendas, and thus it was defined differently by various organisational actors. The environmental movement had created a new maxim: 'reduce, re-use and recycle'. Businesses tended to emphasise only the last of these. The reason is clear: by focusing on consumers and their wastes, two goals would be achieved:

(1) industrial wastes would have more landfill space in the future than if they had to share such space with all the normal municipal wastes; and
(2) attention was shifted from producers to consumers, inducing the latter to become 'environmental' by recycling their solid wastes. This shift also saved industry a great deal of money that would have had to be sunk into capital, new technology, labour and permits.

Many of the community-based environmental groups were thus locked out of the negotiations about recycling operations. While they had developed the concept of recycling, the actual details of recycling *programmes* tended to be worked out behind closed doors by local municipal leaders and industry representatives, especially those from large waste firms. Waste firms and local governments jointly developed large-scale municipal kerbside programmes. Residents would place materials at the curbside, where they would be picked up and transferred to a Materials Recovery Facility (MRF). At the MRF another set of workers would sort and bale materials for resale to firms using recyclables for their production processes.

Operational details varied across cities, especially relating to whom owned the MRF (the city or the waste disposal firm), and who picked up the materials (city workers, or waste firms under contract to the city). A number of factors, however, were remarkably similar across cities:

(1) social movement-based community drop-off centres found this modernised recycling industry politically and economically difficult, and
(2) programmes were guided by a narrow set of economic goals that pushed out the environmental and equity concerns that had traditionally guided social-movement based recycling programmes.

Tables 1 and 2 document some of the important trends in this emerging industry. Table 1 documents the tremendous growth in recycling in the US.

Between 1980 and 1996, the amount of municipal garbage being recycled in the US more than doubled, increasing from 9.6 per cent to 21.9 per cent of the waste stream. In contrast, the amount of waste landfilled decreased from 81.4 per cent to 55.4 per cent. Along with these changes, we also observe a rise in the number of kerbside recycling programmes (see

TABLE 1
MATERIALS RECYCLED, INCINERATED, AND LANDFILLED FROM THE MUNICIPAL SOLID WASTE STREAM IN THE US (IN THOUSANDS OF TONS) AND PER CENT TOTAL GENERATION, 1960–96

| | | | YEAR | | |
|---|---|---|---|---|---|
| | **1960** | **1970** | **1980** | **1990** | **1996** |
| Recycled | 5,610<br>(6.4%) | 8,020<br>(6.6%) | 14,520<br>(9.6%) | 29,650<br>(15%) | 46,610<br>(21.9%) |
| Incinerated | 27,000<br>(30.6%) | 25,100<br>(02.7%) | 13,700<br>(9%) | 31,900<br>(16.2%) | 36,090<br>(17.2%) |
| Landfilled | 55,510<br>(63%) | 87,940<br>(72.6%) | 123,420<br>(81.4%) | 131,550<br>(66.7%) | 116,240<br>(55.4%) |

*Source*: Franklin Associates, Ltd, *The Future of Solid Waste Management and Recycling*, Multi-Client Study, November 19996, Draft.

TABLE 2
INDICATORS OF GROWTH IN THE RECYCLING INDUSTRY AND RECYCLING PROGRAMMES IN THE US, 1990–96

| | YEAR | | |
|---|---|---|---|
| | **1990** | **1995** | **1996** |
| Number of MRFs | NA | 310 | 363 |
| MRF Capacity (in tons per day) | NA | 32,000 | 29,400 |
| Number of Mixed Waste MRFs | NA | 34 | 58 |
| Mixed Waste MRF Capacity (in tons per day) | NA | 20,000 | 34,800 |
| Total MRF Capacity (MRF + Mixed Waste MRF) (in tons per day) | NA | 52,000 | 64,200 |
| Number of Kerbside Recycling Programmes | 2,700 | 7,375 | 8,817 |

*Sources*: USEPA, *Characterization of Municipal Solid Waste in the United States, 1996 and 1997*; and Jennifer Carless, 1992, *Taking Out the Trash: A No-Nonsense Guide to Recycling*, Washington, DC: Island Press.

Table 2). In just six years (from 1990 to 1996), the number of these programmes more than tripled, increasing from 2,700 to 8,817. By 1996, there were 8,817 [*Carless, 1992; US Environmental Protection Agency, 1996, 1997*]. More important to us is the changing form of the recycling programmes. This is more difficult to gauge from existing industry summaries. The rise of large municipal programmes has been so unproblematically accepted that there is no reliable count of municipal versus drop-off programmes. Instead we use a variety of indicators listed in Table 2.

From 1995 to 1996, we observe a steady increase in the number of MRFs being built (53 in a single year). MRFs are places where recyclables are taken to be sorted and baled for resale. MRFs are distinct from drop-off centers because the former often accompany the rise of municipal programs, as places where the waste hauling organisations can take and sort large quantities of materials. They are places of centralisation that are necessary for large-scale recycling programmes. More importantly, we see a substantial rise in *mixed-waste* MRFs, from 34 to 58, with processing capacity growing from 20,000 to 34,800 tons per day. Mixed-waste–or 'dirty'–MRFs are also a product of municipal programmes, which create more 'efficient' operations, by collecting both solid waste and recyclable materials in the same vehicle. They have modernised recycling: large firms using sophisticated machinery to cart and sort high quantities of recyclables, as an integrated part of municipal waste hauling. This is in contrast to the many 'source-separated' recycling systems where garbage and recyclables are sorted by residents at the curbside and picked up by two different trucks.

The conventional narrative has been that recycling – like other environmental initiatives by industry – emerged as waste firms developed an environmental consciousness [e.g., *Szasz, 1994*]. We argue that this narrative is misguided on two accounts. First, the modern history of recycling has little to do with ecological consciousness and more to do with political-economic forces. Second, current organisational practices of waste firms reveal little environmental accounting or stewardship. With this shift to large industrial processing, then, there has been a dramatic change in recycling: a movement away from social and ecological criteria in the decision-making, towards a set of outcomes that are socially regressive. We now turn to the history of Chicago to unpack these dynamics.

*The Early History of Recycling: The Drop-Off Centres*

In Chicago, recycling services were provided by only one organisation for several years – the Resource Centre. This is a drop-off centre in Chicago's Grand Crossing neighbourhood, a community that is 99 per cent African American, with a third of its residents living in poverty. The Resource Center served mostly low-income people who scavenged recyclables out of alleys and garbage cans and brought them to the center, in return for cash. Ken Dunn, a legend in Chicago's environmental movement and community development circles, managed the centre. Dunn came to Chicago in the late 1960s. Like many activists associated with the early recycling movement, Dunn was a Vietnam War protester, counterculture activist, a Peace Corps participant, and graduate student drop-out. As he put it, 'I wanted to do something of value to the community, and being a graduate student at the U. of C. [University of Chicago, Philosophy] was about as far from that vision as you could get!'

By the mid-1970s, the Centre expanded beyond its drop-off operations. It developed collection routes, where residents would leave things on the corner or by garbage bins. This provided much-needed jobs for the growing ranks of low-income minority men who had been disenfranchised and marginalised as Chicago began to deindustrialise. The Centre also had recycling and composting contracts with several neighbourhoods and businesses around the Chicago area. Finally, the Centre branched out into more proactive projects, taking vacant lots and abandoned buildings and 'recycling' them into playgrounds, parks, and gardens. The construction materials extracted from the buildings were even recycled in a process people began to call 'green demolition'.

None of these practices was very lucrative. Their motivations, however, were primarily social and environmental, and secondarily economic. The people behind the Centre wanted to ensure that the recycling programmes were directly supporting the community in two ways:

(1) creating links between environmental cleanup and job creation; and
(2) turning empty lots into productive ones, keeping money circulating within the community.

In the 1980s, Uptown Recycling Inc. (URI) joined the Resource Centre on the North Side of Chicago. Like the Resource Centre, URI's surrounding community was poor and populated by racial minorities. Fully one-fourth of the community's residents were below the poverty line, and many of those were homeless. Both recyclers operated similarly. Local community members, who came from a variety of walks of life, directed them and they were united in viewing recycling as a social movement that was beneficial for the community, the environment, and the local economy. They operated low-tech centres focused on ecological and social criteria. The following is an excerpt from our field notes at the URI site:

> We are standing in an aluminum shed at the Uptown yard surrounded by a can crushing machine, weigh scale, a coffee machine, and a wooden sign leaning up against the wall displaying the prices for a great variety of materials from scrap metal to kitchen sinks. Every couple of minutes very rough-looking people (possibly homeless, definitely under- and unemployed and over thirty years of age) with grocery carts, garbage bags, and children's wagons bring pounds of materials to sell. The site manager, Souma Phosaraj, is Southeast Asian and looks about 50. Souma oversees all monetary transactions. It is important to understand that a recycling 'yard' is just that. It's an outside, open air yard. It has the best ventilation system possible, really. Ventilation is often cited as one of the chief contributors to poor working conditions in MRFs. Here this isn't a problem. However, you are then left at the mercy of the elements [*Fieldnotes, Fall 1996*].

By the mid-1980s, both recycling centres were running smoothly. They were providing employment for workers whom a manager noted, 'would otherwise be considered unemployable.' They had also become important local advocates for progressive community development programmes. They consciously undertook campaigns to raise awareness among constituents, including the disenfranchised individuals who often worked for them. Thus, the centres came to embody three goals:

(1) to divert recyclable materials from the waste stream, thereby reducing Chicago's dependence on landfills and incinerators;

(2) to provide entry-level jobs and a source of income for people in the struggling communities; and
(3) to educate people about the environmental and economic benefits of recycling.

In 1987, the City of Chicago recognised the positive benefits of the community-based recycling centres. The City awarded Uptown Recycling Inc. (URI) a modest 'diversion credit' for its collection routes. Diversion credits were small amounts of money paid to URI, for diverting garbage from the waste stream and landfills. Such credits were meant to give URI the money that the city would otherwise have to pay for landfill usage. With the diversion credit and growing community interest in recycling, both URI and the Resource Centre were fairly stable. For example: URI's weekly household collection programmes grew from collecting around 9–10 tons per month in 1986 to 56 tons per month in 1990. They were collecting recyclables from approximately 19,000 households, with a resident participation rate that reached 80 per cent at one time. URI also published a newsletter with a circulation of 2,000. The staff was usually overwhelmed with the ever-growing requests for speaking engagements, tours of the facility, and general information on residential recycling options in the area.

Interestingly, the people associated with these non-profit recycling centres are proudest of the social goals these organisations achieved during this period. In interviews, people mentioned the ability of the Resource Centre and Uptown Recycling, Inc. to provide employment for people struggling in the deindustrialising economy, and to provide educational campaigns for the community on social and ecological issues. A board member of the Uptown Recycling, Inc. (URI) explains:

> What I'm hinting at is that when Uptown Recycling was getting started, it was very definitely still a [social] movement for these folks because their motivation was based upon a sense of the good, not upon a sense of a cost-benefit analysis or is it cheaper for them to have recycling than to have a larger dumpster. It was a sense of meaning. A qualitative evaluation rather than a quantitative calculation.

In the early 1990s, the City of Chicago began to think more seriously about a larger-scale programme. The City of Chicago needed a system that would bring it into compliance with city and state laws. Illinois law required that Chicago have a solid waste management plan that would achieve a 15 per cent recycling rate by 1994 and 25 percent by 1996. The City was also being pressured by large waste hauling firms to form a partnership to increase recycling opportunities in this large municipal market (more than three million residents). Traditionally, the waste firms hauled trash, leaving recycling to the non-profit, community-based recycling centres. In the late 1980s and early 1990s this began to change. Technological innovations opened up new possibilities for these firms. This was especially true for Waste Management (WMX), the largest waste hauler in the world, which was headquartered in the Chicago area. In particular, new extraction technologies offered the promise of retrieving materials from the raw municipal waste stream. If recyclables could be gathered along with the normal waste stream, then recycling would no longer depend upon people's social consciousness in placing materials on the street. It would be possible to recycle large quantities of recyclables, for the first time making the recycling process potentially financially lucrative.

During this time, a local political conflict provided a catalyst for the City's decision-making process. A coalition of environmental groups called the West Side Association for a Safe and Toxic-free Environment (WASTE) began to challenge the city's efforts to keep open the Northwest Incinerator, which was eventually closed in 1996. They argued that, on social, economic, and environmental grounds, recycling was a sounder option. In turn, this placed even more pressure on the City to recycle, and more pressure on waste firms to develop new waste-disposal practices to replace incineration. Waste Management, Browning Ferris Industries (the world's second-largest waste hauler) and other smaller-scale regional waste hauling firms saw recycling as a golden new political and economic opportunity.

*Modernising Recycling Practices: The Emergence of the Blue Bag Programme*

In 1990, the City of Chicago announced a Request For Proposals (RFP) for developing a comprehensive, city-wide, residential recycling programme. The RFP came as a surprise to people who were familiar with the City's recycling community because it specified that the programme was to be city-wide. By arbitrarily refusing to consider bids for separate sectors within Chicago, it *ipso facto* excluded the full array of community-based recycling centres. Ann Irving, the executive director of a local environmental organisation called the Chicago Recycling Coalition, referred to the RFP process an example of 'bald-faced power playing by a corporation with a monopoly'. Irving and others hinted that the RFP was carefully written to target Waste Management as the ultimate contractor. The Coalition believed that the local recycling community had essentially been locked-out of the proposal. Many activists felt that the non-profit operations like the Resource Center and Uptown Recycling, Inc. had test-piloted a programme, which in turn was then stolen from them. Nobody was surprised when Waste Management was awarded the contract.

Over the following months, the City worked with Waste Management to develop the idea of a 'Blue Bag' programme. The programme required residents to place their recyclables in blue plastic bags. These would be collected along with residential trash (municipal solid waste), in regular garbage trucks. Trucks would compress their loads (to increase efficiency of pick-up costs), and then dump them at new 'Material Recycling and Recovery Facilities' (MRRFs). Both Blue Bags and regular garbage bags would be pulled out and their contents separated manually. Recyclable materials in regular garbage bags would also be pulled out of the stream for hand sorting.

Immediately, people in the recycling community expressed outrage at the lack of social or ecological criteria in the decision making process. The Chicago Recycling Coalition, the non-profit recycling centres, and others in the recycling community, made the following criticisms. First, the Blue-Bag operating system was selected because it was the cheapest way to start a programme. A leading trade publication summarised: 'the primary reason given for adopting the commingled bag/MRRF recycling program is its affordability' [*Solid Waste Management Newsletter, 1990*]. In particular, the Blue Bag programme fit right into WMX's existing structure, with few changes. To quote a former Waste Management manager:

> In 1991 the City went out to look and see how should we recycle and one of the things that they saw is that a lot of places have kerbside programs and they looked at the cost of that. The cost – because you end up sending two trucks down an alley, it was cost prohibitive. So they looked at the Blue Bag programme. Where possible they have externalized and/or minimized costs. For example, the decision to combine municipal solid waste with recyclables in collection and sorting was based on cost minimization. No new trucks, routes, or haulers were needed.[4]

Waste Management focused most of their efforts on cutting costs. The MRRF sites themselves were situated on environmentally-polluted land, because it was cheap. One of the MRRFs was constructed on an old landfill on Chicago's Southeast Side. This is an area of the city that activists commonly refer to as 'the Toxic Doughnut' because it is so inundated with polluting industries that residents are surrounded by toxic facilities on all sides. Finally, the buildings were designed with attention paid only to issues like waste volume and shipping/receiving spaces for trucks. No consideration was made for energy efficiency or ecological design. The buildings had no heat or air-conditioning. In the winter, temperatures were frigid. In the summer, the combination of heat and humidity, rotting garbage, machinery, and people made the building extremely hot.

Second, the decision-making ignored the minimised ecological gains from the programme. The Chicago Recycling Coalition (CRC) led the way in attacking the Blue Bag as a fiscal and ecological nightmare. Ann Irving, then Executive Director of CRC, described to us how the programme was designed solely on economic and political considerations:

> I think the most telling thing about the relationship between the city and WMI was that … the city chose this programme, decided it was going to go ahead with this lengthy process of writing an RFP and during that process there was open discussion about what this programme was going to consist of, but the city was a little cagey as to what it was precisely going to ask for in the RFP. But what it was very up-front about was they were arguing that the contractor would be asked to provide the capital in order to construct the facilities. And that aced out a lot of smaller waste haulers in the area who might have been very interested in doing it. And the city kept arguing that it couldn't accept people like them because they just couldn't be sure that they would have the capital resources necessary to go with these very large facilities, the price of which of course would increase–even double – by the time it was built.

Not surprisingly, the Blue Bag system did not work very effectively. As with regular garbage bags, when Blue Bags were compressed in garbage trucks, they broke. By the time the bags got to the MRRF, the loads consisted of wet trash and contaminated recyclables. Some materials were too contaminated to be saved. None the less, workers were confronted with the task of having to fish out the recyclables from a sea of garbage. A single contaminated load of recyclables can ruin an entire production batch in the remanufacturing process. Contaminated recyclables therefore have much lower exchange-value. For example, there are over 90 varieties of paper. If they are not sorted and a remanufacturer tries to use recycled paper to make a higher-grade paper product, the resulting batch will be defective. An internal memo from the Chicago Recycling Coalition stated this quite forcefully:

> The [Blue Bag] system mixes all recyclable materials together in one bag. Recycling industry representatives say that much of the material will be poor quality and difficult to recycle. FSC Paper, the area's main newsprint buyer, has said that newspaper contaminated with glass shards will damage its machinery. If the city is unable to sell the materials, they will have to be landfilled or incinerated, which defeats the whole purpose of the programme. The city and Waste Management will not be able to sell these low grade materials for top dollar, so the overall cost to taxpayers is likely to be higher.

For cost-saving purposes, Waste Management also refused to either hire trained workers or to slow down the production line. Waste Management uses a temporary job service, Remedial Environmental Management (REM), which effectively operates as a day labour exchange. These oppressive relations of production and the associated low recovery rates have hurt the firm's productivity. Figures released by the City of Chicago and Waste Management as late as fall of 1997 reveal that the percentage of materials (excluding yard waste) that is actually recycled is averaging around five or six per cent.

The following composite description of the MRRF operations reveals a graphic portrait of the social problems generated by the new programme. The composite is based on our interviews with workers and foremen in their homes, who, like the rest of Waste Management's MRF workers, were all African American:

> It is 7 a.m. in the morning. You are a black woman. You are standing in a huge facility (400 yards long) It's freezing cold because there is no heat. You have just walked 1.5 miles because the facility is not reachable by public transportation. And you are too poor to own a car. You are going to spend the next 10-12 hours (often you do not know how long) standing on an assembly line sorting through raw garbage straight from trash cans. You may or may not have protective gloves, so you will have to be careful. Coming down the line could be: hypodermic needles, dead animals, live rats, broken glass, and on the odd day a baby or human body parts. You have seen co-workers splattered with battery acid, and picking up leaking bags marked: 'biohazard'. To quote one of your co-workers: 'I can't remember the first guy who got stuck by a needle ... The guy got stuck by a bloody needle. You don't know whose needle that was. Hopefully, he didn't get infected with HIV, or Hepatitis A or B ...' This worker goes on to tell us that this man was lucky compared to one of his co- workers who picked up a bag of asbestos that came down the line. For this you will be paid $6 an hour, and guaranteed employment for 89 days, at which time you will be fired one day before the 90 days needed for unionization and other benefits to start ... .

This description stands in stark contrast to the drop-off centres. The Blue Bag programme marked an new era in recycling.

### *The Demise of Ecological and Social Roots: Attacking the Drop-Off Centres*

Unfortunately, the City's support for the Blue Bag programme gradually led the administration to withdraw its support for the non-profit recyclers. As Chicago's Blue Bag system became more expensive, it increasingly squeezed out the community-based centres. Beginning 1 January 1996 the

City only provided 25 per cent of its previous support to the non-profits. One year later, the City withdrew 100 per cent of this support.

The non-profit recyclers – URI and the Resource Centre – felt the negative impact of the Blue Bag programme on a daily basis. For example, the non-profit centres started to experience tension with the City of Chicago over the centres' contracts to provide recycling to Chicago's public schools. Officially, under the Blue Bag programme, the choice to use the Blue Bag in a Chicago public school is left to the discretion of that school's principal. Unofficially, however, City employees from the Department of Environment (which oversees the Blue Bag programme) have 'paid visits' to teachers and principals of several high schools. They pressured school officials to cancel their agreements with non-profit recyclers in favour of the Blue Bag.

Finally, because the Blue Bag system was supported by a new municipal ordinance, if Uptown Recycling's residential customers wished to continue recycling, they were legally bound to support the City's Blue bag programme. At the drop-off centre, URI workers complained of dwindling recyclables as the Blue Bag system went on-line. Moreover, a series of ordinances were passed, making it difficult for drop-off centre workers to 'scavenge' alley and trash cans for recyclables. This was a variation of the classical aphorism that 'The rich and poor are equally forbidden to scavenge in alleys'. Finally, workers complained that local law enforcement officers were increasingly giving them a hard time.

The manager of Resource Management, a private suburban Chicago recycling firm, commented on the transition that Chicago's recyclers were experiencing:

> recycling today – because rags, scrap metal, hides were all recycled 100 years ago – but … the post-consumer kerbside pickup, that's fairly recent. I think that type of activity was initiated by not-for profit interests who were interested in making the world a better place to live. [These were people] who had not a profit motive, but were really just doing it because it was something that they believed in. And they did a good job at that level. None of them were really financed properly. What's happened now, is the transition between that time and the time that business got involved because it made economic sense.

By withdrawing support and then coercing others to do the same, the City and Waste Management effectively crippled the non-profits. URI was ultimately forced to close its doors in 1997. The Resource Centre continued to survive but just barely. To quote a manager from the Resource Centre, 'We're not breaking even by any measure of the word …'. In short, their future is highly problematic.

## III. Discussion: Three Critiques of Ecological Modernisation

In this analysis, we challenge one of the core hypotheses of ecological modernisation theory, namely that the design, performance and evaluation of processes of production are increasingly based on ecological criteria in addition to economic criteria. Drawing on data gathered from the US recycling industry, we have raised three critiques:

*Critique No.1*

There is no compelling evidence that the ecological sphere has been emancipated from the economic sphere in decision-making criteria. In this case, it appears that prior social and ecological spheres have been suppressed under a narrow economic agenda. We note the robust character of capitalism has shaped the modernising recycling industry in at least two respects: (1) the ability of market criteria to dominate the agenda, even in the face of strong public support for ecological protection, and (2) the inability of ecological interests to penetrate organisational logics even when market opportunities exist.

This irony is not lost on local activists. Alderperson Toni Preckwinkle, of Chicago's fourth ward, wrote in a local newspaper that:

> Hyde Parkers have had access to quality recycling programs for nearly thirty years. Ken Dunn's pioneering efforts allowed many of us to feel that we got in on the ground floor of an exciting movement. ... After a year and a half of operation, it seems appropriate to consider how effective the Blue Bag program is in Chicago. That isn't as easy as it sounds. A great deal of money has gone into promoting the Blue Bag program. My office has special blue plastic containers, provided by the city, in place by each desk. City employees haul a giant inflatable Blue Bag from festival to festival. The official message about the Blue Bag program is advertised in every medium. However, while I can tell you the reading scores of sixth graders at every school in my ward, I don't have a lot of facts about the Blue Bag program readily at hand. The official statistic is that ten percent of the city's households participate. We don't know what percent of Chicago's refuse stream is impacted by the program. By way of comparison, in Beverly-Morgan Park, the Resource Center reached recycling 26 per cent of the refuse stream with a local participation rate of 70 per cent. If it takes 70 per cent of the citizens to recycle 26 percent of the refuse stream, then ten per cent of the citizens are probably recycling a very small percentage of Chicago's refuse stream. Given the official blessing of the Blue Bag program and the dollars allotted to promotion, this is troubling. Even more troubling is the fact that the

> City of Chicago Department of Environment held up a $482,196 payment due to Waste Management earlier this year, because the company failed to recycle paper, plastic and glass as promised [*Preckwinkle, 1997*].

Ecological criteria were absent in the decision making process about production practices. Interviews with recycling centre managers clearly suggested that there were structural incentives for using more ecological 'words' than 'deeds'. The Blue Bag programme was marketed to the public, prior to its actual implementation, as a 'green' programme. A great deal of money was spent on multi-media marketing and a school-based education initiative throughout Chicago.[5]

During the project development process, however, environmental criteria threatened to raise costs and lower efficiency, making them politically and economically untenable. This difficulty appeared in simple early decisions over truck purchases. No consideration was given to purchasing a cleaner type of truck. It recurred in decisions about production lines. A decision was made to recycle waste in ways that would be efficient but produce a low quality recyclable product. This minimises the capacity of recycled materials to reduce the withdrawal rate of natural resources, since low-quality recyclables require a higher addition of virgin materials in the remanufacturing process in order to maintain material properties. Finally, this same pattern emerged in management decisions to ignore occupational and environmental hazards, which have added more dirty production facilities in communities already burdened with toxic industries.[6] This pattern of decision-making has been repeated by many waste disposal firms across the county. Political and economic considerations thus shaped and drove the City's and Waste Management's adoption of a system that provided standardised, convenient, low-cost, and 'efficient services', capturing economies of scale that could generate profits and create new jobs (albeit hazardous and low-paying ones).

Embedded in this discussion is our critique about technological choice. If and when ecological principles have played a part in this process it is only where they save corporate entities money, which will likely then be reinvested in extracting or polluting processes. This dynamic is again demonstrated by the Waste Management case study. The Blue Bag programme does not take advantage of new technologies that could minimise natural resource withdrawals and additions. Old and dirty trucks are used for pick-up. Moreover, separate recycling trucks could have been used that would have produced more jobs, and cleaner recyclable products (that is, 'source separated' instead of commingled) and higher market returns. The facility eschews technologically efficient sorting, in favour of using a combination of low-income unskilled laborers working on a dangerous high-tech system. This arrangement produces an inexpensive but 'dirty' commingled recycling stream. Technological storing innovations are not used to contain waste waiting to be sorted. This failure serves to attract vermin and exports potential public health problems to the surrounding community. The only place where ecological concerns might have played a role was in the effort to extract as much was as possible from the garbage stream. However, this activity was driven by the overriding economic motive of generating revenue from waste.

*Critique No.2*

The modernisation of recycling appears to lead only minimally to a very narrow set of ecological gains. Many reviewers of our work have commented that, despite our critique, there *are* ecological gains none the less associated with municipal recycling practices, like Waste Management's Blue bag. This is true, but such gains are minimal. Clearly, there is some reduction of natural resource withdrawals. The Blue Bag programme is recovering millions of pounds of recyclables, thereby lessening the extraction of virgin materials for production. These limited gains, however, are well below what more ecologically sound forms of waste disposal could have achieved in this period. Moreover, these gains come at a high social cost.

For example, the Blue Bag recycling system drove existing reuse programmes out of the market-place. We refer to this as a quantitative gain at the expense of qualitative outcomes. Furthermore, modernisation has brought workers into closer contact with environmental hazards, as workers sorting recyclables work in increasingly unsafe conditions, marked by close contact with biohazards and other toxic substances. Thus, there is an increase in environmental risk even where there are some natural resource withdrawal reductions.

As cost became the driving force, Waste Management paid even less attention to working conditions. We interviewed more than two dozen workers and managers who were employed by WMX in the Blue Bag system. Their stories resemble those reported by laborers in the sweat shops, steel mills, coal mines, textile mills, and meat factories of nineteenth-century America, and the contemporary Third World. Health and safety hazards include a myriad number of threats to worker well-being.

Recycling sorting centres are not normally thought of as workplaces that process or produce chemical toxins. But Chicago's mixed-waste 'dirty' MRRFs are much more noxious than 'clean' MRFs. Workers in the dirty MRRFs routinely have to handle toxic substances. Because household hazardous waste is unregulated, recyclable plastic and metal containers that recycling centres collect often contain residues of toxic wastes. As one worker explained, he comes into close contact with 'anything and everything that people just normally throw out in their garbage'. This included bleach, battery acid, paint and paint thinner, inks, dyes as well as razor blades, and homemade explosives.

Recycling organisations are also not organised to process medical wastes. Yet MRRF workers also routinely handle these materials. Several workers became punctured by syringes and hypodermic needles. These are some of the most common and potentially lethal accidents that can occur in materials recycling and recovery facilities [*Powell, 1992*]. Specifically, workers who were stuck by needles were fearful of contracting HIV. An ex-Waste Management manager, later turned whistle blower, offered the following institutional analysis of the rise of medical wastes in household garbage streams:

> Let's take for example, the medical waste issue alone ... in terms of the whole medical field it now has changed. Fewer and fewer people are allowed to stay in hospitals. Most, practically every procedure that they can think of that they could put into an outpatient basis – they're doing it. Which means that people are taking all kinds of hypodermic needles, colostomy bags, and all this stuff home and disposing of it in the garbage. Just say, for example, all the people who are diabetics – all of the people who are forced out of the hospital because their insurance will not allow them to stay any longer, they feel like they can be better taken care of at home. Now they're sending in nurses; there's a whole network that they send out to people's houses. The reason I know this is because my dad just had serious surgery not too long ago. And he was taking all different kinds of injectables and ... he had a colostomy bag for a while. He's fine now, he still has a nurse visiting but he's not injecting anything anymore. But, my point is just think of all the people who have a legitimate use for hypodermic needles – have a legitimate, a hospital-prescribed use for all of these items that are normally disposed of in a hospital setting.

Later discussions with a practicing health professional confirmed that these practices are indeed widespread among hospitals. This outcome is due in large part to the continued restructuring of the US private health care industry. These environmental hazards add a new and disturbing dimension to the discourse around 'the health care crisis' in this nation. Our critique is not of the medical industry, but of the recycling industry. What has made for new biohazards to *recycling* workers is the fact that recyclable materials are *commingled* with raw garbage, which includes these medical biohazards. This is the 'dirty' component of the 'dirty MRRF' nature of Chicago's Blue

Bag programme. Let us be clear: for the sake of reducing the pick-up costs of recyclables, Waste Management and the City of Chicago have eagerly traded off recycling workers' health for the improved profitability of the Blue Bag programme.

Workers also experienced shock and stress on a routine basis. For example, Edward, a former employee, told of a grisly incident that occurred during an evening shift:

> I worked in the primary department. That's where the trucks dump raw garbage right there. One time a dead lady was dumped on the floor in front of me ... One woman [employee] fainted and everybody else was screaming. A couple of guys were just wandering around on the catwalk [a 40 foot structure] looking like they was dazed.

Later, at the same MRRF, two deceased human infants were discovered on the recycling line on consecutive days. Thus, psychological and physical hazards intermingle as people desperate for gainful employment and job security are pressured to continue working in the face of gross health and safety violations.

Finally, Waste Management also saved costs by simply not installing any heating or air conditioning systems in their MRRFs. In Chicago, this means the facility is typically unbearably cold or overwhelmingly hot. Raw garbage, especially in the warm months, generates sufficient odours to make many employees nauseous.

*Critique No.3*

In addition to the first two critiques, above all, we stress here that ecological modernisation has focused narrowly on ecological issues to the neglect of other, equally important components of the social process, such as social equity concerns. The trajectory of the Blue Bag and other recycling systems has negative social equity outcomes. By hiring temporary workers through REM and commingling medical and municipal solid waste with recyclables, Waste Management has constructed a system of hazardous and degrading labour practices. Without a vacation, the worker might earn a gross income of $12,500 a year (assuming periodic overtime earnings)–about half of what a Chicago sanitation worker took home in 1988.

Secondly, the jobs are dead-end. The REM temporary firm's involvement seems to ensure that most of the workers will only be at the facility for a brief period of time. Even if the pay were good, the worker is not employed long enough to get his/her family back on its feet. But, even if they were at the facility, they do not acquire any employable skills. Although the facility is a high-tech structure, most of the jobs are low-tech. Hence, these recycling jobs neither support the community (through wages for families) nor the future prospects of workers (by increasing human capital).

Finally, the MRRF jobs have created ill-will in low-income ethnic communities. The MRRFs have continually used strong-arm coercive management styles to maintain production schedules. Several workers spoke to journalists about the deplorable health and safety conditions in the plants. In response, REM issued a memo to its employees, 'strictly prohibiting' any communication with the media. Workers were explicitly instructed to respond with 'no comment' to any inquiries about working conditions in the MRRFs. They were warned that 'violation of this work rule may result in disciplinary action, up to and including immediate termination of employment'.

Unfortunately, this was only the beginning of a systematic pattern of exploitation by management. Workers regularly complained of being harassed by foremen and managers who rarely let them leave the sorting lines to use the bathrooms. Moreover, managers arbitrarily instituted mandatory overtime. One whistle-blowing ex-manager recalled, '[the managers'] philosophy was to "keep your foot in their ass". That was their verbal philosophy, as communicated to us. That is bound to fail. Nothing new about that.' He went on to describe the extent of the coercive conditions in the plant:

> Yeah, you know that anybody working in those places needs a tetanus shot. You know with all of the dust and bacteria floating around in the air. If you bump your leg on a piece of metal and prick yourself...anything can happen … [they weren't given the shots] … Well it's because of the costs. The thing is that an enormous amount of money changed hands, but all of the workers were circumvented from all that. They were the last thought of part of the puzzle. They had all of these specifications as to how the plant should be built, but they had nothing in regards to workers' safety, training, employee retention, none of that … [man's name] was the site supervisor for REM and when things took a turn for the worse when everybody started to riot at the Medill plant and all the [pay] checks were coming in bad [underpaid, miscalculated], we had armed guards. I don't know if they were policemen or not, but they looked like street thugs. They were sitting around the dining room making sure that workers weren't going to bust any windows out or anything.

Thus, workers at Waste Management were treated poorly and often had to concentrate their efforts on resisting these conditions, rather than on simply being productive employees. There were other barriers facing employees who sought job security. The following quote is from a college course paper written by another ex-manager of Waste Management's MRRFs:

> At the rate of hours we are expected to work, using the hourly scale to estimate pay, moneys not paid range between $18,000 and $23,000 annually. My check stubs indicate a 40 hour work week, however, my actual average work week is closer to 68–70 hours per week. The conditions under which we work include lack of heat, lack of hot or cold water as well as lack of hand washing facilities after using portable toilets ... The majority of female employees have school-age children and are single parents. They seemingly fit the stereotype seen and portrayed in our media as inner-city blacks who only want something for nothing. Although I have found this not to be true, our superiors believe this to be so, consequently, the way upper management treats them is colored by management's personal biases. Our plant is somewhat difficult to reach, even by car. Half of our workers walk 1.5 miles or more through open fields to get to work. There is no public transportation in the area. When hourly workers have to pick-up their children from elementary schools, our supervisors get angry and want to fire or terminate them.

Like many marginalised workers in the current political economy, REM employees face few opportunities for secure employment. Even middle managers of large corporations have no immunity from downsizing [*Rifkin, 1995*]. Recycling workers are especially vulnerable in the labor market, because they are low-skilled and have no collective bargaining power. Without post-secondary education or union representation, they face few opportunities for meaningful or adequately-compensated work. Even though they add value to the discarded recyclables, they themselves rarely gain any real value (human capital, skills).

We note that these sorts of labour and health conditions have been found to be true more generally of the modernising recycling industry in the United States [*Pellow, 1998; Powell, 1992*]. A recent report from the Charles Stewart Mott Foundation stated, 'These jobs tend to be of relatively low quality. With little attention given to job quality or job ladder issues, there is little likelihood that these new jobs will provide an effective avenue for rising out of poverty' [*Mott Foundation, 1998*].

## IV. Conclusion: Proposed Agendas for Future Research

How do we make sense of the trajectory of urban recycling programmes? Hugh Stretton [*1976*], an Australian urban historian and social democratic thinker, noted that programmes of environmental reform in cities were likely to fall within one of three categories.

The first reform was 'The Rich Rob the Poor'. Here, environmental gains were essentially achieved by coercing less-powerful citizens into compliance and participation. He cautioned that this scenario was 'so far to the Right that I do not believe that any society will succeed with it, or even try it' [*1976: 15*]. Stretton termed the second change model 'Business as Usual'. He characterised it as 'a little to the Right of the middle of the capitalist road, and all too probable' [*1976: 15*]. The third pattern he termed 'Trouble and Second Chances', closer to his own (and our own) preferred social democratic reforms. This path has many advocates among non-governmental or social movement organisations concerned about recycling. But it is only infrequently embodied in most municipal programmes. Stretton described it as 'the politics of the possible: the best future that might be practicable in some at least of the capitalist democracies and therefore the one to work for' [*1976: 16*].

Consistent with our earlier work, we argue that much of what passes as progressive change in the United States falls much closer to Stretton's model of 'Business as Usual' as opposed to the 'Trouble and Second Chances' [*Schnaiberg, 1980, 1994; Weinberg, 1997a; 1997b; Weinberg, Pellow and Schnaiberg, 1996*]. For us, the underlying dynamics that have shaped recycling and that characterise recycling practices are typical of the political and economic processes that have characterised the United States in the post-Second World War period.

We have characterised these dynamics as the *treadmill of production*. The underlying dynamic of capitalist production in this model is the attempt by market actors to extract natural resources and convert them into profits through market exchanges. These profits are then reinvested back into the firm through the purchase of new productive physical capital, which in turn can further reduce labour costs while increasing production capacity. The basic organisational strategy is to replace workers – who are expensive and unpredictable – with technology that is consistent, efficient, and productive. At each stage, profits are used to purchase more productive physical capital rather than investing in improving labour, enhancing environmental protection, or increasing social security. At each stage of this acceleration, these financial and technological changes raise the capital-intensification of production. This often results in lower employment capacity, locking organisations into a need for a constant and reliable flow of natural resources to make 'efficient' use of their high-cost new technologies. .

These dynamics typify the change in the post-1945 structure of the political economy in the United States, perhaps more than in most industrial countries [*Longworth, 1998*]. The key driving force for this treadmill has been the growth in exchange value interests in the community's political economy. Exchange values are those associated with market interchanges, and generally revolve around levels of profitability of economic organisations, and/or levels of share prices/dividends for public investors in these organisations. The treadmill paradigm traces the rising influence of agents with exchange values over other community actors with use-values. Thus, market interchanges come to dominate almost every decision made about urban areas [*Logan and Molotch, 1987; Zukon, 1995; Squires, 1994*]. Even those aspects of community life that were not previously related to market activity, like clean air and water, are reduced to market commodities.

Where does this leave ecological modernisation theory? Obviously, our model of the treadmill of production stands in stark contrast to the model of ecological modernisation, at least with regard to the core assumption of an emancipated ecological sector. We believe that the contrasting models call for more empirical work to examine a diverse set of production changes. In particular, we want to argue that research on ecological modernisation has to be based on data, around the actual contexts of decision-making. This is of particular importance because different cultural and political histories may explain a lot about why ecological modernisation is present or absent across different nations.

There is an interesting historical perspective on the question of ecological criteria in production. Some contemporary environmentalists have had access to private firms for the purpose of examining ecological accounting systems and indeed discovered that firms are often very concerned about 'closed loop systems' and waste minimisation. Many environmentalists and scholars mistakenly viewed these concerns as the result of a recently emerged ecological consciousness among industrialists. However, most industries have always been vigilant about 'housekeeping' and waste minimisation, but not for ecological use value: they seek to reduce these impediments to reaping higher economic exchange values. As one steel industry manager told us, 'the way we think about waste is that it's material that you could have used to make more money'. More broadly, we interpret support from the corporate sector for the Blue Bag recycling programme to be embedded in a logic about saving landfill space and other issues related to waste cost minimisation.

This issue is problematic because it also has many practical dimensions for research. We do not have routine access to decision-making processes in firms or the state, or we only have access *post-hoc* and are left to reconstruct motives from the untrained observations of organisational actors who are speculating about the past. Often these actors may not have even been involved in the actual decision making process. Any theorizing about the emancipation of the ecological sphere from the economic sphere must be based on richer cases that can actually document *decisions* made in this context. Over the last few years, a growing list of rich case studies have been produced. These are beginning to allow us to tease out these sorts of questions [*Mol, 1995*] (also Gille, Sonnenfeld, this volume). We would call for an enhanced effort to collect and compare these sorts of case studies. They hold the best promise for not falling into the trap of inferring production decision-making processes while only knowing production outcomes.

These case studies could also focus on certain types of decisions. We have no doubt that there are a growing number of firms who claim to be environmentally concerned.[7] What we do not know, however, are two other things:

(1) to what extent are these claims marketing strategies or organisational cultural beliefs? and
(2) to what extent is the current marketplace favourable for these sorts of ventures?

One place to engage these questions is at points of crisis. Elsewhere we have argued that the pursuit of ecological means within a capitalistic framework will lead to moments of acute crisis for firms [*Weinberg, 1998*]. We can learn much about the organisational culture from how firms respond in those moments. There is a tendency among some ecological modernisationist advocates to argue that ecological modernisation itself is a theory of win-win practices in 'ecologizing the economy' and 'economizing the ecology' (Frijns *et al.*, this volume). In contrast, instead of a sociology of win-win situations, we call for a sociology of zero-sum moments. How do firms, industries and government regimes respond when difficult decisions have to be made? In contrast to a 'win-win' orientation, a zero-sum moments perspective acknowledges the real and inherent conflicts among stakeholders. Ignoring these differences, we argue, has often led to policy making like the Blue Bag system.

The growing debates over ecological modernisation are intellectually exciting. They have brought together a growing international group of environmental social scientists who are debating over core questions of theory and practice. In many respects, we believe the debate has the potential to create the sort of interdisciplinary and international dialogue that the environmental social sciences were premised to support.

## NOTES

1. Many of the early statements about ecological modernisation were not published in English. Instead, we refer readers to contemporary and more accessible statements. People interested in the history of the theory should turn to these sources for thorough statements (see in

particular, Mol [*1995*]; Spaaragen [*1997*].

2. 'Reuse' is the second leg of the triad in the hierarchy of solid waste management proposed by the USEPA several years ago – 'reduce, reuse, recycle'. Waste reuse operations are usually small-scale enterprises like drop-off centers that have reuse components or other socio-ecologically oriented enterprises.
3. While it is true that a greater volume of waste is being recycled by such a system, it is often at the expense of a higher quality product that might have been produced under a system also marked by more humane working conditions. These higher quality systems do exist in the form of non-profit and government-sponsored recyclers that expressly emphasize a balance among the social, economic, and environmental goals.
4. A post-consumer solid waste management infrastructure was already in place because Waste Management had been providing regular waste pick-up service in the area for several years. It had a fleet of trucks and several transfer stations and landfills.
5. Theoretically, we would argue that the robust nature of capitalism explains the actions of recycling firms. Firms rarely gain a competitive advantage by introducing ecological criteria into production decisions; more often, their public relations gains arise from their pronouncements alone. The history of the Blue Bag programme, and recycling more generally, certainly substantiates this point.
6. We note that this is typical of management decisions in the US context, where much of green rhetoric has been formalised into terms like: design-for-environment, sustainable development, responsible care, good neighbour, etc. Most, like the Campaign for Environmentally Responsible Economies (CERES) are voluntary, non-enforceable initiatives promoted by large corporations. Many of these companies are directly responsible for significant habitat destruction and worker exploitation (specifically, many oil and chemical firms fit this category). For example, WMX has told each host community in which it sites a plant that it will be a 'good neighbour' by encouraging local community participation in its affairs through 'community assistance panels'.
7. However, while many oil firms publicly proclaim their environmental concern and have environmental divisions, they quietly operate to reduce environmental protection. Nowhere is this so dramatic as in the follow-up to the Exxon Valdez disaster in Alaska. While Exxon claimed it was cleaning up its act and the environment, it actively litigated to deny damage payments to many groups of impacted workers and citizens [*Hirsch, 1997*]. Moreover, it and other major oil companies had previously worked to reduce environmental regulation during the Reagan administration [*Gramling and Freudenburg, 1997: 81–2*], helping to set the stage for this disastrous ecological accident.

## REFERENCES

Brown, Phil and Edwin J. Mikkelsen (1997 (1990)), *No Safe Place: Toxic Waste, Leukemia, and Community Action,* Berkeley, CA: University of California Press.

Carless, Jennifer (1992), *Taking Out the Trash: A No-Nonsense Guide to Recycling*, Washington, DC: Island Press.

Cohen, Maurie J. (1998), 'Ecological Modernisation: A Response to its Critics', paper presented at roundtable, American Sociological Association, San Francisco.

Franklin Associates, Ltd. (1996), *The Future of Solid Waste Management and Recycling: Multi-Client Study*, Nov., draft.

Gonos, George (1997), 'The Contest over "Employer" Status in the Postwar United States: The Case of Temporary Help Firms', *Law and Society Review*, Vol.31, pp.81–110.

Gould, Kenneth A. (1993), 'Pollution and Perception: Social Visibility and Local Environmental Mobilization', *Qualitative Sociology*, Vol.16, No.2, Summer, pp.157–78.

Gould, Kenneth A., Weinberg, Adam S. and Allan Schnaiberg (1995), 'Natural Resource Use in a Transnational Treadmill: International Agreements, National Citizenship Practices, and Sustainable Development', *Humboldt Journal of Social Relations*, Vol.21, No.1, pp.61–93.

Gould, Kenneth A., Schnaiberg, Allan and Adam S. Weinberg (1996), *Local Environmental Struggles: Citizen Activism in the Treadmill of Production*, New York: Cambridge University Press.

Gramling, Robert and William R. Freudenburg (1997), 'The Exxon Valdez Oil Spill in the Context of U.S. Petroleum Policies', in Picou, Gill and Cohen [*1997: 71–91*].

Hirsch, William B. (1997), 'Justice Delayed: Seven Years Later and No End in Sight', in Picou, Gill and Cohen [*1997: 271–303*].

Kitschelt, Herbert (1986), 'Political Opportunity Structures and Political Protest: Anti-Nuclear Movements in Four Democracies', *British Journal of Political Science*, Vol.16, pp.57–85.

Landy, Marc K., Roberts, Marc J. and Stephen R. Thomas (1990), *The Environmental Protection Agency: Asking the Wrong Questions*, New York: Oxford University Press.

Logan, John R. and Harvey Molotch (1987), *Urban Fortunes: The Political Economy of Place*, Berkeley, CA: University of California Press.

Longworth, Richard C. (1998), *Global Squeeze: The Coming Crisis For First-World Nations*, Skokie, IL: NTC/Contemporary Books.

Melosi, Martin (1981), *Garbage in the Cities: Refuse, Reform, and the Environment, 1880–1980*, College Station, TX: Texas A&M University Press.

Mol, Arthur P.J. (1995), *The Refinement of Production: Ecological Modernisation Theory and the Chemical Industry*, Utrecht: van Arkel.

Mol, Arthur P.J. (1996), 'Ecological Modernisation and Institutional Reflexivity. Environmental Reform in the Late Modern Age', *Environmental Politics*, Vol.5, No.2, pp.302–23.

C.S. Mott Foundation (1998), *Jobs and the Urban Poor*, Detroit, MI.

Moyers, Bill (1992), *Global Dumping Ground: The International Traffic in Hazardous Waste*, Washington, DC: Seven Locks Press.

Pellow, David N. (1998), 'Bodies on the Line: Environmental Inequalities and Hazardous Work in the U.S. Recycling Industry', *Race, Gender and Class*, Vol.6, pp.124–51.

Pellow, David N., Weinberg, Adam S. and Allan Schnaiberg (1995), 'Pragmatic Corporate Cultures: Insights from a Recycling Enterprise', *Greener Management International*, Vol.12, Oct., pp.95–110.

Picou, J.S., Gill, D.A. and MJ. Cohen (eds.) (1997), *The Exxon Valdez Disaster: Readings on a Modern Social Problem*, Dubuque, IA: Kendall-Hunt.

Portnoy, Kent E. (1991), *Siting Hazardous Waste Treatment Facilities: The Nimby Syndrome*, New York: Auburn House.

Powell, Jerry (1992), 'Safety of Workers in Recycling and Mixed Waste Processing Plants', *Resource Recycling*, Sept.

Preckwinkle, Toni (1997), 'Doubts Abound about City's Blue Bags', *Hyde Park Herald*, 4 June, p.4.

Ragin, Charles. (1987). *The Comparative Method*, Berkeley, CA: University of California Press.

Rifkin, Jeremy (1995), *The End of Work: The Decline of the Global Labor Force and the Dawn of the Post-Market Era*, New York: G.P. Putnam's Sons.

Schmitter, Philipe and Gerhard Lehmbruch (eds.) (1982), *Patterns of Corporatist Policy-Making*, Beverly Hills, CA: Sage.

Schnaiberg, Allan (1980), *The Environment: From Surplus to Scarcity*, New York: Oxford University Press.

Schnaiberg, Allan (1986), 'The Role of Experts and Mediators in the Channeling of Distributional Conflict', in A. Schnaiberg, N. Watts and K. Zimmermann (eds.) *Distributional Conflicts in Environmental-Resource Policy*, Aldershot, England: Gower Press, pp.363–79.

Schnaiberg, Allan (1994), 'The Political Economy of Environmental Problems: Consciousness, Coordination, and Conflict', in Lee Freese (ed.), *Advances in Human Ecology*, Vol.3, Westport, CT: JAI Press, pp.23–64.

Schnaiberg, Allan (1997), 'Sustainable Development and the Treadmill of Production', in Susan Baker, Maria Kousis, Dick Richardson and Stephen Young (eds.), *The Politics of Sustainable Development. Theory, Policy and Practice within the European Union*, London and New York: Routledge, pp.72–88.

Schnaiberg, Allan, Weinberg, Adam S. and David N. Pellow (1998), 'Politizando La Rueda De La Produccion: Los Programas De Reciclaje De Residuos Solidos En Estados Unidos', *Revista internacional de sociología*, Nos.19–20, pp.181–222.

Sheehan, Helen E. and Richard P. Wedeen (eds.) (1993), *Toxic Circles: Environmental Hazards from the Workplace into the Community*, New Brunswick, NJ: Rutgers University Press.

Simonis, U.E. (1989), 'Ecological Modernisation of Industrial Society', *International Social Science Journal*, No.121, pp.347–61.

Solid Waste Management Newsletter (1990), 'Chicago Announces 1991 Recycling Plan', Office of Technology Transfer, University of Illinois Center for Solid Waste Management and Research, Vol.4, No.12, Dec.

Sonnenfeld, David (1998), 'Contradictions of Ecological Modernisation: Pulp and Paper Manufacturing in Southeast Asia', paper presented at the American Sociological Assoication Meetings, San Francisco, CA.

Spaargaren, Gert (1997), 'The Ecological Modernisation of Production and Consumption: Essays in Environmental Sociology', dissertation, Wageningen Agricultural University, Wageningen.

Spaargaren, Gert and Arthur P.J. Mol (1992), 'Sociology, Environment, and Modernity: Ecological Modernisation as a Theory of Social Change', *Society and Natural Resources*, Vol.5, No.4, Oct.–Dec., pp.323–44.

Spaargaren, Gert and Arthur P.J. Mol (1995), Book review of Michael Redclift and Ted Benton, *Social Theory and the Global Environment*, *Society and Natural Resources*, Vol.8, No.6, pp.578–81.

Squires, Gregory (1994), *Capital and Communities in Black and White: The Intersection of Race, Class, and Uneven Development*, Albany, NY: SUNY Press.

Stretton, Hugh (1976), *Capitalism, Socialism and the Environment*, New York: Cambridge University Press.

Szasz, Andrew (1994), *EcoPopulism: Toxic Waste and the Movement for Environmental Justice*, Minneapolis, MN: University of Minnesota Press.

United States Environmental Protection Agency (1996), *Characterization of Municipal Solid Waste in the United States*, Washington, DC.

United States Environmental Protection Agency (1997), *Characterization of Municipal Solid Waste in the United States*, Washington, DC.

Weinberg, Adam (1997a), 'Legal Reform and Local Environmental Mobilisation', in Lee Freese, (ed.), *Advances in Human Ecology*, Vol.6, Westport, CT: JAI Press, pp.293–323.

Weinberg, Adam (1997b), 'Local Organizing for Environmental Conflict: Explaining Differences Between Cases of Participation and Non-Participation', *Organization and Environment*, Vol.10, No.2, pp.194–216.

Weinberg, Adam 1998. 'Distinguishing Among Green Businesses: Growth, Green and Anomie', *Society and Natural Resources*, Vol.11, pp.241–50.

Weinberg, Adam, Pellow, David N. and Allan Schnaiberg (1996), 'Sustainable Development as a Sociologically Defensible Concept', *Advances in Human Ecology*, Vol.5, pp.261–302.

Weinberg, Adam, Schnaiberg, Allan and David N. Pellow (1998), 'Ecological Modernisation in the Internal Periphery of the USA: Accounting for Recycling's Promises and Performance', paper presented at the Annual Meetings of the American Sociological Association, San Francisco, Aug.

Zukon, Sharon (1995), *The Cultures of Cities*. Cambridge, MA: Blackwell Publishers.

# Europeanisation and Ecological Modernisation: Agri-environmental Policy and Practices in Finland

PEKKA JOKINEN

This study addresses Europeanisation and ecological modernisation, using the case study of Finnish agriculture. It analyses the assumption that there has been a general transition towards the policies of ecological modernisation by examining changes in institutional arrangements, discourses and policy practices in the last three decades. The findings on Finnish agri-environmental policy are somewhat contradictory. On one hand, especially in the policy discourses, there are signs of transition consistent with the ecological modernisation theory. On the other, from the institutional and practice-oriented perspective, the evidence suggests only some minor changes, even stability in agri-environmental regulation. We find that the Finnish response to EU regulation has been problematic from the viewpoint of agri-environmental policy: the transition period of Finnish agriculture due to the entrance into the European market has retarded, at least temporarily, the policy process of ecological modernisation which began to emerge in the late 1980s.

## Introduction

Western agriculture is a shrinking industry, which in its highly modernised, intensified and rationalised phase has led to various social problems such as chronic surplus production, social effects of structural change, and severe environmental impacts. They seem to result in a continuous need for modifications to policies, which might appear to be rational and effective at the target level, but are usually based on the reconciliation of differing interests and, thus, tend to lead to implementation difficulties and minor changes only at the level of practices. The idea of combining environmental

Earlier drafts of this contribution were presented as a paper at the ISA XIV World Congress of Sociology, Montreal, Canada, 26 July–1 August 1998, RC 24, Session 'Ecological Modernization – Theory and Practice', and at the International Workshop on Ecological Modernisation, University of Helsinki, Finland, 10–12 September 1998. The author thanks his colleagues in these workshops, and Arthur Mol and the anonymous referees of the journal for their useful comments on earlier drafts.

policies with agricultural market and income policies was present in the reform of the European Union's Common Agricultural Policy (CAP) in 1992. This was concretised by the partial shift from price support to direct payments for farmers. Further, there is a separate agri-environmental scheme under the CAP defining the possible agri-environmental measures in member states.

Finland joined the EU in 1995 with Austria and Sweden, and began to adapt to the CAP, including the EU's agri-environmental regulation. However, the impact of Europeanisation on the ecological re-orientation of Finnish agriculture is basically problematic. In recent years, there certainly seems to be an agri-environmental policy shift. As an illustration, 47 per cent of all Finnish state environmental expenditures in 1995 were used as agri-environmental subsidies [*Statistics Finland, 1998b: 9*]. However, a deeper insight into the development of agri-environmental regulation suggests that there are also elements which have actually slowed down the ecological modernisation process of agriculture in Finland, even though Finland has been mentioned as one of the European *environmental policy* 'forerunners', especially on the grounds of well-developed domestic policies [*Andersen and Liefferink, 1997*].

This study addresses Europeanisation and ecological modernisation, using the case study of Finnish agriculture. With regard to the assumption of a general transition towards the policies of ecological modernisation (EM), the paper examines changes in institutional arrangements, discourses and policy practices of Finnish farm pollution control in the last three decades. The empirical findings of this article are based on qualitative analysis of official documentary material on Finnish agri-environmental policies and of interviews[1] with present Finnish agri-environmental policy-makers. In addition, material is drawn from a couple of earlier studies [*Jokinen, 1995; 1997; Niemi-Iilahti and Jokinen, 1999*].

## Ecological Modernisation as a Sociological Theory and a Framework for Environmental Policy Analysis

Environmental social scientists are used to addressing three interrelating basic meanings for the notion of 'ecological modernisation' [e.g., *Mol, 1995*]. Firstly, it is known as a sociological theory linking environmental questions with the general-level theoretical discussion on industrial/post-industrial society and the conditions for its environmental reform [*Spaargaren and Mol, 1992*; also *Blowers, 1997; Cohen, 1997*]. At this level, EM seems to manifest itself as an alternative social theory, for instance, to the risk society theory of Beck [e.g., *1992*], to neo-Marxist approaches [e.g., *Schnaiberg and Gould, 1994*], and to green political theory [e.g., *Goodin, 1992*]. Secondly, in environmental social science studies, EM has been applied as a theoretical framework for environmental policy analysis [e.g., *Hajer, 1995; Weale, 1995*]. Further, according to Mol [*1995*], ecological modernisation appears also more practically and normatively as an environmental policy strategy that is supposed to be successful.

As Mol and Sonnenfeld show (this volume), it is clear that with the growing number of studies based on the ecological modernisation perspective, the theoretical discussion is expanding and taking in an ever wider range of issues. However, from this it follows also that it is more difficult to pinpoint the exact theoretical and conceptual locus of EM. In spite of the theoretical diversification, the concept is essentially tied up with theories of modernity [*Spaargaren and Mol, 1992; Mol, 1996*]. It is considered fundamental that the process of ecological change ('emancipating ecological rationality') in society proceeds via a continuous, active and reflexive institutional reorganisation, utilising modern science and technology and economic dynamics. Thus, this 'modernisation out of the environmental crisis' seems to imply 'learning processes' for most of the central institutions of modern society [also *Hajer, 1996; Macnaghten and Urry, 1998*].

As mentioned above, in environmental social science studies the concept of ecological modernisation has also been used in analysing and describing the 'new' paradigm of environmental policy from a historical and empirical point of view, and its interpretations seem to be increasingly variant. There certainly are some very narrow versions to be found, for instance, that by Schnaiberg [*1997*]. He has drawn a parallel between ecological modernisation and the notion of 'industrial ecology', the latter being defined as 'a form of re-engineering to reduce the environmental disruption generated by treadmill production' [*ibid.: 76*]. However, if ecological modernisation is defined in this way, it undoubtedly follows that the role of different actors in environmental policies (that is, the level of 'environmental politics') is disregarded. More interesting is the conceptualisation by Baker *et al.* [*1997*], which has set the concept of ecological modernisation in the context of sustainable development. This being the case, EM is equated with a 'strong sustainable development'. It is characterised by, for instance, the principle that environmental protection is a precondition of economic development, and, interestingly, by *more* state intervention and an environmentally regulated market, high local self-sufficiency and a strengthened redistribution policy.

The often cited narrative by Hajer [*1995*] reports that a broader perspective on environmental problems, labelled the paradigm of EM, emerged in the early 1980s and began to dominate the conceptualisation of environmental problems and policy strategies in Western Europe [also *Blowers, 1997*]. The paradigm of EM started from the assumption that environmental problems are caused by institutionalised failures of industrial society, however, in contrast to the radical environmentalist thinking of the 1970s, it assumed also that it is possible to control the problems by means of existing political, economic and social institutions. It is considered particularly important for the success of this paradigm that environmental policy is not seen as a zero-sum-game: the successful environmental policy and the maintenance of economic growth could be realisable at one and the same time. Another main conclusion in several studies is that actors above the immediate national decision-making level (for example, the OECD, the UN, and the EC/EU) have contributed strongly to this re-orientation of environmental regulation [*Hajer, 1995; Weale, 1995; Gouldson and Murphy, 1996*].

According to empirical studies, the conceptual and institutional shift to the policy of ecological modernisation is manifesting itself in many different ways. Hajer [*1995*] notes that a significant change in environmental policy-making has appeared. The EM thinking aims to introduce alternative and innovative policy measures and utilise the evident synergy between environmental policies and economic development. Thus, the goal is increasingly to integrate environmental policy with other policy sectors, especially those related to production (for example, agricultural policy and energy policy). Economic policy measures have also gained increasing importance, and along with the general societal trend of deregulation in the 1980s, responsibility for environmental control has partly shifted from the state to the industries themselves. Frouws and Mol [*1997*] have pointed out, however, that economic means of environmental policy should be understood as complementing rather than replacing traditional state regulation. In other words, the scope of environmental policy has at once expanded and deepened. Further, the new environmental policy paradigm includes avoiding conflicts (typically between the state and environmental movements) and encouraging a larger number of people (particularly from environmental organisations as well as local residents) to take part in policy-making.

The EM perspective has not often been applied to agri-environmental policy analysis. An exception, however, is the study by Frouws and Mol [*ibid.*] examining the ecological restructuring of the central institutional factors (that is, science and technology, economic dynamics, state policy, and social movements) of Dutch agriculture. At a superficial level, they have found a lot of progress in agriculture's institutional practices. This is illustrated by new policy measures and environmentally benign technologies, eco-labels for food products, participatory regional planning practices etc. However, the authors also emphasise that 'there is still a long way to go'. The political modernisation of agri-environmental policy is in its initial phases, and a 'social contract' between the farmers and society is the visionary direction taken in order to strengthen the legitimate ecological restructuring of agriculture [cf. *Glasbergen, 1992*].

### *Neo-corporatist Structures in Agriculture – a Counter Force for the Policies of Ecological Modernisation?*

For the studies on environmental politics and policies, and, thus, also for the ecological modernisation perspective, the issue of policy actors and networks is basic [e.g., *Mol, 1995*]. Further, relating to the topic of this study, it is evident that agricultural policy should not be interpreted as just another policy sector (as regards the form of policy-making and articulation of interests). Instead, it has often been seen as an illustrative example of corporatist structures and practices, and there has even been discussion about the 'exceptionalism' of agricultural policy [*Grant, 1995*]. The theoretical view of corporatist policy-making and interest mediation has been further elaborated in the policy network approach by means of the concept of the 'policy community' [e.g., *Rhodes and Marsh, 1992; Daugbjerg, 1996*]. The ideal type of policy community refers to a close and institutional relationship between pressure groups and the state in the process of policy-making. Essentially, this relationship is insulated from parliamentary control and from the public.

The essential qualities of a policy community relate to the consensual mode of policy-making processes and to the output of these processes at the programme level. Epstein [*1997*] has also paid attention to the insularity of the policy community as well as to the stability and continuity in its membership. Further, he points out the 'win-win' game played in the policy community: both the state actors and the interest group actors are engaged in the exchange of policy resources. In order to maintain the closed and consensual policy-making process, the community members favour an incremental approach in their policy programmes. Similarly, Epstein suggests, the depoliticisation of issues and a programme sectorisation serve the objective of excluding dissident actors from decision-making. In sum, the application of the concept of policy community to agricultural policy studies refers to the mutual interests of the agricultural state agencies and the farmers' unions (that is, farmers' interest organisations). The agricultural policy community is an autonomous, stable and closed-circuit institutional structure which can be used to explain the features of agri-environmental policy [e.g., *Garner, 1996*].

It is an interesting issue whether the differences between agricultural policy and other policy areas is gradually eroding [e.g., *Just, 1994; Grant,*

*1995*] and, similarly, whether we will increasingly observe new positions of actor groups in agri-environmental policy-making [e.g., *Frouws and Mol, 1997*]. However, we might conclude so far that the ecological modernisation perspective and the agricultural policy community notion (which emphasises, above all, the neo-corporatist style of policy-making) seem to propose rather dissimilar understandings of agri-environmental regulation. In the first place, the openness of agri-environmental policy-making is necessary for the process of ecological modernisation; the policy community perspective suggests that the intention of the agricultural policy community is the opposite. According to the ideal type of EM policy-making, environmental and agricultural policy goals should be integrated with each other. The agricultural policy community approach, again, appears to suggest a strict sectorisation and detachment of these issues (or, alternatively, the dilution of environmental factors into agricultural policies). Further, if we rely on the EM approach, we can suppose that even radical changes in agri-environmental policies, or alternatively, rather stable policies on the basis of the policy community approach will take place. Finally, the restructuration of policies by EM should produce pro-environmental solutions, whereas environmental regulation arising from the agricultural policy community is typically supposed to lead to agri-technical and defensive[2] solutions, which primarily support the income and production goals.

## The Formation of Finnish Agri-environmental Policies

### *Agricultural Modernisation and the Environmental Record*

The modernisation of Finnish society has resulted in the somewhat late but then rapid transition of the industrial structure into a service-dominated society and also in the depopulation of rural areas, which peaked in the late 1960s. Thus, the role of farming has gradually changed, and in pure economic terms, agriculture's significance nowadays seems to be rather minor (Table 1). This change, especially from the 1960s up to the 1990s, has been strengthened by national agricultural policy which has advanced rationalisation and strongly increased productivity. The number of farms has decreased and, in the present context of the European Union's agricultural policy, it is expected to decrease further. Meanwhile, the average farm size is growing. However, two core features of the industry have been preserved. First, it is firmly based on family-farming which has been supported by rural policy stressing the countryside's vitality. Secondly, Finnish agriculture has traditionally been oriented towards maintaining self-sufficiency so the products are mainly channelled into domestic consumption.

TABLE 1

THE CHANGES OF STRUCTURE IN FINNISH AGRICULTURE 1960–95

| Year | Per cent of total GDP | Per cent of total employment | Number of farms | Arable land (1000 ha) |
|---|---|---|---|---|
| 1960 | 10.7 | 28.7 | 331263 | 2654 |
| 1970 | 6.9 | 20.3 | 297527 | 2667 |
| 1980 | 4.3 | 10.8 | 224721 | 2563 |
| 1990 | 3.2 | 6.9 | 199385 | 2545 |
| 1995 | 1.7 | 6.1 | 169707 | 2146 |

*Sources*: Statistics Finland, Statistical Yearsbooks of Finland.

Finnish production is based mainly on animal husbandry, with dairy farming and beef production accounting for nearly half of the total agricultural product. In 1995, only eight per cent of the country's total land area was agricultural land, and there were 99,960 active farms (that is, engaging in agricultural production) with an average arable area of 22 hectares [*MAF, 1996c: 22–3*]. However, the country-level figures do not tell the whole truth, because within the (EU's fifth largest) country the regional differences are great. The intensive crop production is concentrated in Southern and Western Finland, where in numerous rural municipalities field area accounts for 30–50 per cent of all land area. There, for instance, the average farm size (30 ha) and the proportion of large farms (13 per cent of farms larger than 50 ha) are higher than in other parts of the country [*ibid.*].

Agriculture's environmental record and the policy pressure put on it are of course bound to certain endogenous factors, notably the absolute mechanisation, chemicalisation and intensification of the modernised production. However, they depend also on its relative position (that is, on the proportions of total pollution caused by various human activities). In the case of Finnish water pollution management, the emphasis has traditionally been put on the pulp and paper industry [e.g., *Joas, 1997*]. In practice, however, up to the 1970s, the Water Courts issued generous water pollution permits for Finnish industry in the name of economic goals and raising the standard of living. The last two decades have witnessed a rather drastic change: due to the stringent environmental policy norms and environment-technological responses to them, industrial emissions into waters have been reduced substantially [e.g., *Statistics Finland, 1998b: 26–7*].

As can be seen in Table 2, nowadays Finnish agriculture contributes a major share to water pollution. In particular, the eutrophication of coastal and inland surface waters due to nutrient losses from field cultivation is seen as a severe problem [e.g., *MoE, 1995; Finnish Government, 1998a*]. In summer 1997, the most widespread and prolonged toxic blue-green algae blooms ever recorded on the Finnish sides of the Baltic Sea[3] and also in inland waters brought the eutrophication issue strongly to the Finnish public consciousness too.

TABLE 2

MAIN SOURCES OF WATER POLLUTION* (TONNES PER YEAR) IN FINLAND IN 1996

| | Phosphorus | Nitrogen |
|---|---|---|
| Industry | 290 | 3.760 |
| Communities | 250 | 14.380 |
| Fish farming | 150 | 1.180 |
| Cattle farming | 300 | 2.900 |
| Crop production | 1.500–3.500 | 25.000–35.000 |

*Excluding natural alluviation and airborne pollution (deposition).

*Source*: Statistics Finland [*1998a: 41*].

Although Finnish agriculture as a whole can be seen as relatively extensive compared to the most intensive areas of West European agriculture, it also exerts other pressures on the environment. There are, for instance, some local problems of groundwater pollution caused by nitrate leaching from fields and the ammonia and greenhouse gas (methane) emissions from manure of animals [e.g., *MoE, 1995*]. Further, in the 1990s, the notion of biodiversity has established novel legitimacy for nature conservation [*ibid.*]. Increasing attention is thus also paid to the reduction of species and habitats, which certainly is a long-run negation of the agricultural modernisation process. For instance, the seminatural grasslands and pastures, which are important habitat types for maintaining biodiversity in agricultural land, have disappeared along with the rationalisation and specialisation of production. It has been estimated that in Finland about 20 per cent of threatened species are living in the agricultural environment [*MoE , 1997a: 67*].

Above, we have discussed theoretically-based characteristics of 'ecological modernisation' in agriculture. In order to clarify empirically the formation of Finnish agri-environmental policy, its major developments and the points of change, we turn to the analysis of agricultural pollution identification and problem definition and, therefore, of the consequential mode of policy and regulation. According to the traditional definition in policy studies, the term 'environmental policy' has referred to the goals and means used to prevent and remedy ecological problems. Our belief is that disputes over the need for, the contents of or control over these goals and means need to be examined through a sociological study of the evolution of environmental policies [e.g., *Jokinen and Koskinen, 1998*]. The starting point for environmental policies is the variety of definitions of and ideas about environmental problems assumed by different major actors. Further, these definitions and ideas provide the basis of the actors' environmental-political solutions to the control of environmental problems. In sum, we examine a field where actors with differing interests, representing the state, different economic sectors, and various organisations seek to control the content of agri-environmental policy [*ibid.; Weale, 1995*].

The examination is organised in three periods [also *Eckerberg and Niemi-Iilahti, 1996; Jokinen, 1995; 1997; Niemi-Iilahti and Jokinen, 1999*]. They are called (1) 'agricultural pollution as a non-issue' (early 1970s–1987), (2) 'agriculture equal to other polluters' (1988–94), and (3) 'environmental issues attached to the maintenance of agriculture' (1995– ). The periods in question have been distinguished from each other on the basis of discursive and institutional changes in agri-environmental policy. However, it can not be considered an autonomous sub-policy with clear-cut boundaries. Instead, as has been emphasised by the policy network theorists cited above, its formation is tied to the power of and interaction between agricultural and environmental sectors and, thus, to more comprehensive developments in agricultural and environmental policies.

### *From the 1970s up to the Late 1980s: 'Agricultural Pollution as a Non-issue'*

As in the Western world in general, the late 1960s can be described in Finland as the phase of increased and radicalised environmental consciousness and activism. In addition to addressing the increasingly severe pollution issues, one of the environmentalists' main goals was the development of national environmental administration and legislation, and especially the establishment of a separate ministry focusing on environmental issues [*Rannikko, 1996*]. However, the institutionalisation of Finnish environmental policy started rather slowly in the 1970s. The first Report of the Environmental Protection Committee in 1970 and the foundation of the National Water Board (within the authority of the Ministry of Agriculture and Forestry) in the same year were important initial steps [e.g., *Joas, 1997*]. It makes no sense to talk about Finnish agri-environmental policy beginning before the 1970s, although several sub-areas of agricultural policy, such as production, income and structural policies, can be identified as originating immediately after the Second World War [e.g., *Granberg, 1995*].

The chemicalisation of Finnish agriculture had increased strongly from the late 1950s on [e.g., *Miettinen, 1998*]. For instance, the use of herbicides increased about eight-fold from 1960 up to the peak year of 1980 [*Statistics Finland, 1994: 45*]. Similarly, the nitrogen input in fertilisers increased fivefold from 1960 up to 1990 [*ibid.: 44*]. The health impacts of agricultural pesticides were to some extent discussed, but early environmental policy was, however, far more focused on the side effects of rapid modernisation and the increasing standard of living. In other words, industrial plants, urban areas and consumption habits were regarded as the main causes of increasing pollution and waste. Instead, in the 1970s, several *ad hoc* committees focusing on water management issues – such as the First National Programme of Goals for Water Pollution [*National Water Board, 1974*] – considered farming primarily a non-polluting industry, which utilises natural resources in a sustainable way [e.g., *Jokinen, 1995*]. Thus, the main characteristic of the first period of Finnish agri-environmental policy, lasting even until the late 1980s, was the dominating view of 'agricultural pollution as a non-issue'.

In agricultural policies this phase can be regarded as an institutionalised 'secure age' for the agricultural policy community: the prices of products were annually negotiated inside the agricultural sector between the Ministry of Agriculture and Forestry (the MAF) and the farmers' union. Thus, income issues and productivity goals were under the control of actors involved in the corporatist agricultural policy-making. Further, the MAF administered both the agricultural productivity and agri-environmental issues before the foundation of the Ministry of the Environment (the MoE) in 1983. As regards agri-environmental problems, it is fair to generalise that the 'agrarian' discourse held by the agricultural policy community actors dominated agri-environmental discourse and the problem definition arena in the 1970s and in the early 1980s [*Jokinen, 1997*]. As a result, there were no significant policy initiatives connecting environmental protection goals to agricultural policy issues (for example, to the emerging surplus production problem).

Thus, taken together, during this period the main environmental policy discourse was defensive towards farm pollution problems and little environmental policy pressure was put on farming. From the view point of institutional practices, agri-environmental regulation remained very minor despite a small number of voluntary-based economic and informative measures. The former measures refer, for instance, to loans for farm water management and the latter to consultation and farming-technical recommendations on environmental protection.

### *From the Late 1980s up to the Entry into the EU: 'Agriculture as Equal to Other Polluters'*

However, around the middle of the 1980s, identifiable signs of a somewhat changing water pollution problem definition appeared in Finland. The first of these was the work by the *ad hoc* Committee of Farm Water Management in 1983 which reported that, in the southern and western parts of Finland, some rivers are polluted by agriculture alone [*KM, 1983: 66*]. Further, the permanent Committee of Water Management, which was preparing the Programme for Water Pollution up to 1995, in its report in 1986 adopted the same starting point as regards agricultural load on waters [*KM, 1986: 42* ]. This emerging change of course reflected, firstly, the 'objective' side of the matter, which was the increasing natural scientific evidence of the environmental impacts of the intensified farming. However, secondly, it also foreshadowed a changing notion of the relative position of agriculture compared to other polluters. Further, along with the foundation of the MoE in 1983 the dynamics of agri-environmental policy-making was becoming gradually more complicated: a new player had entered the policy field with a basic view on farm pollution greatly differing from that held by the MAF.

Thus, from the middle of the 1980s, the foundations of the second phase of the Finnish agri-environmental policy were taking shape: we find that from the late 1980s up to the middle of the 1990s agriculture was regarded as 'equal to other polluters'. The first official, target-level indication of this new notion can be found in the Programme of Goals for Water Pollution up to 1995 [*MoE, 1988*]. This programme was approved by the Government in 1988, and its general target was that communities, industry, and agriculture had to cut down equally their pollution of water. Thus, for the first time, concrete objectives for the reduction of agricultural load on waters were set. For instance, the ambitious goal was to reduce the phosphorus load by even 50 per cent [*ibid.*].

For the agricultural sector, the new goals of water pollution management appeared rather hard, and the farmers' interest organisation did not basically accept the goals but regarded them as excessive and unjust [*Jokinen, 1995*]. In the late 1980s, the organisation questioned the whole legitimacy of the findings, which were suggesting that agriculture was the major contributor to the euthrophication problem. To clarify this disputed issue, there was pressure to obtain scientific evidence of farm pollution. An extensive research project, named 'Agriculture and Water Quality', was founded, and its legitimacy was strengthened by its joint finance (by, for example, the MAF, the MoE, the farmers' union, and the major Finnish fertiliser producer company). The main finding of the study was that more eutrophication of water had been caused by agriculture than by industry and communities combined [*Rekolainen et al., 1992*].

On the basis of the findings of the 'Agriculture and Water Quality' study, a significant mile-stone of Finnish farm pollution management was reached: the MAF and the MoE together prepared the Environmental Programme for Rural Areas [*MoE, 1992*]. This programme was important in that, for the first time, agricultural and environmental policy-makers together approved a programme based on a relatively uncontested identification of the farm pollution problem. On the other hand, however, by adapting to the preferences of agricultural policy players, the programme did not introduce any new regulation ideology. The policy was still based on 'soft' informative policy tools, such as recommendations of and consultation on economically-motivated cultivation practices (for example, avoidance of over-fertilisation) having positive environmental effects as well. Thus, the diversification of regulation towards 'harder' regulatory or economic measures was ignored.

In sum, the notion of a worsening farm pollution problem (as expressed by the two committees mentioned above) had led to the introduction of rather ambitious targets in the Programme of Goals for Water Pollution up to 1995. Moreover, the strongest dispute over agriculture's contribution to the load on waters was diminishing in the early 1990s along with the new, established scientific evidence. However, as regards the mode of Finnish agri-environmental policy, the main principles were rather stable in the first and second periods discussed so far: the activities causing agri-environmental problems were regarded as primarily belonging to the autonomous area of farming-technical decision-making at the individual farm level, so trust in the voluntary action of farmers dominated. The goal to decrease the effluent escaping into waters by 50 per cent by 1995 was far from realised [*MoE, 1994*], no doubt in large part due to this lack of a comprehensive regulation strategy.

By which factors then should we explain the continued emphasis on 'soft' regulation in Finnish agri-environmental policy? What was the contribution of various policy agents to this approach? We have assumed here that agri-environmental policy consists of the interaction between the rather different policy elements, which is determined by the relative strength of agricultural policy actors and environmental policy actors. Certainly, the dominance of the 'agricultural policy community' can be regarded as the most essential general factor when the first and second phases of the Finnish policy-making up to the 1990s are explained [*Jokinen, 1997*]. The MAF and the farmers' union had a mutual interest in maintaining their traditional control over agricultural policy, which also resulted in agreement on principles of agri-environmental policy. In short, the policy community defined the severity of water pollution problems as moderate, emphasised their controllability, and was definitely reluctant to accept any regulative interventions for environmental policy reasons.

The evident counter-force for the agricultural policy sector in agri-environmental policy issues consists of environmental state agencies and environmental NGO's (or the like). Buttel [*1995*] has suggested that we are, indeed, witnessing the second great transition of the century in world agriculture, that is, the 'environmentalisation' of agriculture. Environmental movements and organisations are considered a main element of this transition because of their reshaping influence on policies and practices of agriculture. In fact, agriculture seems to have become one of the major targets of environmentalists in several Western countries [e.g., *Lowe et al., 1997; Frouws and Mol, 1997*].

Finland has been a different case: the environmentalists have not shown any major interest in farm pollution. The Finnish Nature Conservation Association, being the most institutionalised part of the environmental movement, has at times expressed its concern about farm pollution, for example, in its official policy statements [*Jokinen, 1997*]. Further, in the late 1980s and in the 1990s, it has been typical for a representative of the Association to be nominated as a member of governmental working groups preparing agri-environmental programmes, which has become the main way for environmentalists to try to influence the policy outcomes. However, compared with the great activism the Finnish environmental movement has shown, for instance, towards forestry issues [e.g., *Rannikko, 1996*], the response to farming's environmental impacts has been very mild. This particular neglect by the environmentalists has also been reflected in public opinion. Although in the late 1980s more eutrophication of waters was caused by agriculture than by industry and communities combined, a survey finding shows that 58 per cent of Finns in 1989 regarded the chemical and the pulp and paper industries as the major water polluters [*Suhonen, 1994: 141*]. An important conclusion can be drawn: given the considerable passiveness of Finnish environmentalists in farm pollution issues, the seriousness of these issues has in practice been emphasised by the environmental state agency (that is, the MoE) alone.

Finally, however, the overall demarcation line between the first and second phases of agri-environmental policy is also related to a more general change in Finnish environmental policies: namely, in the late 1980s, the optimistic notion of 'sustainable development' penetrated different sectors of society, agriculture being no exception [e.g., *Macnaghten and Urry, 1998; Jokinen and Koskinen, 1998*]. A rapid discoursive re-orientation appeared: along with the general 'greening' of the Finnish society (that is, a transformation in environmental values and discourses), the agricultural policy-makers also adopted an environmental discourse in the early 1990s, a discourse whose concepts they were able to utilise in major environmental discussions [*Jokinen, 1997*]. The MAF [e.g., *1992*] began, for instance, with the further support of the organisation of farmers' advisory services (the Association of Rural Centres), to construct a new environmental challenge with the idea of 'agri-ecological services' (for example, activities sustaining rural landscape resources). In addition, increasing investment into environmental research (for example, 'agri-ecology' and environmental economics) appeared in the research institutes under the MAF's steering. Similarly, the shift in discourse was rapid in the farmers' interest organisation: in contrast to the strong denial of any agri-environmental problems up to the late 1980s, it began to emphasise the national importance of the 'sustainability' of agriculture, which, however, should have been subsidised [*Jokinen, 1995*]. This new attitude of the farmers' organisation towards agri-environmental issues was strengthened by the foreseen economic change due to Finnish membership in the European Union.

### *Adapting to the European Market: 'Environmental Issues Attached to the Maintenance of Agriculture'*

Finland's accession in 1995 to the European Union was a fundamental event, especially for the agricultural sector. Overall, the entry into the common agricultural market immediately influenced the economic operation model of the (formerly protected) domestic agriculture and changed the basic parameters of food production. The market prices of agricultural products fell on average by 40 per cent of the 1994 level [*Kettunen, 1996*]. The Finnish Government decided it was essential to moderate this income reduction, again, by providing domestic income support for farmers.

Joining the EU also marked a turning point between the second and third phases in Finnish agri-environmental policy as the main principles of the policy were reshaped in two respects. First, by the introduction of the novel EU-based agri-environmental schemes and to the diversification of agri-environmental regulation instruments. Secondly, agri-environmental policy was tightly linked to the political and economic adaptation of Finnish agriculture to the EU's agricultural policy. The latter gives grounds for labelling the present (the third) period as 'environmental issues attached to the maintenance of agriculture'.

The three important EU schemes relating to farming are the Nitrates Directive,[4] the 'Natura 2000' nature conservation network,[5] and the Regulation 2078/92.[6] In addition to these new transnational schemes, domestic environmental policy has also been developed further: according to the new Programme of Goals for Water Pollution up to 2005, agriculture, like other industries, is required to cut down further its pollution of water [*Finnish Government, 1998a*]. The ambitious goal of this programme is to reduce the phosphorus and nitrogen loads from agriculture 'by at least 50 per cent' (from the average level of 1990–93). However, like its predecessor [*MoE, 1988*], the programme is limited to the principal targets and does not introduce any special measures that are obviously needed for its progress.

As they are declarative in nature, the new pollution reduction goals introduced in the national Programme of Goals for Water Pollution have aroused no substantial criticism, for instance, by the farmers' organisation. The reverse is true of the EC Nitrates Directive, the basic purpose of which is to reduce and prevent water pollution by nitrates from agricultural sources. Individual countries have to designate areas 'vulnerable to the leaching of nitrates' [e.g., *Brouwer and Lowe, 1998*]. In Finland, the implementation of the Directive was prepared in 1996–97 by a working group of state officials, the majority of them from the MoE and a minority from the MAF [*MoE, 1997b*]. It is remarkable that the Nitrates Directive is the first substantial regulative (referring to permits, rules, prohibitions, etc.) agri-environmental scheme introduced into Finnish policies. Consequently, it is not surprising that its substance became an issue of conflict between the MoE and MAF, which was strongly supported by the farmers' organisation.

At the core of this conflict was, firstly, the extent[7] of Finnish areas identified as vulnerable to nitrates. The MoE pushed for a large area including 18 per cent of the total land area and 40 per cent of all farms, whereas the MAF did not want to define any areas as vulnerable. Secondly, the MAF was suspicious of the environmental protection requirements the MoE was proposing under the Finnish nitrates rule. Thus, the MAF demanded light measures in order to ensure that farmers can not be obliged to carry out, without monetary compensation, such measures which otherwise are substantially subsidised by the Finnish Agri-environmental Programme. The final decision on the implementation of the Directive can obviously be seen as rather favourable to the MAF's targets for it comprises lighter measures than those in the Agri-environmental Programme and no 'vulnerable' areas [*Finnish Government, 1998b*].

The European 'Natura 2000' nature conservation network has been founded in order to conserve natural resources through a programme of habitat surveillance, evaluation and conservation. In Finland, its implementation appeared unexpectedly to be the most disputed environmental issue of the 1990s. The preparation work was done in 1996–97 by the MoE in a rather unsuccessful way: it received approximately 14,000 written complaints from land-owners, whose mobilisation was impressively supported by the farmers' organisation. The political arguments against the programme typically emphasised, for example, the 'illegality' and 'interventionism' of the programme, and the Ministry's planning strategy was criticised because it ignored negotiations with the land-owners concerned.

However, compared with the volume of the conflict and the huge public debate, the programme (though covering 12 per cent of the country's total area) was far from dramatic: in the end, 97 per cent of the areas in the Finnish 'Natura 2000' were already established conservation areas, which had been included in earlier programmes and decisions [*Finnish Government, 1998c*]. Rather than the objective impact of this novel programme, an obvious explanation for the severity of the conflict is that the Natura process unleashed both the Finnish farmers' traditionally critical stand to nature conservation and the strong political opposition to the EU [*Jokinen, 1997*]. This conflict was strengthened by the exclusion of the representative party of agricultural interests, the Finnish Centre, from the wide government coalition of the Left, Social Democrats, the Greens, and the Conservatives, which was formed after the parliamentary elections in 1995. From a narrow sector perspective, unlike the Nitrates Directive issue, the conflict over 'Natura 2000' was of course not directly connected to farm pollution control. In practice, however, it was a fundamental issue for the whole legitimacy of environmental policies and also an essential part of the re-strengthened tension between Finnish farmers and environmental authorities in the late 1990s.

The most important new scheme of the present policy phase is the extensive agri-environmental subsidy system, named the 'Finnish Agri-environmental Programme for the years 1995–99' [*MAF, 1996a*]. It is based on the EU Regulation 2078/92, which, however, does not determine the form and content of national agri-environmental programmes. Instead of obligatory activities, it makes financial support available for different measures [e.g., *Baldock and Lowe, 1996; Brouwer and Lowe, 1998*]. The main outlines of the Finnish programme could, thus, be adapted to the 'national agricultural interest', which was defined in 1993–94 by the Finnish Government (formed by the rural party, the Finnish Centre and the conservative National Coalition Party) in the negotiations on Finland's Accession Treaty with the EU. It follows that the present Finnish agri-environmental subsidies must be understood as a part of the general support strategy aimed at compensating farmers for the fall in producer prices to the common market's price level. The more exact 'national agri-environmental interest' took shape in a particular planning process by sector policy-makers: in keeping with the general objectives of the Government, the programme was prepared in 1994 by a working group of agricultural and environmental state officials. This process will be dealt with below.

There is considerable variation in the ways this particular EU Regulation has been interpreted and implemented by single states [e.g., *Baldock and*

*Lowe, 1996; Curry and Stucki, 1997; Brouwer and Lowe, 1998*]. In the Finnish programme, we can identify three general principles, which are not derived from the EU rule [*Niemi-Iilahti and Jokinen, 1999*; also *Miettinen, 1998*]. Firstly, it is mainly directed towards water pollution problems caused by arable cultivation. Secondly, unlike, for instance, the Environmentally Sensitive Area schemes in certain Western European countries, the Finnish programme is not spatially targeted: it is applicable throughout the country, it is open to all farmers, and by 1996 about 90 per cent of the agricultural land was already under its influence [*MAF, 1996a*]. Finally, as a necessary result of the former, the total cost of the programme is remarkably high.

More specifically, the new agri-environmental subsidy system costs a total of 270 million ECUs[8] (about 1.6 billion FIM or US$300 million) per year. It is a significant financial source of Finnish environmental policy: in 1995-98 these subsidies accounted for 42–47 per cent of all state expenditure on environmental protection [*Statistics Finland, 1998b: 9*]. Furthermore, the financial significance is valid also from the viewpoint of Finnish agricultural policy: environmental subsidies are equal in value to the EU's Common Agricultural Policy support and the Less Favoured Areas support. In 1998, Finnish farmers received approximately 1.6 billion FIM as CAP support and also 1.6 billion FIM as LFA support. Finally, as agricultural subsidies accounted in 1998 for 41 per cent of the total income of Finnish farmers and agri-environmental subsidies for 18 per cent of all these agricultural subsidies, the strong agricultural policy function of the agri-environmental support is clear.

The Finnish agri-environmental programme 1995–99 'aims at compensating the farmers for the costs or income losses caused by measures to protect the environment or landscape management, and at securing the income of farmers in the changing conditions' [*MAF, 1996a: 15*]. Its main environmental goal[9] is to decrease the total nutrient leaching to waters, by about 30–40 per cent in 5–10 years [*MAF, 1996b*]. A retardation of eutrophication is expected, especially in the coastal areas of the Baltic Sea, and in some inland areas, the water quality is even supposed to improve. In practice, the farmers must take certain environmental measures[10] in order to obtain annual payments, varying between 40 and 900[11] ECUs per field hectare. In the south-western region especially (which had to be left outside the support for Less Favoured Areas) the farmers consider environmental subsidies an important source of income; indeed, 20 per cent regard environmental support as indispensable for the future[12] of farming [*Siikamäki, 1996*].

### *The Differing Framings of the Farm Pollution Problem as the Basis for Present Policies*

It has been shown that new agri-environmental schemes and regulation instruments have been brought into use in Finland. However, tension and conflicts have arisen because of the implementation of EU rules (after the earlier rather consensus-seeking period from the late 1980s up to the middle of the 1990s). These conflicts have appeared especially between the environmental central agency and the agricultural central agency together with the farmers' organisation. In order to clarify further the present process of Finnish farm pollution control, we now draw on interviews with Finnish agri-environmental policy-makers.

Lowe and Ward [*1998*; also *Lowe et al., 1996*] have found that inside the process of agri-environmental pollution control, there are two conflicting notions of pollution. According to one, farm pollution is a technical problem, in other words, a 'form of rule breaking'. By contrast, it is characteristic of the other notion that farm pollution is seen as something discreditable, in moral terms, as a 'type of environmental crime'. The distinction between these competing discourses points to a central factor of policy-making: the definitions and notions of environmental problems provide the basis for differing environmental policy demands by the actors. Thus, Lowe and Ward suggest that the former definition (pollution as a 'problem for farming') leads to agriculturally-led solutions, whereas the latter definition (pollution as a 'problem of farming') implies environmentally-led solutions. Further, not surprisingly, they argue that agricultural actors, referring to agricultural policy-makers and advisors, and to farmers, seem to have adopted the notion of 'relative pollution' (that is, showing understanding for the production context in which the pollution arises). Environmental pressure groups, the wider public, and environmental regulatory officials, again, tend to consider farm pollution as an 'environmental crime'.

In the Finnish case, the 'relative pollution' claim was put forward by several of the agricultural policy-makers interviewed. The claim consists of two elements, namely, the priority given to production issues and the critique of the dominant pollution problem definition. Referring to the latter, a number of critical comments were expressed about the official information in environmental statistics concerning the domestic causes of water pollution. As one interviewee[13] put it:

> It is true, per se, that the load on waters is mainly produced by agriculture. However, there is a misinterpretation in the environmental statistics, since the actual impacts on water quality have been ignored. The farm pollution (unlike pollution from point sources) is produced

> throughout the country. It follows that it becomes, indeed, weak in large waters.

Also Finland's responsibility for international water pollution was offered as an issue for public debate. This is illustrated by the following quotation (an agricultural state official): 'If we speak, let's say, of the pollution of the Gulf of Finland, we should also ask how large is the Finnish share of it ... I think, indeed, our business is well-managed.' It must be noted that, despite its international emphasis, even this view of the exaggeration of Finnish farm pollution is primarily directed to domestic environmental policy players. Namely, they have often justified their demands for a tightened agri-environmental regulation by the eutrophication problem in the Baltic Sea.

The other subject, which is interpreted here as a claim for 'relative pollution', is the priority given to production activities. According to this way of thinking, some farm pollution has been and will be unavoidable, and, therefore, environmental actors should not have an ultimate say in the way the practices of agricultural production are carried out. The farmers' interest organisation (here echoed by its representative) in particular wants to link agri-environmental issues primarily to the decrease in the number of farms: 'As a result of the ending of agricultural activities, the real environmental threat in the countryside will be the deterioration of human landscape.'

In contrast to agricultural policy actors, environmental policy actors seem to consider the environmental impacts of modernised agriculture a serious problem, and to quote Lowe and Ward [*1998*] again, even an 'environmental crime'. However, it must be emphasised that in the Finnish context we use 'environmental crime' as a strong metaphor referring to a form of problem framing, totally separate from any juridical meaning. In any case, this kind of frame gives priority to nature and was echoed by those interviewed, above all, for water quality and aquatic habitats. Thus, agriculture's environmental impacts, which were exemplified above by water pollution and the loss of biodiversity, are characterised as absolute: 'indeed, it is catastrophic that the bottom of the sea is thoroughly lifeless on the coasts'. Therefore, the only logical consequence can be the cutting down of farm pollution: ' ... it makes no sense to restore the water system if the load on waters is not reduced.'

In sum, paralleling Lowe's and Ward's [*1998*] findings, our study shows that, in Finland, agri-environmental policy players and experts tend to frame farm pollution problems in two differing ways. The representatives of the agricultural policy sector regard these problems above all as 'a problem for farming'. However, they set these problems and their solutions in a wider context which reflects a concern for the survival of the domestic agriculture and for the vitality of rural areas since Finland's entrance to the European agricultural market. On the other hand, for the environmental state officials and the representatives of the major Finnish nature conservation organisation, farm pollution primarily signifies 'a problem of farming'. Therefore, the agri-environmental issue is discussed in the context of environmental protection and nature conservation.

These differing ways of framing farm pollution clarify the different stands the policy players take on the present phase of policies and especially on the new agri-environmental subsidy system. Many of those interviewed at the central agency level had been involved in the preparation of the subsidy programme. Generally, they seem to agree that there is a notable agricultural policy element in the agri-environmental scheme: environmental support is linked to the total process of adapting Finnish agriculture to the common market. However, it does not follow that there is an agreement on the desirability of the state of affairs. Differing stances were taken on the conditions and scope of support. Agricultural actors mainly expressed a true satisfaction with the strong income element. They seemed to think that since the most important goal is to preserve domestic agricultural production, the outstanding environmental subsidies are simply rational. By contrast, environmental policy actors were rather dissatisfied with some of the policies included in the agri-environmental support. Generally speaking, they tend to think that 'environmental support' should promote nothing but environmental protection goals. Thus, the minor role reserved for organic farming and for more developed environmental protection techniques was regarded as the weak point in the present Finnish subsidy policies. The environmentalists emphasised that the massive financial support conflicts with the idea of environmental incentives and the polluter pays principle.

### *Present Friction between Agricultural and Environmental Policies Slowing down Institutional Reforms*

Due to the EU's Common Agricultural Policy, the opportunity for the Finnish agricultural policy community to formulate *agricultural* policy has diminished sharply. Instead, the Ministry of Agriculture and Forestry now focuses on the implementation of CAP (although to implement EU policies at national level does not mean being completely powerless). The farmers' union is adapting to the new role of a traditional-type pressure group. From the institutional perspective, it is a characteristic of the present period of the Finnish agri-environmental regulation that the integration of sector policies has increased. Agricultural and environmental policy-makers are obliged to co-operate more and more on farm pollution issues, at central, regional and local agency levels. However, in *agri-environmental* policy, agricultural policy-makers do have a firmer hold on policy resources than the environmental policy-makers: the MAF is responsible for the large agri-environmental programme, and it is administered, implemented and controlled by the regional agricultural and rural policy districts and by municipal rural officers subject to the supervision of the MAF [*MAF, 1996a*].

Agricultural policy-makers seem to be satisfied with the present way of integrating environmental goals to agricultural policies. According to them, the main lines of Finnish agri-environmental policy-making were well established already in the early 1990s and, thus, 'Europeanisation' has not brought any major innovations. Rather, the EU's agri-environmental subsidy scheme is regarded as an important channel giving the option to support the already existing environmentally-friendly farming practices. This idea of success is illustrated by the following quotations of agricultural state officials:

> Indeed, concerning the goals of policies, the Finnish response to the EU Regulation 2078/92 must be seen as a continuation to the first national Environmental Programme for Rural Areas (in 1992 by the MAF and the MoE, PJ) and to the associated practical recommendations ('Good Cultivation Practices' in 1993 by the MAF, PJ).
>
> Farmers adapted to the environmental conditions for the support really rapidly. The programme reconciles Finnish and the EU's agricultural policies with Finnish and the EU's environmental policies. Further, we had to make farmers accept the programme, even on a voluntary basis. Thus, the programme must be seen as a great achievement.

Unlike agricultural policy-makers, environmental policy actors appear to adopt a rather ambivalent attitude towards the main principles of the present agri-environmental policy. The Finnish subsidy model is regarded both as a policy success and as a matter for criticism. The positive assessment can be understood from the international policy viewpoint. It is acknowledged that, in contrast to many western countries, in Finland there is a remarkable policy response to the urgent agri-environmental problem (that is, water pollution). However, from the domestic point of view, they think that the effectiveness of the policy should still be greatly improved. In brief, environmental policy actors argue that in order to realise[14] the present water pollution reduction targets the conditions (see note 10) to qualify for major agri-environmental support should be stricter and more targeted.

At first sight, the main problem for the agri-environmental policy makers we interviewed seems to be reaching a mutual understanding, at the principal level, on the function of agri-environmental subsidies. There are two contradictory basic options: whether to have as many farmers as possible involved in the subsidy scheme or to have as effective agri-environmental measures as possible under subsidisation. The former must necessarily be seen as a partial income support and, thus, typically an 'agriculturally led' solution. In the Finnish case, the major part (80 per cent) of the agri-environmental subsidies (of the scheme in 1995–99) are explicitly legitimised by the thinking that the rather light measures with rather good financial compensation will result in a large number of farmers joining and, thus, in an improvement of the environmental awareness and attitudes of most farmers[15] [*MAF, 1996a*; cf. *Miettinen, 1998*]. Environmental policy actors, however, consider such subsidies 'ineffective with respect to their costs'. The latter option, subsidies on condition of targeted and developed environmental protection, can rather be called an 'environmentally led' solution. In the Finnish case, a minor part (20 per cent) of the funding is reserved for this type of support. Agricultural policy actors, again, regard this support as 'too scattered because of the many measures included in it'.

Basically, these differing views on the economic tools of agri-environmental regulation reflect rather profound difficulties in developing agri-environmental regulation, which throws together differing definitions of the farm pollution problem as well as organisational rationalities and interests. In the Finnish case, we find that both the agricultural policy sector and the environmental policy sector are increasingly interested in farm pollution issues and, thus, are contributing to the development of agri-environmental policies. However, the traditional interest conflicts have not disappeared.

Finally, we can also find minor signs recently of some kind of opening up of the agri-environmental policy-making process. This is illustrated by the planning practices of the Finnish Agri-environmental Programme, which was prepared in 1994 solely by agricultural and environmental state officials [*MAF, 1996a*]. In 1995, a follow-up working group for the programme was set up, and also representatives from farmers' organisation, farmers' advisory service organisation, and of nature conservation association were appointed members of this group [*MAF, 1998*]. Further, in 1998, a working group responsible for the preparation of a 'Finnish Agri-environmental Programme for the years 2000–2006' was set up, and it also consists of the several interest representatives mentioned above. However, in the Finnish case, the voice of the consumers' organisation is still not heard in agri-environmental policy processes.

## Findings: The Transition towards Ecological Modernisation?

Concerning the assumption of a general transition towards 'the policies of ecological modernisation', our findings on the case of Finnish agri-environmental policies are somewhat contradictory. On one hand, from the late 1980s on, there are certain signs of transition consistent with the ecological modernisation theory. On the other hand, from the institutional and practice-oriented perspective, our evidence also suggests only some marginal changes and even stability in agri-environmental regulation, notwithstanding the present transnational context of regulation.

First, an undeniable change has appeared in agri-environmental policy discourses. Along with the relatively uncontested identification of the farm pollution problem and with the general 'greening' of social and political discourses in the late 1980s, the agri-environmental issue is no longer conceptualised as a 'non-problem' but rather as a challenge by different actors, including agricultural policy-makers. Secondly, we can also find some institutional and practical reaction to this environmental challenge. With reference to one of the core criteria of ecological modernisation, that is, the integration of policies, especially due to the present large agri-environmental programme, farm pollution issues have been institutionalised in Finnish agricultural policy-making more profoundly than has been the case before. It is another matter that, notwithstanding the partial move away from the purely sector-driven approach, the weight given to agricultural and environmental issues is still certainly far from equal. Since the mid-1990s a clear transition from informative to economic regulation has appeared. Thus, increasingly, ecological factors are involved in production activities, although signs of the 'greening' of consumption so far are limited to the expansion of organic farming. In recent years, a slight opening up of agri-environmental policy-making practices can also be noted, which is illustrated especially by the planning and follow-up processes of the Finnish Agri-environmental Programme 1995–99 (and of the coming programme for 2000–2006). All in all, in the Finnish case, there are signs of change that may allow a more effective agri-environmental policy and a more environmentally friendly farming.

However, there are also indications of slowness in the process of ecological modernisation. First, as is typical for all European farm pollution control, the polluter pays principle and its consequences are still ignored. Secondly, from the viewpoint of technological innovations, it can be argued that new 'clean' agricultural production methods have not been introduced to any great extent. Rather, from the 1980s on, the long-term trend in Finland has been to emphasise the importance of 'good cultivation practices', which are regarded *also* as environmentally-safe (in other words, to improve slightly the existing production practices and, accordingly, to 'eco-label' them). The technological advance implied in the major measures of the present Finnish Agri-environmental Programme 1995–99 is modest too: for instance, the technical standards and zoning solutions can be considered neither innovative nor strict (see note 10). Finally, though the agri-environmental planning practices have been partially opened up in the 1990s, the policy process itself is mainly running according to the rules of the traditional actors. Thus, concerning both farm pollution and a more environmentally-friendly production, the Finnish environmental and consumer NGO's seem to remain almost as passive as they have always been.[16]

What effect, then, has 'Europeanisation' had on the ecological re-orientation of Finnish agriculture? As we have emphasised above, agri-environmental policy consists of differing policy elements. In the early 1990s, when Finnish membership of the EU emerged, especially in the environmental state agency, much trust was placed in European farm pollution control [*Jokinen, 1995*]. It was expected in the Ministry of the Environment that the coming of new transnational rules would enable the use of legal and economic tools in addition to communicative measures, which would also considerably strengthen the position of environmental policy actors in farm pollution control practices. However, after having examined the formation of the present practices of Finnish agri-environmental regulation, we draw a different conclusion: rather, at least temporarily, the transition period of Finnish agriculture due to the entrance into the European market has somewhat retarded the policy process of ecological modernisation, which had begun to emerge in the late 1980s.

This is because the concern that the (still powerful) agricultural interests feel for the survival of domestic agriculture has reinforced the 'agricultural policy community' type of thinking and the problem definition in those policy issues which have remained within national discretion. The implementation of the EU rule (Regulation 2078/92) resulted in the use of the agri-environmental scheme (The Finnish Agri-environmental Programme) as an important part of the income policy. It was also employed by the rural party-dominated Government to soften the Finnish farmers' attitudes towards environmental policies and, above all, their strong political opposition to the EU. Thus, as a result of 'Europeanisation', the national agricultural policy actors have become more influential in agri-environmental regulation even at the expense of environmental policy actors, but the (potentially even strong) internal 'greening' of agricultural policy, which should also lead to institutional changes, has been trampled on by the income and production policy objectives.

## Conclusions

This study has focused on ecological modernisation and the case of the Finnish agri-environmental policy of the last three decades, even if the present phase with the 'Europeanisation' of policies from the mid-1990s has been examined in a more detailed way. We have found that the Finnish response to EU regulation has been problematic from the view point of agri-environmental policy. It follows, thus, that there is a broad implication for the case of European agriculture and the environment. The EU's Common Agricultural Policy currently has an ambivalent effect on the agri-environmental issue. Some of the agri-environmental measures in CAP have their positive impacts. They are, however, a very minor part of the total structure of the CAP. In our case, Finnish agriculture must be seen as relatively peripheral compared to the West European highly intensive farming system, and immediately after losing her national agricultural policy, Finland seems to have belonged to the periphery also from the viewpoint of European agricultural *policies*. According to our view, the Finnish stance to the Common Agricultural Policy because of its economic effects has been mainly reactive,[17] which, again, has retarded the policy innovations promoting the ecological re-orientation of agriculture.

Thus, it seems to be clear that the relationship of Finnish agriculture to the CAP and the European market is a separate issue which is of fundamental importance from the environmental point of view (however, at the turn of the millennium, the CAP itself is being reshaped as well). It is an open issue whether we should interpret the present time as a simple 'adaptation phase' or, as we suggest, as a more profound and complex transition phase of Finnish agriculture leading to new structures and a new policy set-up. In any case, from the 'horizontal' angle, the pro-environmental changes of agriculture should be based not only on external pressures but also on internally motivated factors. Further, from the 'vertical' angle, the need for more locally inspired and negotiated innovative action, instead of responses to 'top-down' rules and demands, is a challenge for the future of Finnish agriculture.

As regards the ecological modernisation position and especially the role of the state and the market, an issue pervading our study is whether we should see the West European agri-environment issue and, further, the Finnish agri-environment issue, as a different case compared with various production- and consumption-related environmental issues. We propose that the answer is, at least to some extent, in the affirmative. First, it seems that European agriculture will necessarily continue to be a highly state-centred (referring to the Common Agricultural Policy), protected and not growth-oriented (compared with industries such as industry, energy, and transport) industry even in the future, and this certainly challenges any strong trust in market dynamics, irrespective of whether or not the trust is well-grounded *per se*. Second, the Finnish agri-environmental policy case must be understood also in its Nordic context. The Nordic model of environmental policy has been characterised as 'cooperative and consensual' in a corporatist and 'rule-deferential' culture [*Christiansen and Lundqvist, 1996*]. Thus, an implication of this case for the EM perspective is that we should consider further (and rather carefully) the thesis according to which the institutional role of the state in environmental policies is in deep transition.

## NOTES

1. The interviews in 1996–97 were with government officers and experts at the central, regional, and local state agency levels, and representatives of the farmers' interest organisation and of the nature conservation association. These semi-structured theme interviews were conducted by Pekka Jokinen and Anita Niemi-Iilahti [e.g., *Niemi-Iilahti and Jokinen, 1999*].
2. Agri-environmental problems are either mainly denied or admitted 'conditionally' in exchange for compensatory agri-environmental subsidies.
3. The problems of eutrophication and blooming algae in the Baltic Sea have already led to the signing of several international agreements on the protection of marine environment (the 1974 Helsinki Convention; the 1988 Baltic Sea Declaration of the Ministers of the Environment; the 1992 Helsinki Convention). For instance, the 1988 Declaration demanded a 50 per cent reduction in the pollution load by 1995. Among others, Finnish agriculture and fish farming have failed to achieve the target.
4. 91/676/EEC.
5. 92/43/EEC.
6. The full name of the Regulation is 'Agricultural Production Methods Compatible with the Requirements of the Protection of the Environment and the Maintenance of the Countryside'. For its details, see, for example, Baldock and Lowe [*1996*].
7. The practices have varied in West European countries. Some, such as Austria, Denmark, Germany, and the Netherlands, have defined their whole territory as vulnerable, whilst Ireland has decided that it has no areas vulnerable to nitrates [*Brouwer and Lowe, 1998*].
8. In 1999: approximately 1 ECU = 6 FIM = 1.1 $.
9. More precisely, according to the Finnish Government decision, the programme has four general environmental objectives. It aims at (1) reducing pressures on the environment, especially in surface waters, groundwater and air, and at reducing hazards caused by the use of pesticides (2) preserving biodiversity and managing agricultural landscape (3) protecting wildlife habitats and endangered species of flora and fauna (4) producing agricultural commodities in an extensive and environmentally friendly manner [*MAF, 1996a: 15*].
10. The environmental support is available for each farm irrespective of its contribution to water pollution, The subsidies are further divided into two main parts; the 'basic support' and the 'special support'. About 80 per cent of the subsidies are reserved for the former and 20 per cent for the latter. In order to qualify for the basic support, farmers must meet certain conditions [*MAF, 1995*]. For instance, an Environmental Management Plan must be drawn up for each farm, the use of fertilisers per hectare must not exceed the prescribed limits, filter strips 1–3 metres wide and covered with vegetation must be established along water bodies, and at least 30 per cent of the arable land of the farm must be covered with vegetation in winter time. The conditions for the special support are clearly stricter, though also more heterogeneous, than those for the basic support. The special support measures include,

among others, organic farming and conversion to it, and riparian zones which will be completely removed from production under a 20-year set-aside.

11. The payments are highest in the southern and western parts of the country and they also vary greatly with the (combination of) measures taken.
12. In 1999 the preparation of the 'Finnish Agri-environmental Programme for the years 2000–2006' was in progress.
13. All the following quotations have been translated from Finnish to English by Pekka Jokinen.
14. Certainly, the changes in the load on waters can be verified only after a long period (from five up to even twenty years), and even then the effects of separate measures – not to say of separate policies such as agricultural policy and environmental policy – can hardly be identified. Very simply, natural scientific uncertainties leave room for differing interpretations by policy-makers.
15. Already in 1996, about 90 per cent of the field area was under the influence of the programme, which, again, led the follow-up working group for the programme [*MAF, 1996b: 30*] to conclude that the large commitment of farmers to the programme provides evidence of the positive change in farmers' environmental attitudes. However, it seems to us that it much more likely indicates the need for and attractiveness of the subsidies than an extremely rapid change in environmental attitudes.
16. Actually, it is far from easy to explain the remarkable passiveness of Finnish environmentalists (excluding the rising animal rights activism) even though the environmental impacts of agriculture have gone up on the 'official' severity scale of environmental problems. Certainly, the structural small-scaleness of agriculture and the subsequent unclearness about the environmental responsibility of individual farmers can be seen as one of the main reasons for the absence of activism [*Jokinen, 1997*]. Among the obvious policy factors are also the MAF's interest in integrating environmentalists into the formal 'partnership' in the agri-environmental policy process with the very minor representation given to them. Further, the effect of the strategic emphasis Finnish environmental organisations have laid on forestry issues must be realised. However, additional study is needed in order to examine further cultural, social, political and environment-related reasons.
17. This is so, despite the fact that nowadays Finland is trying to push through the EU the notion of a 'Northern Dimension' paying attention, for instance, to the short duration of the growing season in Finland (but explicitly apart from agri-environmental issues).

## REFERENCES

Andersen, M.S. and J.D. Liefferink (1997), 'Introduction: the Impact of the Pioneers on EU Environmental Policy', in Andersen and Liefferink (eds.) [*1997: 1–39*].

Andersen, M.S. and J.D.. Liefferink (eds.) (1997), *European Environmental Policy: The Pioneers*, Manchester: Manchester University Press.

Baker, S., Kousis, M., Richardson, D. and S. Young (1997), 'Introduction. The Theory and Practice of Sustainable Development in EU Perspective', in Baker, Kousis, Richardson and Young (eds.), *Sustainable Development: Theory, Policy and Practice in the European Union*, London: Routledge, pp.1–40.

Baker, S., Kousis, M., Richardson, D. and S.C. Young (eds.) (1997), *Sustainable Development Theory, Policy and Practice in the European Union*, London: Routledge.

Baldock, D. and P. Lowe (1996), 'The Development of European Agri-environmental Policy', in M. Whitby (ed.), *The European Environment and CAP Reform. Policies and Prospects for Conservation.*, Guilford: CAB International, pp.8-25.

Beck, U. (1992), *Risk Society: Towards a New Modernity.*, Wiltshire: Sage.

Blowers, A. (1997), 'Environmental Policy: Ecological Modernisation or the Risk Society?', *Urban Studies* Vol.34, Nos.5–6, pp.845–71.

Brouwer, F. and P. Lowe (1998), 'CAP Reform and the Environment', in F. Brouwer and P. Lowe (eds.), *CAP and the Rural Environment in Transition. A Panorama of National Perspectives,* Wageningen; Wageningen Pers, pp.13–38.

Buttel, F.H. (1995), 'Twentieth Century Agricultural-environmental Transitions: A Preliminary Analysis.', *Research In Rural Sociology and Development*, Vol.6, pp.1–21.

Christiansen, P.M. and L. Lundqvist (1996), 'Conclusions: A Nordic Environmental Policy Model?', in P.M. Christiansen (ed.), *Governing the Environment. Politics, Policy and Organization in the Nordic Countries*. Århus: The Nordic Council, pp.337–63.

Cohen, M.J. (1997), 'Risk Society and Ecological Modernisation. Alternative Visions for Post-Industrial Nations', *Futures*, Vol.29, No.2, pp.105–19.

Curry, N. and E. Stucki (1997), 'Swiss Agricultural Policy and the Environment: An Example for the Rest of Europe to Follow?', *Journal of Environmental Planning and Management*, Vol.40, No.4, pp.465–82.

Daugbjerg, C. (1996), *Policy Networks under Pressure: Policy Reform, Pollution Control and the Power of Farmers*, Aarhus: Aarhus University.

Eckerberg, K. and A. Niemi-Iilahti (1996), *Implementation of Agri-environmental Policy in the Nordic Countries*, Vaasa: University of Vaasa.

Epstein, P.J. (1997), 'Beyond Policy Community: French Agriculture and the GATT', *Journal of European Public Policy*, Vol.4, No.3, pp.355–72.

Finnish Government (1998a), *Decision on the Programme of Goals for Water Pollution up to 2005*, Helsinki (in Finnish).

Finnish Government (1998b), *Decision on the Implementation of the EC Nitrates Directive*, Helsinki. (in Finnish)

Finnish Government (1998c), *Decision on the Natura 2000 Network in Finland*, Helsinki (in Finnish).

Frouws, J. and A.P.J. Mol (1997), 'Ecological Modernization Theory and Agricultural Reform', in H. de Haan and N. Long (eds.), *Images and Realities of Rural Life*, Assen: Van Gorcum, pp.269–86.

Garner, R. (1996), *Environmental Politics*, Guildford: Harvester Wheatsheaf.

Glasbergen, P. (1992), 'Agro-environmental Policy: Trapped in an Iron Law: A Comparative Analysis of Agricultural Pollution Control in the Netherlands, the United Kingdom and France', *Sociologia Ruralis*, Vol.32, No.1, pp.30–48.

Goodin, R.E. (1992), *Green Political Theory*, Padstow: Polity Press.

Gouldson, A. and J. Murphy (1996), 'Ecological Modernization and the European Union', *Geoforum*, Vol.27, No.1, pp.11–21.

Granberg, L. (1995), 'A Break in Agriculture. From Peasant Values to Welfare Rationality', in L. Granberg and J. Nikula (eds.), *The Peasant State*, Lapin yliopisto, Rovaniemi, pp.71–87.

Grant, W. (1995): 'Is Agricultural Policy Still Exceptional?', *The Political Quarterly*, Vol.66, No.3, pp.156–69.

Hajer, M.A. (1995), *The Politics of Environmental Discourse. Ecological Modernization and the Policy Process*, Guildford: Oxford University Press.

Hajer, M.A. (1996), 'Ecological Modernisation as Cultural Politics', in S. Lash, B. Szerszynski and B. Wynne (eds.), *Risk, Environment and Modernity: Towards a New Ecology*, London: Sage, pp.246–68.

Joas, M. (1997), 'Finland: from Local to Global Politics', in M.S. Andersen and J.D. Liefferink (eds.), *European Environmental Policy: The Pioneers*, Manchester: Manchester University Press, pp.119–60.

Jokinen, P. (1995), 'The Development of Agricultural Pollution Control in Finland', *Sociologia Ruralis*, Vol.35, No.2, pp.206–27.

Jokinen, P. (1997), 'Agricultural Policy Community and the Challenge of Greening. The Case of the Finnish Agri-environmental Policy', *Environmental Politics*, Vol.6, No.2, pp.48–71.

Jokinen, P. and K. Koskinen (1998), 'Unity in Environmental Discourse? The Role of Decision-makers, Experts and Citizens in Developing Finnish Environmental Policy', *Policy and Politics*, Vol.26, No.1, pp.55–70.

Just, F. (1994), 'Agriculture and Corporatism in Scandinavia', in P. Lowe, T. Marsden and S. Whatmore (eds.), *Regulating Agriculture*, London: David Fulton Publishers, pp.31–52.

Kettunen, L. (1996), 'Adjustment of Finnish Agriculture in 1995', in L. Kettunen (ed.), *First Experiences of Finland in CAP*, Vammala: Vammalan Kirjapaino, pp.7–25.

KM 1983:66 (1983), *Ad hoc Committee Report on Farm Water Management*, Helsinki: VAPK (in Finnish).

KM 1986:42 (1986), *Permanent Committee on Water Management: A Proposal for the Programme of Goals for Water Pollution up to 1995.*, Helsinki: VAPK (in Finnish).

Lowe, P., Ward, N., Seymour, S. and J. Clark (1996), 'Farm Pollution as Environmental Crime', *Science as Culture*, Vol.5, No.4, pp.588–612.

Lowe, P., Clark, J., Seymour, S. and N. Ward (1997), *Moralizing the Environment: Countryside Change, Farming and Pollution*, London: UCL Press.

Lowe, P. and N. Ward, N. (1998), 'Field-level Bureaucrats and the Making of New Moral Discourses in Agri-environmental Controversies', in D. Goodman and N. Watts (eds.), *Globalizing Food: Agrarian Questions and Global Restructuring*, London: Routledge, pp.256–72.

Macnaghten, P. and J. Urry (1998), *Contested Natures*, London: Sage.

Miettinen, A. (1998), 'Finland', in F. Brouwer P. and Lowe (eds.), *CAP and the Rural Environment in Transition. A Panorama of National Perspectives*, Wageningen: Wageningen Pers, pp.323–44.

Ministry of Agriculture and Forestry (MAF) (1992), *Agriculture 2000, the Revised Programme*, Helsinki: VAPK (in Finnish).

Ministry of Agriculture and Forestry (MAF) (1995), *EU Guide for the Countryside* (in Finnish).

Ministry of Agriculture and Forestry (MAF) (1996a), *The Agri-environmental Programme in Finland.*

Ministry of Agriculture and Forestry (MAF) (1996b), *Agri-environmental Programme 1995–99: An Interim Report of the Follow-up Working Group* (in Finnish).

Ministry of Agriculture and Forestry (MAF) (1996c), *Farm Register 1995*, Helsinki: Hakapaino (in Finnish).

Ministry of Agriculture and Forestry (MAF) (1998), *Agri-environmental Programme 1995–99: A Final Report of the Follow-up Working Group* (in Finnish).

Ministry of the Environment (MoE) (1988), *The Programme of Goals for Water Pollution up to 1995*, Helsinki: VAPK (in Finnish).

Ministry of the Environment (MoE) (1992), *Environmental Programme for Rural Areas*, Helsinki: VAPK (in Finnish).

Ministry of the Environment (MoE) (1994), *Environmental Programme for Rural Areas. An Interim Report of the Follow-up Working Group*, (in Finnish).

Ministry of the Environment (MoE) (1995), *Finnish Environmental Programme up to 2005*, Forssan Kirjapaino (in Finnish).

Ministry of the Environment (MoE) (1997a), *Environmental Policies in Finland*, Helsinki: Edita.

Ministry of the Environment (MoE) (1997b), *Proposal for the Implementation of the EC Nitrates Directive in Finland*, Helsinki: Edita. (in Finnish).

Mol, A.P.J. (1995), *The Refinement of Production: Ecological Modernization Theory and the Chemical Industry*, Utrecht: Van Arkel.

Mol, A.P.J. (1996), 'Ecological Modernization and Institutional Reflexivity: Environmental Reform in the Late Modern Age', *Environmental Politics*, Vol.5, No.2, pp.302–23.

National Water Board (1974), *The Principles of Water Management up to 1985*, Helsinki: VAPK (in Finnish).

Niemi-Iilahti, A. and P. Jokinen (1999), 'Three Years' Experience of EU Integration: The Case of the CAP and Agri-environmental Subsidies in Finland', in M. Joas and A. Hermanson (eds.), *The Nordic Environments: Comparing Political, Administrative, and Policy Aspects*, Aldershot: Ashgate, pp.233–59.

Rannikko, P. (1996), 'Local Environmental Conflicts and the Change in Environmental Consciousness', *Acta Sociologica*, Vol.39, No.1, pp.57–72.

Rekolainen, S., Kauppi, L. and E. Turtola (1992), *Agriculture and Water Quality*, Helsinki: Luonnonvarainneuvosto (in Finnish).

Rhodes, R.A.W and D. Marsh (1992), 'New Directions in the Study of Policy Networks', *European Journal of Political Research*, Vol.21, Nos.1–2, pp.181–205.

Schnaiberg, A. (1997), 'Sustainable Development and the Treadmill of Production', in S. Baker, M. Kousis, D. Richardson and S.C. Young (eds.), *Sustainable Development: Theory, Policy and Practice in the European Union*, London: Routledge, pp.72–88.

Schnaiberg, A. and K.A. Gould (1994), *Environment and Society. The Enduring Conflict*, New York: St. Martin's Press.

Siikamäki, J. (1996), *Suomen maatalouden ympäristötukijärjestelmän sisältö ja toiminta* (The Contents and the Implementation of the Finnish Agri-Environmental Programme), Helsinki: MTTL.

Spaargaren, G. and A.P.J. Mol (1992), 'Sociology, Environment, and Modernity: Ecological Modernization as a Theory of Social Change', *Society and Natural Resources*, Vol.5, No.4, pp.323–44.

Statistics Finland (1994), *Environment Statistics*, Helsinki: Tilastokeskus.

Statistics Finland (1998a), *Statistical Yearbook of Finland 1998*, Hämeenlinna: Tilastokeskus.

Statistics Finland (1998b), *Finland's Natural Resources and the Environment 1998*, Helsinki: Tilastokeskus.

Suhonen, P. (1994), *Mediat, me ja ympäristö* (Media, Us, and the Environment), Tampere: Hanki ja Jää.

Weale, A. (1993), 'Ecological Modernisation and the Integration of European Environmental Policy', in J.D Liefferink, P.D. Lowe and A.P.J. Mol (eds.), *European Integration and Environmental Policy*, London: John Wiley.

*Transitional Economies*

# Ecological Modernisation as Cultural Politics: Transformations of Civic Environmental Activism in Lithuania

LEONARDAS RINKEVICIUS

This article elaborates on the notion of ecological modernisation as cultural politics [*Hajer, 1996*] by exploring structural tensions and institutional learning among civic, bureaucratic, and economic actors in Lithuania, from 1957 to the present. More particularly, it tests the eco-modernist claim that a shift has occurred in environmental activism from a grass-roots, radical and conflict orientation in the 1970s and 1980s, to a professional, reform and consensus orientation in the 1990s. The study finds that both types of environmentalism were present in Lithuania throughout the time studied. It suggests that the environmental movement's critical, grass-roots, nonconformist aspect is likely to remain along with a more professionalised and dialogue-oriented component, even while such manifestations continue to evolve in form.

## Introduction

This study analyses changes in the manifestation of civic environmental activism in Lithuania – one of the transitional economies in Eastern Europe which recently departed rapidly from Soviet totalitarianism, re-established itself as an independent state, and is now seeking to become an open, democratic society – in four distinct historical periods:

- state socialism under Soviet domination until 1986;
- *perestroika*, 1986–88;
- national liberation, 1988–89; and
- re-establishment of an independent state and transition to a market economy, 1990–present.

An important part of research related to this contribution has been carried out in the framework of the EU-funded project, Public Participation and Environmental Science and Technology Policy Options (PESTO) (contract IC20 CT 960040). The author would like to express gratitude to the European Commission for making this work possible, and to the project co-ordinator, Professor Andrew Jamison, for his insightful comments on earlier versions.

Examining discursive strategies, practices and cultural tensions, the study argues that the organisational form and ideological orientation of environmental activism in Lithuania shifted from latency under Soviet rule, to becoming a social movement in the late 1980s, to developing in more professional directions in the 1990s. The analysis draws data from semi-structured interviews, secondary sources, and participant observation.

The dynamics in Lithuania's environmental movement are contrasted with those in Western Europe and North America which have been well documented and theorised. Comparisons are made along the lines drawn by Western social scientists *inter alia* those developing ecological modernisation theory.

As noted in the ecological modernisation literature [*Hajer, 1995; Mol, 1995; Spaargaren, 1997*], environmentalists in the 1970s were often antagonistic in their attitudes and self-identities *vis-à-vis* other actors, especially industrialists and public authorities. Similarly, the notion of 'appropriation by the movement's opponents' articulated by Eder [*1996*], implies a continuation of tensions between civic and economic or bureaucratic domains. On the other hand, recent empirical findings and theoretical accounts point to the gradual alleviation of tensions between environmental organisations and actors in other domains. For instance, the ecological modernisation literature has illuminated how environmental activists are involved on the boards of industrial enterprises and on governmental committees and are engaged in new, dialogue-based public-policy interfaces [*Hajer, 1995, Mol, 1995*]. Activists now also play important roles in consensus-oriented, environmental partnerships and institutional experiments epitomised by the emblem of Local Agenda 21.

Jamison [*1996*] finds that social movement organisations pursuing civic goals inspired by environmental concerns are deploying new strategies and engaging academics, marketing professionals and even political elites (as suggested, for instance, by use of the image of the Duke of Edinburgh in a World Wildlife Federation advertisement). Such examples indicate that contemporary environmental organisations are acquiring a new character, in contrast to that of grass-roots organisations of the 1970s. Evidence of new institutional developments, discursive strategies and practices suggests a trend among contemporary environmental activists to replace confrontation in various domains of environmental science and technology policy. At the same time, Jamison [*1996*] and other environmental social scientists anticipate the emergence of new grass-roots environmental activism [also Szerszynski *et al.*, 1996].

This study shows that there was neither clear-cut conflict nor consensus among Lithuanian environmental activists during the periods studied. Ecological modernisation in Lithuania is often portrayed as a mixture of collaborative efforts and tensions among the green movement and public authorities as well as industrialists and academics. Here it is argued that Lithuanian environmental activism has been typified from the beginning by the presence of *both* a dialogue- and consensus- oriented trend – often subsumed as a feature of ecological modernisation [*Mol, 1995*]; and a grass-roots, more critical, more radical stance – typical of Western environmentalism in the 1970s.[1]

The notions of 'policy domains' and 'cultural politics' are important conceptual tools in this study and are introduced briefly in the next section. Subsequent sections examine in some detail the development of environmental activism in Lithuania during the four periods studied. The analysis concludes with a discussion of the implications of the Lithuanian case for ecological modernisation theory.

## Policy Domains and Cultural Politics

Analysis of the environmental movement and public engagement in environmental policy deals with the interplay of actors in civic and other policy domains. In their work regarding science and technology policy, Jamison and Baark [*1990*] distinguish three policy domains or cultures: bureaucratic, entrepreneurial and academic. This analytical framework has recently been expanded to include a fourth 'ideal type', namely the civic.[2] (See Figure 1.)

FIGURE 1
CULTURAL TENSIONS BETWEEN DIFFERENT POLICY DOMAINS

| Domain ⇨ | Bureaucratic | Economic | Academic | Civic |
|---|---|---|---|---|
| Important traits ⇩ | | | | |
| Doctrine | *order* | *growth* | *enlightenment* | *democracy* |
| Steering mechanism | *planning* | *commercial* | *peer review* | *assessment* |
| Ethos | *formalistic* | *entrepreneurial* | *scientific* | *participatory* |

*Sources*: Jamison [*1997*], Rinkevicius [*1998*].

Each policy domain is characterised by different *doctrines*, or objectives and ideals. The doctrine of order prevails in the bureaucratic domain, for instance, in contrast to the doctrine of growth in the economic domain. The domains differ as well in the nature of their *steering mechanisms*. In the bureaucratic domain, rules may be based on cumbersome planning procedures. The economic domain is based on rapid and dynamic commercial transactions institutionalised as routine practices.

Each policy domain is characterised also by dominant *ethos* or value-orientations. Most civic-oriented activity is based on faith in democratic participation and policy openness to public scrutiny. In the bureaucratic domain, formalism prevails, excluding those outside formal decision-making structures. In the entrepreneurial ethos, economic actors respect effective and efficient, smart and flexible kinds of activities.

Following some of the sociological literature [Eisenstadt, 1968; Abercrombie *et al.*, 1994], policy cultures/domains might be more generally described as social institutions or institutional spheres, because those structural entities imply a conjunction of prevailing guiding principles, beliefs and norms as well as routine dominant practices and forms of organisation.

### *Cultural Politics*

The term, 'cultural politics', has been defined as *cultural tensions between discourse coalitions in the course of environmental policy-making process* [*Hajer, 1996*]. Jamison and Baark's [*1990*] distinction between policy domains, and focus on inter-domain tensions is helpful in elaborating the notion of cultural politics. Such cultural tensions and politics, I would like to suggest, are derived not only from divergent ideologies followed by eco-modernists and those who perceive ecological modernisation as a defence of '*status quo*' or view it as a 'technocratic project'. The cultural politics of ecological modernisation can be also seen as induced by, or growing out of, diverging doctrines, principles, daily practices and value-orientations institutionalised in different policy domains, and thus leading to different 'policy cultures'.

From such a perspective, ecological modernisation can be sociologically analysed and interpreted as a controversial process of cultural politics whereby the diffusion of eco-modernist ideology depends on people's attitudes, beliefs and practices which are structurally shaped by different policy domains. Eco-modernists suggest that cultural politics leads to new dialogue-seeking and consensus-building experiments in society, which in turn alleviate tensions through inter-domain learning and reach closure regarding the objectives and trajectories of sustainable development.

Transformations of environmental activism are one aspect of this complex process of change in structure and agency. Research into the cultural politics of environmental activism is one of the projects of the further development of ecological modernisation theory. We turn now to analyse the development of environmentalism in Lithuania during the Soviet era.

## Latent Environmentalism Under the Soviet Regime Until 1986

The Soviet era in Lithuania, from the 1950s through the 1980s, can be characterised as a period of the colonisation of other spheres of society by the bureaucratic domain. This pattern is identifiable in most social structures, although official ideology emphasised the development of collective, democratic rule by working people. Oppression by the totalitarian state and its bureaucracy had the most significant impact upon the civic domain in the countries of the former Soviet bloc.

Political culture of Lithuanian society under Soviet rule might be characterised as 'double-faced', having both a facade and a latent side [*Palidauskaite, 1996*]. Some scholars portray Lithuania in the 1960s through the 1980s as a country with a 'practically totally demolished civic society' [*Vardys, 1993*]. Environmentalism, a particular arena of civic activism, was not an exception from this general pattern. The Soviet regime blocked the emergence and penetration of environmentalism into public space and discourse[3] until well into the *perestroika* years.

In Western Europe and North America, by contrast, it was the extensive public environmental debates and active social movements – the civic domain – which emerged in the 1970s that carved out the way for spreading environmentalist attitudes and values in the society at large. Social movements created a temporary public space [*Jamison, 1996*] for spreading the seeds of environmentalism. Based on the ecological critique, they widely propagated the idea of an alternative society, and articulated a new set of values and cognitive interests [*Eyerman and Jamison, 1991*], although it is still debated which factors were most important in consolidating and mobilising the newly emerging movements [*Yearley, 1994*].

Until the mid-1980s, Lithuanian society found itself in a state of stagnation, at least on the face-side of social life. In spite of this, important institutional changes took place indicating the existence of a *latent form of environmentalism*. The core feature of latent environmentalism were the brave civic stands by particular individuals in public administration and economic enterprises, expressed in pro-environmental actions. Such actions were not noted in public discourse or debates, as they were not covered by the news media. Latent environmentalism differed from Western environmentalism, which expressed ecological anxiety through social movements and movement organisations. Latent environmentalism was an important part of what might be termed a 'double-faced' policy culture, formally following directives from the top in Moscow but actually adjusting and ameliorating them at a local level in Lithuania.

Latent environmentalism yielded particular fruits: Lithuania was the first republic in the USSR to establish the State Nature Protection

Committee in 1957. Lithuania was also the first among the republics of the Soviet Union to pass the Law on Nature Protection in 1959. The Nature Protection Society was established in 1960. The latter was 'public', however, only to the very limited extent, because the Soviet regime allowed only strictly defined ideological and organisational boundaries. Those legal and institutional developments were achieved not by Soviet public authorities, but rather by outstanding individuals, through informal networking and lobbying, and personal contacts. While occupying important posts in the state bureaucracy, such individuals shared civic pro-environmental orientation and ethos.

One such personality was Viktoras Bergas, chairman of the State Nature Protection Committee. He led this committee from its establishment until well into the 1970s. Most above-mentioned and other institutional and legal developments [also *Rinkevicius, 1997*] in the sphere of nature protection in Lithuania were to a large extent the result of activism by Bergas and a few other officials. They were able to generate ideas, to bridge informal networks among key persons in the Soviet bureaucracy, and mobilise necessary resources. This type of actor is similar to what we elsewhere called *policy entrepreneurs* [*Giuliani, Rinkevicius and Sverrisson, 1998*].

Many people who belonged to the Soviet administrative structures in Lithuania (top leaders of the Communist Party, the Council of Ministers and the Government of Lithuania, municipal administration) were born before the Second World War. At least part of their or their parents' socialisation took place in the pre-war rural communities that formed a nucleus of Lithuanian society. Many of those who stood behind environmentally-informed institutional innovations in Soviet times shared ecocentric world views stemming from traditional Lithuanian communal pantheist culture respecting the work in, and being *with*, nature [*Kavolis, 1994; Rinkevicius, 1997*].

This tradition, more precisely a blend of pantheism and Christianity, stemmed in Lithuania from the fifteenth to sixteenth centuries. Ecocentric and preservationist, idealistic and romantic with respect to nature, it is similar to the Arcadian tradition described by Worster [*1977*]. This tradition continued through the two decades of Lithuania's independence between the two world wars, although it was gradually altered by modernisation. Even today, some of the traits of traditional pantheism can be observed among elder rural Lithuanians.

Of course, it was dangerous for any individual even occupying a high post to resist the fulfilment of five-year plans, including the drastic expansion of Soviet industry. However, even without public pressure and long before Lithuania's environmental social movement emerged, individuals took risks to suspend construction or request reconstruction of particular plants which were very important for the Soviet economy. In the Siauliai region, for example, environmentally conscientious actions by water quality inspectors such as Kastytis Cechavicius affected the dairy in Luoke, the Vairas bicycle factory, the Elnias tannery in Siauliai, the Akmene cement factory, the Paventes sugar plant, and other companies in the 1970s and 1980s.[4] In most cases, it was the first time that environmental issues had been integrated into processes of technological reconstruction or upgrading of industrial enterprises. Issues that previously had been largely neglected by industrialists were brought on to the agenda of the day by environmentally and civilly oriented public servants.

Through the 1970s and 1980s, it gradually became a rule that environmental inspectors, along with other public administrators, were entitled to assess each case of industrial development and approve or disapprove changes in every production site. Some companies, for example, the Luoke dairy and the Elnias tannery, had to redesign and reconstruct their facilities or particular pieces of equipment.

Some companies were even encouraged to develop prototypes of what today has come to be called industrial ecology. For instance, a closed-loop water treatment and reuse system developed at the Akmeme cement factory included treatment not only of its own sewage, but also sewage from households in the entire nearby city of Naujoji Akmene. All industrial and household sewage was collected, biologically treated and reused for industrial production at the same factory. This system was gradually improved during the 1970s and 1980s and regularly awarded the gold medal at the All-Union Exhibition of the Achievements in National Economy (Moscow).

Gradual integration of environmental concern into the daily praxis of industrialists and design engineers could be found in various instances in Soviet Lithuania. However, it was not social structures, for example, regulatory institutions, market forces, or social movements, but rather civic-oriented individuals who were trying to pursue environmentally friendly industrial development integrating ecological rationality into modernisation of the country's economy. Construction of new companies and reconstruction of existing ones was increasingly combined with construction of environmental facilities. When industrial entities of national importance such as the Mazeikiai oil refinery were developed, regional environmental inspectors spent four days a week on-site to ensure that they would be built with maximum environmental precaution. Such cases could be regarded as the first instances of dialogue among actors from different domains – bureaucratic, economic, and academic.

Algirdas Brazauskas, Secretary of the Lithuanian Communist Party's Central Committee, which was in charge of industrial development, regularly visited the site of Mazeikiai refinery. He assessed the construction works together with an environmental inspector, local engineers and designers. Mazeikiai's environmental facilities were completed nearly six months ahead of its production facilities. This case has often been referred to by Communist Party leaders and others as a 'good example' of environmentally friendly industrial development in the former Soviet Union.

The ways in which decision-makers as well as common citizens influenced industrial and technological projects indicate that both shared a pro-environmental perspective and civic stand. According to Kastytis Cechavicius in Siauliai and Rapolas Vasiliauskas in Kaunas, who were actively involved in the environmental inspection of industrial sites in the Soviet period, it was not only civic stand and nonconformity which were necessary to overcome structural impediments and achieve environmental improvements in industry. Professional competence in technological and legal aspects also were very important in communicating a message to state bureaucracies that suspending or improving particular industrial projects was necessary from an economic as well as environmental point of view. According to the aforementioned interviewees, inspectors had to be brave, nonconformist, professionally competent, and also 'clean', that is, not having even a hint of corruption in their past.

Some actions of such individuals can be interpreted as 'antimodernist', because they reflect a tendency not to seek environmentally benign industrial development but rather to sustain Lithuanian society based on premodern agrarian social structures, traditions and values. This can be illustrated by the citation from memoirs about Bergas, chairman of the State Nature Protection Committee:

> He bravely maintained: there is no need for the cotton mills, oil refineries – Lithuania [is an] agricultural country. When it was planned to construct an oil plant nearby Jurbarkas, he was very resistant ... : it will flood meadows and pollute the river Nemunas – where [will] such a policy lead? [*Zemulis, 1996: 10*].

The civic stand of ecologically and socially concerned individuals like Bergas and others within the Soviet-dominated administration and elite, I would argue, was one of the main reasons behind Lithuania's current concentration of food processing, textiles, wood and other 'light' and 'renewable' industries rather than Soviet-style heavy industry.

Due to this 'double-faced' policy culture – formally following directives from the top in Moscow but actually adjusting and ameliorating them at a local level – Lithuania avoided the significant natural and social bifurcation visible in other former Soviet republics, including Belarus, Latvia and the

Ukraine. Major industries in Lithuania were not concentrated in the one or two biggest towns as in Latvia and Estonia. Instead, many new industrial enterprises emerged in local or regional towns, giving employment to former peasants from surrounding villages and preventing the immigration of large numbers of workers from Russia and elsewhere in the former Soviet Union. In this way, disturbances of social community were mitigated and peoplele's alienation from previous social ties and the natural environment was less painful.

Adjustment of industrialisation and urbanisation policies to local natural and social conditions, on the other hand, is not only a result of *latent* policy culture characterised by *informal* communication among elite decision makers and local community members. Official Communist Party and Lithuanian government documents demonstrate the existence of official, facade policies reflecting the precautionary approach with regard to environmental and social disturbances related to the development of industry and other sectors of Lithuania's economy.

One example is Decree No. 303 by the Central Committee of the Lithuanian Communist Party and the Council of Ministers on the *Limitation of Industrial Development in Large Cities,* issued in 1981. This and related documents officially communicate a policy aimed at stopping industrial construction in the two major cities of Vilnius and Kaunas; reducing it in the cities of Klaipeda, Kedainiai, and Mazeikiai; and giving priority for industrial development in the rural areas and small towns of Alytus, Telsiai, Taurage, Ukmerge, and Plunge. Communist Party leaders, governmental decision-makers and planners affiliated with research institutes expected that this policy would prevent the merging of industrial cities into agglomerations.

As with most policy documents and decisions under Soviet rule, this decree was not accessible to the general public and there were no public discussions or scrutiny. The contents of this decree as well as other documents indicate, however, that those who developed such policies had their own ways to exchange information and receive feedback from rural communities. The Communist Party functionaries, governmental decision-makers or researchers and planners used different channels of communication including visiting one's parents living in rural areas, helping rural relatives in fieldwork and household activities, meeting informally with former classmates and friends still living in native villages. Weekend hunting and fishing expeditions were especially popular among Soviet elites and Communist Party functionaries.

These and other communication channels helped Soviet decision-makers learn about local social and natural circumstances. Decisions related to industrialisation and urbanisation were thus not allowed drastically to disturb existing communities, which helped soften the disruption of people's roots and relationship with their social and the natural environment.

These are just a few examples of the complex process of Lithuania's modernisation under Soviet rule, 'balancing' prevailing technocratic ideology, bureaucratic 'command-and-control' decision-making style and deliberate, precautionary policy, taking into account local social values and environmental concern. It is still a largely under-researched area with the exception of a few texts in native language by elder generation social scientists in Lithuania [e.g., *Grigas, 1995*].

Under Soviet rule, the civic domain in Lithuanian society was weak and peripheral compared to other policy domains. It began to gain strength in the mid-1980s, however, with resultant changes in interfaces between policy-oriented institutions and the general public. These changes are addressed in the following section.

## First Tribunes: Environmental Activism in the Era of Perestroika, 1986–88

A group of intellectuals (intelligentsia) were the first public tribunes in Lithuania in the mid-1980s to disseminate signals relating to the risks involved in short-sighted and irresponsible science and technology decisions. In the fall of 1986, during the early beginnings of *perestroika* and *glasnost,* a protest letter was published in the *Literaturnaya Gazyeta* in Moscow, signed by more than 20 outstanding writers, poets, actors, painters, composers, architects. Three of them were Russians, the others Lithuanian. They warned about the plans of Moscow technocrats to start oil extraction near the Curronian peninsula, a Lithuanian pearl on the coast of the Baltic Sea. A few weeks later, this protest letter was reprinted in the major Lithuanian cultural newspaper, *Literatura ir Menas*.[5]

This was the first publicly visible attempt to initiate public debate and influence environment-related science and technology decisions in Lithuania that otherwise were likely to be made by governmental authorities in a traditional way – without any public assessment or scrutiny. It was the first time that Lithuanian actors in the civic domain deployed the mass media to disseminate an important message that induced social-environmental anxiety. The letter was publicised six months after the Chernobyl catastrophe that had added to growing public anxiety about the advancement of science and technology in the former Soviet Union.

The discourse which developed with the publishing of the protest letter involved various types of symbols, emblems, and arguments. Some arguments were historical, cultural, mythological, poetic, moral, and deeply emotional – again analogous to the Arcadian tradition described by Worster [*1977*]. The way the policy of extracting oil on the Baltic coast was developed implied socio-cultural irresponsibility and alienation between the 'centre and periphery' in the Soviet empire. The letter was imbued with an intrinsic, intuitive sense of risk similar to that addressed by 'risk society' theory [*Beck, 1992*]; it focused on inherent threats in science and technology development not susceptible to precautionary, preventative public scrutiny and control.

In addition to the Arcadian, ecocentric language, this coalition also tried to phrase their arguments in a language with which actors in the bureaucratic and economic domains were familiar: rational, quantitative, based on weighting economic and environmental costs and benefits:

> When 5 thousand tons of crude oil will spill off at once, the damage as follows shall be made for the state: 1) fines for pollution will amount to 155 million roubles, 2) the fisheries will suffer from some 40 mln. roubles damage, 3) damage for recreational coastal zone will amount to 38 mln. Roubles, 4) clean-up of the polluted territory will cost 49 mln. Roubles … All together such a spill-off would cost a state about 284 mln. roubles.[6]

This was probably the first time that environmental concerns in Lithuania were expressed differently from the cultural and moral terms prevailing in the 1960s and 1970s such as Bergas' 'it will flood meadows and the Nemunas river [*sic*] it will pollute the Baltic Sea'. By contrast, the message was based on very concrete facts, and was spelled out in a language that one would now call cost-benefit analysis – assigning it to the vocabulary of ecological modernisation. This indicates that the cultural politics of environmentalism in Lithuania induced by a coalition of well-known intellectuals were intended not simply to criticise environmentally irresponsible public policy and publicly exhibit distinct pro-environmental self-identity of those who signed the letter. On the contrary, the language chosen and the ways in which the message was communicated were further indications that civil actors were attempting to find a constructive solution than signs of an attempt to take a radical, alternative, non-constructive stand which would confront or neglect the objectives and practices of other actors.

One peculiar aspect of this letter is the fact that among all the intellectuals who signed it there was only one scientist. To argue about environmental damage in cost-benefit terms it was necessary to involve some expert scientists, especially those who have access to important information which at that time was usurped by the central bureaucracy. The empirical evidence available to date suggests the existence of informal communication channels among environmentally concerned actors encompassing various societal domains – the bureaucratic, the economic, and the academic.

The protest letter contained rather sharp statements of mistrust of the command-and-control system based upon directives from Moscow. The quest for decentralisation and local control echoes an ideology promulgated by environmentalists in the Western countries during the 1970s [*Cotgrove, 1982; Pepper, 1994*]. The call for decentralisation in environmental science and technology decision-making implied the call for more general decentralisation from Moscow, and therefore it was a very brave step even for a coalition of outstanding intellectuals.

> We thought such *foolishness* and *open crime* against nature is impossible in the *Republic of Lithuania* [my italics - L.R.]. First of all because deep, nature respecting customs blossom here for many years, and the Lithuanian nation is especially unanimous and concordant in implementing elementary unwritten laws of nature protection, however, in the given case not everything depends on the republic.[7]

This letter not only calls for precaution and deliberation in environmental science and technology policy, but also expresses the need to increase national sovereignty. It reflects growing controversy in the relations between the 'centre and republic', as it was called in those times. However, the call for sovereignty was articulated in a rational (although sharp) eco-modernist language: emphasising the relevance of decentralisation for saving natural resources and avoiding environmental disasters which would be too costly to mitigate.

The protest letter resulted in the temporary suspension of the technocratic plans to drill oil in the Curronian peninsula. None of the members of coalition that signed the letter was physically or otherwise punished by the repressive structures of the Soviet regime, perhaps because it was already the period of *glasnost*. In the mid-1990s, in spite of active protest by the Lithuanian government and civic society at large, the plans were renewed by the Russian government and implementation started on Kaliningrad part of the Curronian lagoon.

This 1986 letter was probably the first case where environment-related information was used for disseminating the politically significant call for decentralisation and sovereignty of Lithuania. Lithuanian environmentalism has included a mix of ecological concerns and ideas of national sovereignty ever since. Even under *glasnost*, the strong totalitarian boundaries kept policy-making relatively closed to democratic public scrutiny. Those boundaries were breached two years later.

## Civic Environmentalism and National Liberation, 1988–89

In 1988–89, the reawakening of civil society and steps towards the reconstitution of the national sovereignty of Lithuania were closely intertwined with the awakening and mobilisation of the Lithuanian green movement. In the first period of the movement's development, a few years before the collapse of the Soviet Union and about two decades later than the onset of contemporary environmentalism in the Western countries, Lithuanian environmentalists aimed to appropriate and control the sphere of nature protection, in conjunction with protest against Soviet 'imperialism'. Lessons learned in that process had important implications for the subsequent development of a mix of dialogue and confrontation in Lithuanian environmentalism.

### *Roots of Lithuania's Green Movement*

During the years of the 'Singing revolution' of 1988–89, the dialogue among the environmental activists and actors in other domains was of a different kind as compared to the earlier-mentioned latent communication channels in Soviet times. In the late 1980s, the nucleus of the Lithuanian green movement was formed by a younger generation that did not belong to the state administration or regulatory structures, at least during that movement's awakening and mobilisation. The movement was led by individuals such as Saulius Gricius, Zigmas Vaisvila, and Saulius Piksrys, all well into their twenties, but seldom more than 30 years old. Many young people came to the green movement from so-called nature tourism clubs and similar organisations.

During Soviet times in Lithuania, nature conservation as well as historical and cultural heritage preservation activities were institutionalised in nature tourism clubs, folk music groups and other organisations usually based at universities and research institutes. Such clubs were quite popular among students and young employees of Vilnius University, Kaunas Polytechnic Institute, Kaunas Agricultural Academy, the Academy of Medicine, Institutes of Physics, Chemistry, and other institutions. Students and other young people expressed environmental and cultural cognitive interests through such clubs by organising summer expeditions, collecting folk songs and tales from elders in Lithuanian villages, organising conservationist work in nature reserves and parks, and working in summer work camps for students.

The first major green clubs in Lithuania – a local prototype of what is called NGOs in the current terminology – were formed in the three largest cities of Lithuania, namely 'Atgaja' in Kaunas, 'Zemyna' in Vilnius, and 'Zvejone' in Klaipeda. Such clubs started with a few enthusiasts and grew to 30–50 people. Smaller clubs of a similar orientation mushroomed in other cities as well as smaller towns and villages. Such clubs made the core of emerging green movement, and the former activists and members of the clubs of nature tourism, folklore, and preservation of historical monuments formed many of them.

Even seemingly 'innocent' activities like nature preservation were met with suspicion by the Soviet regime. For example, the name, 'National Park', was perceived to have political connotations by the upper hierarchy of the Communist Party in Vilnius and Moscow. Expeditions to such parks by young volunteers were met with suspicion as well. Therefore, when the first green clubs and other NGOs showed first signs in 1987–88, the Communist Party and other ruling structures 'guided' them in such a way that they would be institutionally affiliated with the Communist Youth League – the *Komsomol.*

As a consequence, the first green clubs in Lithuania were formally associated with *Komsomol*, whereas ideologically only very few of the new green grass-roots activists had anything in common with the communist doctrine. This was just one of the first organisational compromises from both the side of environmentally oriented civic actors and the state bureaucracy. It allowed the new grass-roots environmental groups to legitimise obtaining a formal status of public organisations in the Soviet system during the years of late *perestroika* and *glasnost.*

Thus, through the hobby-like activities in tourism and culture clubs the future green activists gained important experience pursuing politically 'delicate' goals and avoiding open confrontation with the Soviet regime. This is an important dimension of explaining why Lithuanian greens from the very beginning pursued unorthodox ways of combining protest and dialogue.

### *The Alternative Character of Early Lithuanian Environmentalism*

There is a lack of agreement among analysts of Western environmentalism regarding the social antagonism of those movements. Some ecological modernisation theorists [*Hajer, 1996; Mol, 1995; Spaargaren, 1997*] note the 'alternativist', or in other words antagonist, character of Western environmentalism at its early phase of development. This is expressed in attempts by environmental movements and their activists to search for rapid change of dominant values and institutions, calls for alternative lifestyles and technologies, and the radical ways in which environmentalists expressed their identity among other members of Western societies. This self-distinction and antagonism is most notable in the attitudes of environmentalists with regard to industrialists, the political-economic elite and the state bureaucracy [*Cotgrove 1982; Hajer 1995*]. By contrast, there are other, quite different, views and interpretations of Western environmentalism in the 1970s, for instance, seeing the movement, its intellectuals and their influential writings as articulating 'a positive programme of social ecology, rather than a depiction of doom and crisis' [*Jamison, 1996: 229*].

Initially, the environmental movement in Lithuania propagated an ideology of 'alternative technologies and alternative life styles'. However, the term 'alternative' did not have the same meaning as in the West in the 1970s: the common denominator for 'alternative' in Western environmentalism was a disillusion in dominant values and institutions prevailing in industrial societies. By contrast, in Lithuania radical environmentalist critiques protested the Soviet totalitarian regime's 'imperialist' ideology, both in the environmental and political sense. In 1988–89, the discourse of the Lithuanian environmental movement was therefore shared by much broader circles of society compared to Western environmental movements in the 1970s.

In Lithuania, environmental activism of this period was most often geared towards the protest against, and critique of, the Soviet totalitarian system, although critique was wrapped into discourse about particular social, economic, ecological issues. Therefore, a certain antagonism has always been present in this civic activism. However, it was a different antagonism from that of Western environmentalists who were criticising the material affluence and capitalist industrialisation and addressed their critique to political elite and industrialists. In Lithuania, material welfare and the capitalist mode of production have not formed an ideological platform on which environmental activism would be based. Shortages of material goods were so common that the critique against affluent society which was so important in Western environmentalism in the 1970s [*McCormick, 1989*] was irrelevant in the Soviet Union.

Regarding worldviews and cosmology, the following statement illustrates the 'alternative' character of Lithuanian greens. Written by one of the movement's intellectuals, it expresses an Arcadian approach to human-nature relations and frames a distinctive discourse coalition:

> Rational logical cognition is to be complemented by intuitive-meditative way of grasping the essence. According to dogmatic scientists, the Greens are getting here into an ocean of mysticism, but in fact they just span the boundaries of categorical thinking ... We learn about the right livelihood not only from science and parties, but also from stars in the sky, folk songs, Ciurlionis' paintings,[8] and the eyes of a child [*Karalius 1990: 16*].

The self-distinction of the greens among other members of society on a discursive level is strengthened by the ways they frame their agency.

> In general, I also see the agency of the Greens as a controlling one – controlling not only the Government or particular organisations, but also nature protection in general ... The Greens will co-operate but also control the situation as no other 'public' controlling committee[9] has ever done before [*Balbierius, 1989: 4*].

Such statements by the leaders and ideologists of the Lithuanian green movement reflect the direction of change in relations between greens and actors rooted in other domains that this social movement was promulgating in the early period of its development in the late 1980s. I will discuss below a claim whether after an upswing of the social movement environmentalism is later 'appropriated by the movement's opponents' [*Eder, 1996: 203*]. The statement above highlights an opposite tendency in the first period of the green movement's development in Lithuania: environmentalists aimed to appropriate and control the sphere of nature protection.

This strong self-confidence among the greens and shared environmental concern expressed through various remarkable actions of public protest, strongly affected science and technology decisions. The most significant were stopping the construction of the third bloc of Ignalina nuclear power plant, suspending the construction of Kruonis hydro-accumulation power plant, and suspending the development of an oil import terminal on the Baltic coast of Lithuania.[10]

The lessons of 'diplomacy' and dialogue learned in Soviet times when establishing first environmental organisations had important implications for the ways in which the movement disseminated its message in the public during the years of national revival. Some illustrations of this mix are given in the next section.

### *Early Observations of a Mix of Dialogue and Confrontation*

The first big rallies organised by the greens in the summer of 1988, particularly in Kaunas, attracted a large number of people from various circles of society. Those rallies raised a curiosity, cognitive interest, and most importantly, awareness among people that public critique against the system was feasible, and the repressive structures were no longer as threatening as they used to be. Public manifestations organised by the greens spontaneously evolved into protest actions against the Soviet regime. Environmental critiques provided a very widely shared set of metaphors and cases pointing to the shortcomings of the state socialism and setting a strategy for radical change. The broad circles of Lithuanian society shared this because it fulfilled people's long-sought but latent aspirations of national liberty and democracy.

Following the successful summer rallies, the first general assembly of the Lithuanian green movement was organised, in October 1988. Delegates representing all the green clubs and some academic organisations, as well as individual participants gathered on the premises of the Kedainiai chemical plant, one of a few heavy industry companies in Lithuania. Geographically, the factory is located in the middle of Lithuania; politically and in terms of media coverage, it is relatively peripheral.

Why did the greens convene their first assembly in the province and not in the major conference hall in the big cities of Vilnius or Kaunas where the most active and publicly visible green clubs have been set up? And why did this assembly take place at the Kedainiai chemical plant, which had been heavily criticised by the greens during the first big ecological protest march organised in the summer of 1988? Addressing such questions will help to explain the cultural politics of environmentalism in Lithuania and the official as well as latent ties of civic actors and those in the economic and bureaucratic domains.

The central authorities in Vilnius preferred to minimise the public influence of this new rapidly growing social movement promulgating the 'alternative' values, technologies, and life styles. Therefore, the public authorities informally suggested that the greens convene their assembly in Kedainiai. Kedainiai was acceptable because organising the greens' assembly outside of major cities was less likely to attract overwhelming public attention and provoke radical actions by environmentalists. Understandably, such an option was not the most favourable for environmentalists, especially in terms of public visibility, and the ability to involve as many adherents as possible. In spite of this, the greens took this compromise step rather than fostering confrontation with the bureaucratic structures.

The repressive Soviet structures on which the totalitarian regime was based were still strong and empowered in Lithuania in 1988. The threat of direct oppression in the case of open confrontation, understandably, was an important factor shaping the strategies of the environmental movement. The choice of the venue for its first assembly in Kedainiai can be interpreted as one example of this unequal confrontation.

The threat of oppression by the Soviet regime was, however, not the only reason for a shift from an open confrontation towards adjustment and dialogue of the greens with the public authorities. Another important factor was that the latter were no longer monolithic. Many actors in the bureaucratic domain were influenced not only by the still powerful regime, but also by new value-orientations and aspirations of liberty and democracy brought to the public arena, in part, by the environmental movement. This civic orientation among some officials in the Communist Party structures, the national Government, municipal authorities and top management of industrial enterprises was very important in paving the way for their latent dialogue with the greens.

The choice of Kedainiai was probably a kind of compromise reached through such informal 'dialog' between the environmental activists and actors in the bureaucratic and economic domains. Recent interviews with the leaders of the greens, Saulius Piksrys and Linas Vainius,[11] confirm that such assumptions regarding the motives for dialogue and compromise seeking are adequate. This and other cases indicate not only the existence of an informal communicative interface between the greens, public administrators and industrialists. They also reveal a tendency that industrialists, for example, top management of the Kedainiai chemical plant, confronted by the pressure of environmentalists, preferred to help (or to please) the latter actors rather than to continue confrontation.

The industrialists not only allowed the greens' assembly to be held at the premises of their factory, some of the staff members of Kedainiai factory attended that meeting. Although they were not among grass-roots activists, they shared certain pro-environmental attitudes with the rest of the audience. Mr Pilkauskas, one of the top managers of Kedainiai company, admitted that they were aware of the need to shift towards environmentally benign production, but the Soviet economic system did not allow them to make the improvements and changes they would like to. So helping the greens might be also interpreted as a reflection of industrialists' ecological consciousness and civic stand, and not merely as a way to reduce the threat of grass-roots critique. This company allowed a rock concert on its territory on the top of a hill made out of industrial waste (phospho-gypsum) during the Peace March organised by greens in the summer of 1989. Moreover, Kedainiai's plant subsidised the protest actions of the greens by providing them with substantial funding (*c.* 10,000 roubles).

Another example of a mixed co-operation-confrontation, pressure-dialogue between the greens and industrialists was the financial assistance (10,000 roubles), transportation (coaches and trucks), and other support provided by the Jonava' fertilisers plant *Azotas* for the ecological protest march in 1988. Paradoxically, during this march and on some later occasions the greens organised quite radical protest demonstrations against this plant, they actively disseminated information on serious negative environmental impact caused by *Azotas* upon humans and nature. Thus, again the industrialists, who were fiercely criticised by the greens in the first place, were on the other hand providing material (and discursive) support for environmentalists. As interviews with the leaders of green movement[12] indicate, industrialists were indicating that they shared pro-environmental attitudes, and had made some efforts to mitigate the negative environmental impact of their productive activities. However, they admitted that their efforts were insufficient in the extant Soviet political-economic system that was too cumbersome and ineffective. Helping the greens by various means was explained by industrialists as the way to show their environmental awareness, although they knew in advance that their companies would be criticised by the greens.

In sum, the new social-environmental movement which emerged in Lithuania in 1988–89 played an important role in Lithuania not only by carving out a temporary public space [cf. *Jamison, 1996*] and publicly communicating an important message, but also by consolidating the national liberation movement. Through an environmentalist critique, it expressed genuine ecological concern as well as protest against the dominant Soviet regime. With regard to environmental policy, this movement opened up new avenues for changing, democratising the public-policy interface. It disseminated signals promising the emergence of new types of dialogue, a belief in the constructive power of public participation, a promise of reshaping the country's economic and technological development in new, environmentally sound, ways that are based on community, and shared responsibility.

## Environmental Politics in the Period of Transition, 1990–Present

Since 1990, civic environmental activists in Lithuania have continued to seek constructive dialogue with actors in bureaucratic and economic domains, even while maintaining radical, non-conformist pressures on the same.[13] Environmental politics during this period have been affected by the ups and downs of economic and political transformation of Lithuania from a Soviet republic to an independent, market-oriented society. It has resulted in a professionalisation of environmentalism in Lithuania, new organisational forms for the environmental movement, 'multiple identities' and identity crises for some of the movement's activists and leaders, and resurgence of new grass-roots activism.

### *The End of Euphoria*

The stream of euphoria, self-confidence and faith in public participation has gradually slowed down among common citizens as well as environmentalists. Radical actions by environmental activists became marginal; they no longer mobilised a broad spectrum of actors representing society at large. Environmental movement leaders have even explicitly stated that the movement would refrain from drastic impact upon various sectors of economy, primarily because of the transitional difficulties that Lithuanian society was facing [*Zemulis, 1995: 11*]. This position was strengthened when Moscow started an economic and political blockade of Lithuania after it declared its independence on 11 of March 1990.

Public policy analysts, journalists, scientists and society at large did not forget, however, the radical actions inspired and led by the green activists during the 'Singing revolution', and associated those actions with the economic downswing in Lithuania during the first years of transitional period. Such 'bad memories' were most prevalent during the difficult years between 1990–94 when society was facing a shortage of heat, rapid growth of prices for electricity, food, shelter, and difficulties in fulfilling basic needs. There were voices blaming the greens for their actions' negative impact upon the fragile Lithuanian economy and raising social anxiety. Although many goods have been in short supply, Lithuanian society has not experienced the shortage of heat, electricity, affordable food and shelter during the last decades of Soviet rule. Rapidly growing difficulties in fulfilling basic needs made the greens one of the social groups that the rest of Lithuanian society in transition could blame without deeply analysing the real underlying causes of the worsening economy.

This might be one of the reasons why, amongst the wider public, there have spread relatively negative attitudes towards public (lay) participation in environmental science and technology policy. The positive impact of the environmental activists in mobilising society and bringing to the public arena ideas of democracy and sustainable development were gradually forgotten [*Stoskus, 1996: 12*]. Only the negative side was remembered, especially the greens' impact upon Ignalina nuclear power plant, Kruonis hydro-accumulation power plant, Butinge oil import terminal on the Baltic coast. The impact of environmentalists upon those and other environmentally sensitive policy issues was one of the main factors that increased people's negative attitudes and expectations regarding public participation in the public policy process [*Rinkevicius, 1998*].

Another lesson from Lithuanian environmentalism of this period is that civic activism, collective identity, social mobilisation and engagement in policy-making was significant *only during the period of structural shift* that was existentially important for society at large. Those changes embodied the fulfilment of long-sought aspirations, that is, the shift of Lithuanian society towards liberation from the totalitarian regime and reestablishment of a sovereign democratic state.

### *Professionalisation of Environmentalism in Lithuania*

Later, since 1993–94 or so, one can observe a tendency of institutional strengthening in the state environmental authorities, municipal authorities, industrial enterprises, academic institutions and newly emerging business enterprises. Various kinds of foreign aid programs were launched to support pollution prevention and cleaner production in Lithuanian industry; environmental authorities have developed and more actively enforced economic policy instruments, the National Environmental Strategy has been developed, other important policies adopted [*Rinkevicius, 1998*].

There are indications that, as in the West, the economic, bureaucratic, and academic domains have gradually overshadowed the civic domain of environmental activism in Lithuania. I would refrain from using in this case the term 'environmentalism being appropriated by the movement's opponents' [*Eder, 1996: 203*], for two reasons. First, I would interpret appropriation not as a process of taking over the initiative or key role from civic-oriented actors by those rooted in other domains, but rather as gradual diffusion of environmental consciousness among public administrators, industrialists, and academics. Second, the relative increase in the environmental activity and visibility of actors in the bureaucratic, economic and academic domains is closely related to the process of professionalisation of the environmental movement.

The Lithuanian greens have professionalised in the process of environmental politics and policy-making, and this process resembles the trajectory towards greater professionalism among the Western environmental movement activists and organisations as described by Jamison [*1996*]. However, a quite high degree of professional competence and constructive input has *always* been present in the discourse and actions of the Lithuanian environmental movement *from its onset* in 1988.

Since 1988–89 and into the 1990s, actors in the environmental movement have been not merely civic-oriented lay persons, but rather people with good educational backgrounds encompassing a broad spectrum of disciplines in the technical, natural, and social sciences. In 1988–89, the Ignalina nuclear power plant became a key issue on the public agenda. At that time, the greens' discursive strategies and actions regarding this issue were formed by the leader such as Vaisvila, who holds a Ph.D. in nuclear physics and had several years' experience of working at one of the top Soviet research institutes in this field. He and other activists who led the public debates in Lithuania about Ignalina in the late eighties were professional physicists, radiologists, energy specialists, mathematicians, biologists.

In 1996–99, a similar combination of grass-roots activism and professional competence was observable. In this period, the actions of the greens aimed at closing down the Ignalina plant were shaped by close collaboration of their leaders, Vainius, Braziulis and Piksrys, with academic specialists from the Lithuanian Energy Institute, for example, Professor Matas Tamonis and his colleagues. The latter group of academics has been hired by the greens to carry out a comprehensive ecological, economic, and technical research project in order to assess the real implications of phasing out Ignalina. This study has shown that the closedown of Ignalina is not as socially painful as cheap energy proponents portray it.

Collaborative efforts and professional dialogue have stretched out not only among environmental activists and academics, but also bankers and financiers including the European Bank for Reconstruction and Development (EBRD). In 1996–99, it was one of the very first instances in Lithuania that the greens would work together and support the international financial institutions in their policies. In this case, coherent actions and dialogue helped both the Lithuanian greens and the EBRD in their attempts to receive the commitment of the Lithuanian Government to close down Ignalina.

The intertwined radical protest and constructive input by environmentalists regarding the reduction of production of contaminated food can serve as another illustration of this complex profile of the greens: integrating radical critique and professional input. This critique by the greens has been addressed against the public authorities and agri-business structures whose activities were aiming at fulfilling and exceeding the five-year plans by any means, and caused over fertilisation and thereby contamination of vegetables and milk products:

> The greens of Lithuania are calling everybody [sic] to boycott milk products ... Only unanimous action can shake the agri-industrial lions ... Our two-week fast can make the leaders of agri-industrial monster descend down on earth from their thrones. The fast in the name of children! [*Karalius and Lekevicius, 1990: 25–6*].

This quotation illuminates the radical, non-conformist approach of the greens, and confrontational ways in which this message was disseminated in society. The approach based on civic disobedience, namely the boycotts and pickets, echoes the early New Environmentalism that emerged in Western countries in the 1970s [*McCormick, 1989*]. However, the above-mentioned ultimatum contained more than radical nonconformist statements and slogans. The greens also constructively proposed: 'food products for sale should contain certificates about the concentration of hazardous substances'; 'price differentiation depending on the environmental quality of products', 'production of high-quality fertilisers', 'application of agro-technical methods'.[14]

Thus, besides the radical critique, the greens were generating and disseminating ideas that are now ascribed to the ecological modernisation discourse. Its important aspects are the faith in individual ecological consciousness; in the environmentally responsible behaviour of consumers and actors in industry and public administration given sufficient information and economic motivation; belief in the need for, and feasibility of, a change in dominant technological and economic institutions.

There are indications that green messages have been taken seriously by policy-makers, academics and industrialists. For example, a warning concerning the urgent need to construct wastewater treatment plants in major cities gradually reached and was incorporated by the bureaucratic domain. The government of Lithuania declared the construction of sewage treatment plants in five major cities – Vilnius, Kaunas, Klaipeda, Siauliai and Palanga – to be a national priority. It became one of the key areas of the National Public Investment Program (PIP) adopted in 1994. The Ministry of Environmental Protection as well as several municipalities (in Kaunas, Klaipeda, Siauliai and others) have allocated a substantial part of available funds as well as the major share of foreign environmental aid for the design and construction of the wastewater treatment facilities.

*New Forms of Organisation and Activity*

Organisational forms of environmentalism in Lithuania have changed since 1990, from the grass-roots green clubs towards new professionalised institutions with an increased role for environmentalists as brokers between different social domains. For example, those who earlier formed a core of the Zemyna green club[15] and the ecological centre, Alternatyva, in Vilnius, for example, Arturas Abromavicius and Aidas Vaisnoras, have established an environmental consulting firm. In the mid-1990s, this firm grew into a major environmental consulting and engineering company, the Baltic Consulting Group. It is a joint venture involving some of the strongest Nordic counterparts, such as the Danish company, Kroeger; the Swedish firm, Rust; and the Nordic environmental financing consortium, NEFCO.

Through the Baltic Consulting Group, the former movement activists got actively involved in designing sewage treatment plants and assessing contamination of the former Soviet military sites as well as designing and implementing technologies for their rehabilitation.[16] Thus, those areas that once were a major focus of the green critique and protest, later became the areas where former members of the green movement found their new professional identity and employment.

Environmental action has been gradually changed by ecological, economic and technological innovation, and 'reflexive rationalisation' to use Ulrich Beck's [*1994*] term. For instance, since 1994 the Lithuanian green movement together with the Lithuanian National Television channel, LTV, and the Ministries of the Environment and Urban Development, has conducted a contest among households called, 'My House: Ecology, Economy, Conservation'. This popular competition is sponsored by various local and Western industries and commercial enterprises promoting, for example, heat isolation and insulation materials, environmentally friendlier products and technologies that enable more rational usage of natural resources by households.

Individualisation and rationalisation are gradually overweighing the mass participation in the protest actions. The 'alternative' and nonconformist ideology is gradually shifting towards an emphasis on daily energy and material saving. For instance, environmentalists in Kaunas, namely the green community, Atgaja, have established a special consulting centre that is actively propagating energy conservation in households and bio-transportation, especially bicycling. This community is also very active in developing in Kaunas a collaborative project on Agenda 21. Together with the Kaunas' Municipality, the greens are forming a nucleus for new social partnerships and participatory experiments. Leaders of the greens, most often their vice-chairman Linas Vainius, are invited to the sessions of the Parliament committees, board of the Lithuanian Environmental Investment Fund, environmental strategy advisory board of the Ministry of the Environment, and other public as well as private organisations.

The greens in Klaipeda, members of Zvejone club, are actively involved in academic activity and professional environmental research. Through various professional educational and research projects, for example, preparing projects of ecological education for schoolchildren, they contribute to solving ecological problems on the coast of the Baltic Sea. This work is performed not in a traditional 'alarmist' way but with some professional pedagogical skills and seeking funding for such projects in an innovative and entrepreneurial fashion which is a characteristic feature of ecological modernisation. This entrepreneurship is expressed, for example, in the ways in which various 'stakeholders' are incorporated in the joint projects, for example, the Coalition Clean Baltic. Entrepreneurship is also reflected in comprehensive, professionally made applications drafted by Zvejone' activists who succeed to obtain funding *inter alia* from major foreign donor agencies, for example, the Swedish International Development Agency (SIDA).

Albeit very fragmented, such examples demonstrate the gradual shift in the goals and forms of activity of the greens as a social movement towards increased professionalism. Those examples not only indicate the professionalisation of the movement but also illuminate new trends in its search for a new niche of green activism in the emerging 'environmental marketplace' [*Eder, 1996*] that the movement had once created. This change of the forms and kinds of environmental activism can also be interpreted as a more general tendency of ecological modernisation.

### *Multiple Identities and Identity Crises*

In 1989, the greens undertook several trials to foster 'diplomatically' central and local authorities to make the final decision regarding the type and location of the municipal sewage treatment plant in Kaunas that was regarded as one of the ecologically hottest spots in Lithuania. Dissatisfied by the slow and bureaucratised response, the activists from the local green club, *Atgaja,* organised a hunger strike until the public authorities would select a site for the waste water treatment plant.

The leader of this protest action was Saulius Gricius, chairman of the green club, Atgaja. At the time the greens held a hunger strike, Gricius was not just a movement activist and leader, he was also recently employed as a senior inspector in the Kaunas Regional Nature Protection Committee (an equivalent of the state environmental protection agency). Piksrys, another key actor in the green movement and participant of hunger strike, was also a recent employee, an environmental inspector in the same Nature Protection Committee.

Those are not the only examples of a 'multiple identity' among environmental activists and movement leaders. In 1990, the green movement split into a political Green Party and movement organisations. The Green Party, led by Vaisvila, a 30-year-old physicist and active critic of the Ignalina nuclear power plant, has won six seats in Seimas, the first democratically elected Parliament of Lithuania, which soon (on 11 March 1990) declared the reestablishment of Lithuania's independence. By contrast, the green movement has chosen a different strategy and aimed at municipal level rather than parliamentary elections, justifying this strategy by the objective to be involved in daily, practical solving of most acute environmental problems.[17] Two years later after the hunger strike in Kaunas, its organiser Gricius became a vice-mayor of the city of Kaunas. He was the first vice-mayor in Lithuania in charge of environmental protection and *inter alia* of the municipal wastewater treatment issues which just a few years ago was the object of criticism and pressure by the environmental movement led by this same person.

How could one interpret such cases when the greens became state officials at the parliamentary, governmental and municipal institutions while maintaining a 'double identity' as civic environmental activists? One might read this as environmentalism being mainstreamed at an early stage of its development. This might lead to the conclusion that, in the case of Lithuania, environmentalism – in some form – is not antagonistic to the state, i.e. this reads more as a victory, albeit partial, for environmentalism.

However, later events have shown this image of mainstreamed environmentalism to be premature. Gricius shortly thereafter (in 1992) committed suicide. Close friends have witnessed that, during his last weeks and days, Gricius was deeply frustrated. He was disappointed at how the environmental ideals and activism were confronted by the brutal reality embedded in the bureaucratic and economic structures seeking short-sighted benefits by misusing the institutional shortcomings of the transitional society.

In 1994, when the left-wing Democratic Labour Party gained a majority in the parliamentary elections, another leader of the greens and actor with a 'multiple identity', Piksrys, lost his position as a senior inspector at the Kaunas Regional Environmental Protection Department. The reason was ideological divergence with the direction taken by the new governmental leaders, especially the new Minister of the Environment. Yet another example is Vaisvila, Chairman of the Green Party, who in 1992 became Vice Prime Minister of Lithuania. He soon left that post due to frustration about the course of public policy and internal politics which had not been as he had expected at the start of the 'Singing revolution'.

Such cases illustrate that the 'multiple identity' of actors who have led the movement was often a cause of internal and external tensions and frustrations, rather than a way to mainstream the environmentalism. Those people chose a very uneasy way – acting on the one hand as movement activists, that is, civic 'outsiders' with respect to public authorities or business enterprises, and at the same time being legitimate 'insiders', that is, employees of public administration or other non-NGO organisations. Based on available empirical evidence, I read such cases of movement activists and leaders becoming servants in different institutions of public as instances of *individual career trajectories* rather than a coherent process of environmentalism being mainstreamed in Lithuania at a very early stage.

### *Signs of New Grass-roots Mobilisation*

In the 1990s important developments support the claim that the grass-roots movement has not ceased to exist. There are reshaped and newly emerging groups that take a radical stance regarding environmental science and technology policy. A new grass-roots group mobilised in Klaipeda City in 1991–93. Focused on the issue of oil import terminal on the Baltic Seacoast, this group was primarily formed by Klaipeda's inhabitants and led by a local female artist-painter. Many of those individuals who formed the nucleus of this newly emerged social grouping were not among those who formed the green club, Zvejone, in Klaipeda in the late 1980s.

This new group has organised numerous radical protest actions against the construction of the oil terminal. Sometimes the 'old' movement activists have backed up their new grass-roots colleagues in protest actions. The movement 'veterans' from Kaunas and Vilnius, however, have more often been involved in organising professional assessment of the new terminal design and location options, lobbying at the level of the Lithuanian government, and networking with the greens, academics, and governmental officials in the neighbouring Latvia.[18]

The new local grass-roots in Klaipeda have intended to oppose any option regarding the oil import terminal, whereas the council members of the Lithuanian green movement have tried not simply to oppose, but rather to find a constructive, ecologically, economically and technologically viable option for the terminal. Their proposal was to lay a pipeline between the Mazeikiai oil refinery in Lithuania, and the already existing Ventspils harbour in Latvia. They expected that both the economic and ecological objectives could be thereby harmonised.

Lithuanian public policy makers neglected this option and decided to construct a terminal in Lithuanian territory. This decision was most notably influenced by the vested interests of the newly emerging political-economic oligarchy which seems to have some links with criminal structures from Lithuania and the former USSR [*Rinkevicius, 1997*]. This example, ongoing as this volume went to press, illustrates that objectives and mode of activity of the newly-emerged grass-roots are as radical and nonconformist as those of actors who formed the core of the environmental movement in the late 1980s.

Another illustrative case of new grass-roots mobilisation was a big public rally held in Vilnius in 1997. The radical critique was aimed against intensive deterioration of timber resources after lands and forests in Lithuania have been returned to the private ownership. This protest action was organised not by newly-emerged grass-roots activists, but by the core of existing green movement. This shows that professionalisation does not preclude the radical, non-conformist character of the movement, and the two seemingly divergent traits of its character may coincide and enrich each other.

The greens and concerned citizens did not demand the prohibition of felling the forest. Rather, the stance taken by those people was conservationist, not preservationist. It was more rooted in the utilitarian paradigm of the 'rational use of natural resources' following Pinchot's tradition in the early 20th century [*Worster, 1977*]. However, some nostalgia of harmonious rural community, living and working in, or together with nature, was present in this discourse too. Therefore this kind of environmental discourse and activism can be seen as a cultural mix involving ecocentric, nature-respecting attitude and value-orientation as well as a conservationist, utilitarian, environmental management-like stream of environmentalism [cf. *O'Riordan, 1981; Pepper, 1994*].

Another series of newly emergent grass-roots actions, guided by nature-respecting attitudes and value-orientations, have centred on the issue of the expansion of an airport in the Curronian peninsula. A very small airport for emergency cases has been developed in this sacred nature spot in the Neringa town in Soviet times. In the mid-1990s, the economic-driven objectives to expand this airport for developing the tourism industry have been pursued by the Lithuanian government. The green movement has initiated multidisciplinary group of professional ecologists and economists to evaluate the potential impact of such an airport upon the fragile nature of Neringa.

The conclusions drawn by the scientists as well as a special commission launched by the Ministry of the Environment recommended a suspension of the development. In spite of this recommendation, the government has allocated significant resources to continue the project of expansion. This led the greens to organise a very drastic picket in front of the National Government building in May 1999. The slogans used by the greens were as radical and nonconformist as in 1988–89. Some of the leaders of this rally were the same people, for example, Piksrys and Vainius, who led the grass-roots movement in the late 1980s. However, there were a lot of younger generation activists who joined the grass-roots protest action perhaps for the first time. The magnitude of the collective action was not as great as those big rallies in the late 1980s with up to 70,000 people. In this case, it was only 20–30 people protesting. However, the greens' radical activity and public visibility as well as their impact on the policy-making process was significant. The Minister of the Environment promised to delay the construction works.

The new grass-roots groups seem to mobilise on a single-issue basis. In any case, such empirical examples indicate that grass-roots environmental activism has not ceased to exist in Lithuania, that new social groupings and new forms of radical action can still be expected to emerge in Lithuania as in the West [*Szerszynski, 1996; Jamison, 1996*]. It is too early to predict whether this process will go in parallel with the mainstreaming of environmentalism, or whether another 'alternativist' movement will emerge out of it. In either case, the emergence and cultural politics induced by such new grass-roots activism could be interpreted as a part of the dialectics of ecological modernisation or ecological transformation of contemporary societies.

## Conclusions

Several conclusions can be drawn from the discussion and empirical evidence provided above. Lithuanian environmentalism is a complex phenomenon. It is not only an expression of antagonist mistrust and critique against dominant social values (materialistic, hedonistic, individualistic) and social structures (furthering centralisation, hierarchy and expert domination). Since it began, the movement has brought modernist elements into ecological discourse such as cost-benefit analysis, scientific knowledge, public information and constructive dialogue with actors in the bureaucratic, economic and academic domains. The discursive and action strategies of the greens reflect a mixed type: composed of rational, techno-economic (eco-managerial) as well as romantic, idealistic (Arcadian) beliefs and values.

Civic environmental activism in Lithuania relies not only on the radical public protest actions typical of traditional social movements, but also on informal and formal networks among the greens and industrialists, academics, and public authorities. This is exemplified by various cases when the greens communicate and collaborate with actors in other policy domains. Actors sharing a 'double identity' also exemplify this: environmental activists occupy important posts in the public authorities and other public and private organisations. This tendency of 'double identity' was visible also in the era of Soviet rule.

Are rationality and idealism/romanticism compatible in ecological consciousness and in the development of ecological-social activism? The co-existence of such mixed value-orientations is visible throughout the development of the Lithuanian green movement. Our analysis suggests that the ecocentric doctrine does not intrinsically contradict the belief in techno-economic reforms and institutional learning advocated by the doctrine of ecological modernisation. Both ideological streams seem to exist in the discourse and actions of Lithuanian greens. This movement cannot be characterised as either solely antagonistic or consensus-oriented with regard to actors in the economic, bureaucratic and academic domains. The actual processes of social and institutional change indicate the existence of a mixture of both tendencies, that is, a cultural politics, where, depending on the case and the context, one tends to prevail over another.

The process of 'appropriation', in the sense as Eder [*1996*] describes it (environmentalism being appropriated from the movement by its opponents), is not so clear-cut in Lithuania. There are empirical examples indicating that actors in the bureaucratic, economic, and academic domains have gradually taken the lead in the course of environmental reforms. It seems that, if any, this shift in initiative was first preceded in Lithuania by the greens aiming for leadership in the societal greening. On the other hand, in spite of certain single cases there is no clear general tendency that the state bureaucrats or economic entrepreneurs would take over responsibility and leadership in environmental reforms from the environmental activists.

What is visible is the shift in the balance towards eco-modernist types of policy innovation, increasing public environmental education and information, deployment of economic instruments in environmental regulation, promotion of the greening of industry. Although actors in the economic, bureaucratic and academic domains are increasingly active in such eco-modernist policy innovations, the movement organisations and

NGOs are by no means marginal or peripheral in the process of the ecologisation of Lithuanian society.

Civic environmental activists and NGOs are expected to play an increasingly important role not only in critically assessing the policies and practices of actors in the economic, bureaucratic and academic domains, as well as raising public environmental awareness. Based on our research into the Lithuanian environmentalism, it could be expected that movement organisations, NGOs and new charismatic leaders will take important initiatives in bridging the new networks among actors in different domains, and experimenting with the new forms of public participation in environmental science and technology policy.

## NOTES

1. I suggest calling this phenomenon a 'cultural mix'. The process of its making and the transformations of environmentalism in Lithuania can be referred to as 'cultural politics', following the conceptual approach suggested by Hajer [*1996*].
2. Applied in the PESTO project, involving the UK, Italy, Netherlands, Sweden, Norway and Lithuania.
3. See Eder [*1996: 204*] for discussion of the concepts, 'public space' and 'public discourse'.
4. Interviews with Kastytis Cechavicius (May 1999) and Rapolas Vasiliauskas (April 1999) who in Soviet times worked as inspectors at the Siauliai and Kaunas regional agencies of the Ministry of Water Resources and Melioration and was one of those brave environmental administrators.
5. 'Kiek gali kainuoti vieno munduro garbe?' (How much could the honour cost?), *Literatura ir Menas,* 15 Nov., 1986.
6. *Literatura ir Menas,* op. cit.
7. *Literatura ir Menas,* op. cit.
8. Ciurlionis (1875–1911) – the most-famous Lithuanian painter and composer whose significance for Lithuanian society and culture might be compared with the one of Sibelius in Finland or Grieg in Norway.
9. The term 'public controlling committee' refers to a particular social institution within the Soviet command-and-control system that has performed the role of public inspection in various public spheres.
10. Those cases were briefly discussed in my earlier paper in *Pesto Papers 1* [1997].
11. Interviews conducted in April, 1999.
12. Ibid.
13. In Lithuania this pattern is characterised by numerous cases, particularly the suspending of construction at Kruonis hydro-accumulation power plant, boycott of agricultural and food processing industry, stoppage of the construction of the third block of Ignalina Nuclear Power Plant, etc. Those cases, as mentioned earlier, are briefly reviewed in my article in *Pesto Papers I* [1997].
14. For entire text, see Karalius and Lekevicius [1990].
15. It was famous as an intellectual centre generating green ideology in Lithuania, mobilising the public to confront *inter alia* the construction of the third bloc of the Ignalina nuclear power plant, etc.
16. Interviews with Arturas Abromavicius, director of the Baltic Consulting Group (May 1996), and Alvydas Karalius, Vilnius Municipality, both of whom are the former leaders of the green club, Zemyna (Oct. 1997).
17. Personal communication with Saulius Gricius (May 1990).
18. This story line has been discussed in my earlier paper in *Pesto Papers* I [*1997*].

## REFERENCES

Abercombie, N., Hill, S. and B.S. Turner (1994), *The Penguin Dictionary of Sociology*, London: Penguin Books.

Balbierius, A. (1989), 'Lietuvos Zalieji: Tabu Isvarymas' (Lithuanian Greens: the Expulsion of Taboo), in *Mokslas ir gyvenimas*, No.1.

Beck, U. (1992), *Risk Society: Towards a New Modernity*, London: Sage Publications.

Beck, U., Giddens, A. and S. Lash (1994), *Reflexive Modernization: Politics, Tradition and Aesthetics in the Modern Social Order*, Cambridge: Polity Press.

Cotgrove, S. (1982), *Catastrophe or Cornucopia: The Environment, Politics and the Future*, Chichester: Wiley.

Eder, K. (1996), 'The Institutionalization of Environmentalism: Ecological Discourse and the Second Transformation of the Public Sphere', in Lash, Szerszynski and Wynne [*1996*].

Eisenstadt, S.N. (1968), 'Social Institutions: the Concept' and 'Social Institutions: comparative study', in D.L. Sills (ed.), *International Encyclopedia of Social Sciences*, Vol.14, New York: Macmillan/Free Press, pp.409–29.

Eyerman, R. and A. Jamison (1991), *Social Movements: A Cognitive Approach*, Cambridge: Polity Press.

Giuliani, M., Rinkevicius, L. and A. Sverrisson (1998), 'Making Participation Happen: The Importance of Policy Entrepreneurs', in A. Jamison (ed.), *Technology Policy Meets the Public (Pesto Papers 2)*, Aalborg University Press.

Grigas, R. (1995), *Tautos Likimas* (Nation's Destiny), Vilnius: Rosma.

Hajer, M. (1995), *The Politics of Environmental Discourse. Ecological Modernisation and the Policy Process*, Oxford: Clarendon Press.

Hajer, M.A. (1996), 'Ecological Modernisation as Cultural Politics', in Lash, Szerszynski and Wynne [*1996*].

Jamison, A. (1996), 'The Shaping of the Global Environmental Agenda: The Role of Non-Governmental Organisations', in Lash, Szerszynski and Wynne [*1996*].

Jamison, A. (1997), 'Introduction', in A. Jamison and P. Ostby (eds.), *Public Participation and Sustainable Development: Comparing European Experiences (Pesto Papers 1)*, Aalborg University Press.

Jamison, A. and E. Baark (1990), *Technological Innovation and Environmental Concern: Contending Policy Models*, Research Policy Institute, Lund University.

Januskis, V. (1990), *Gamta ir Mes: Ekologines Problemos* (Nature and Us: Ecological Problems), Vilnius: Mokslas.

Karalius, A. (1990), 'Zalioji Ideologija: Kodel Zalieji Zali?' (Green Ideology: Why the Greens are Green?), in *Kulgrinda*, Vilnius: Alternatyva.

Karalius, A. and E. Lekevicius (1990), *Lietuvos Zaliuju Keliamu Problemu Apzvalga* (An Overview of the Issues Raised by the Greens), Vilnius: Lithuanian Information Institute.

Kavolis, V. (1994), *Samoningumo Trajektorijos* (Trajectories of Consciousness), Vilnius: Vaga.

Lash, S., Szerszynski, B. and B. Wynne (eds.) (1996), *Risk, Environment and Modernity: Towards a New Ecology*, London: Sage Publications.

McCormick, J. (1989), *The Global Environmental Movement*, London: Belhaven Press (Pinter Publishers).

Mol, A.P.J. (1995), *The Refinement of Production: Ecological Modernisation Theory and the Chemical Industry*, Utrecht: Van Arkel.

O'Riordan, T. (1981), *Environmentalism*, London: Pion.

Palidauskaite, J. (1996), 'Lietuvos Politines Kulturos Raida Valstybingumo Atkurimo ir Itvirtinimo Laikotarpiu' (Development of Political Culture in Lithuania), Ph.D. dissertation, Kaunas University of Technology.

Pepper, D. (1993), *Eco-socialism. From Deep Ecology to Social Justice*, London: Routledge.

Pepper, D. (1994), *The Roots of Modern Environmentalism*, London: Routledge.

Rinkevicius, L. (1997), 'Lithuania: Environmental Awareness and National Independence', in A. Jamison and P. Ostby (eds.) *Public Participation and Sustainable Development: Comparing European Experiences (Pesto Papers 1)*, Aalborg: Aalborg University Press.

Rinkevicius, L. (1998), 'Ecological Modernisation and Its Perspectives in Lithuania: Attitudes, Expectations, Actions', Ph.D. dissertation, Kaunas University of Technology.

Spaargaren, G. (1997), *The Ecological Modernisation of Production and Consumption. Essays in Environmental Sociology*, Wageningen: Landbouw Universiteit.

Stoskus, K. (1996), 'Kada Lietuvoje Gali Atsirasti ir Ekologine politika?' (When Could Environmental Policy Emerge in Lithuania?), Vilnius: *Lietuvos rytas*, 31 June.

Szerszynski, B., Lash, S. and B. Wynne (1996), 'Introduction: Ecology, Realism, and the Social Sciences', in Lash, Szerszynski and Wynne [*1996*].

Vardys, V. (1993), *Lietuvos Politine Kultura ir Laiko Reikalavimai (Lithuanian Political Culture and Requirements of the Time)*, Kaunas: 'I Laisve' foundation.

Worster, D. (1977), *Nature's Economy: A History of Ecological Ideas*, Cambridge: Cambridge University Press.

Yearley, S. (1994), 'Social Movements and Environmental Change', in M. Redclift and T. Benton (eds.), *Social Theory and the Global Environment*, London and New York: Routledge.

Zemulis, F. (1995), 'Zaliuju Vadas Tvirtina, jog Zalieji Buvo, Yra ir Bus' (The Leader of Greens Maintains that There Were, Are and Will Be the Greens), Lietuvos rytas, 20 May.

Zemulis, F. (1996), 'Pirmasis Lietuvos Gamtosaugos Vadovas Tapo Legenda' (The First Lithuanian Environmental Manager Became a Legend), *Lietuvos rytas*, 20 Jan.

# Legacy of Waste or Wasted Legacy? The End of Industrial Ecology in Post-Socialist Hungary

ZSUZSA GILLE

In recent decades in the most developed countries, new approaches to environmental problems, including industrial ecology and ecological modernisation, have become prevalent. Based on interviews and archival research, this analysis argues that during state socialism Hungary had already instituted a production-centred, preventative approach to industrial wastes akin to industrial ecology. Ironically, after the collapse of state socialism, instead of keeping catching up with the West in this respect, the waste institutions and policies were disassembled with the slogan of privatisation and marketisation. As a result, the present dominant environmental discourse in Hungary argues that the environmental goals of the transition from state socialism can and should be achieved without state assistance, simply following market considerations. The resulting environmental policies represent a return to the domination of remedial, end-of-pipe technologies handed down from the West.

## Introduction

'Values on the order of tens of thousands of dollars among wastes'. 'The waste reuse proposals so far accepted in the R.M. Works amount to 'a saving of more than 34 million.' 'New proposals to reuse waste materials are born *en masse*'. When and in what country do you think these headlines were printed? They could have been from the US from the 1980s or 1990s, reporting for example about Dow Chemical, whose Louisiana Division implemented waste reduction programmes [*Nelson, 1994*]; Hewlett-Packard, which has been experimenting with 'design for reuse' [*Paton, 1994*]; or General Motors, Ford, Chrysler [*Klimisch, 1994*] and AT&T [*Sekutowski, 1994*] which started designing their products 'for recycling'. But they are not from the US, nor are they dated in the 1980s. These headlines are from the early 1950s and from a then-state socialist country,

This study examines changes in the concept of waste in Hungary from 1948 to the present.

Hungary! Why and how such waste reduction and reuse projects emerged in, of all places, 'wasteful' state socialism, and what happened to them after 1989, is the subject of this analysis.

In the past few years, industrial ecology has emerged as a new approach to resolving the most developed countries' environmental problems. Its proponents have argued that, given sufficient freedom, business is capable of integrating environmental concerns into the production process, and of self-monitoring its environmental impact, with the help of such tools as voluntarily assumed environmental quality standards, green accounting, and eco-design, among others. As a result, industry's relationship with the state and with regulation becomes more active, horizontal and co-operative. As Socolow says:

> In industrial ecology, industry becomes a policy-maker, not a policy-taker. Industry demonstrates that environmental objectives are no longer alien, to be resisted and then accommodated reluctantly. Rather, *these objectives are part of the fabric of production*, like worker safety and consumer satisfaction [*Socolow, 1994: 13*, emphasis added].

Or in the words of Robert M. White, the president of the USA's National Academy of Engineering, 'the objective of industrial ecology is to understand better how we can integrate environmental concerns into our economic activities' [*White, 1994: v*].

Proponents of ecological modernisation theory also agree that a transition in environmental discourse and practice has begun in the most developed countries, with the Netherlands and Japan leading the way [*Hajer, 1995; Mol, 1995; Spaargaren and Mol, 1992*]. Their argument is that while before environmental protection was primarily focused on how 'safely' to displace hazards from production and environmental politics was concentrated on the distribution of hazards, now the intention is to keep emissions and wastes within the sphere of production. It is in production that emissions can be reduced or prevented, and it is in production that by-products can be re-used or recycled. Such an internalisation of environmental externalities is now seen as consistent with efforts to increase efficiency and even as inducive to technological innovation.

While industrial ecology and the project of ecological modernisation may be seen as revolutionary approaches in the West, a strikingly similar social experiment was initiated in state socialist countries half a century ago. That social experiment, while not flawless and originally not framed in terms of environmental protection, had progressive elements that have been dismantled and forced into oblivion after the collapse of socialism. I revisit this historical example for two reasons. First of all, it makes possible a less stylised, historically better contextualised, and more empirically grounded understanding of the relation between ecology and the post-socialist transition in Eastern Europe. Second, this social project arguably has some lessons to offer for present efforts of production-centred environmentalism.

For evidence I rely on my research on the history of the concept of waste in Hungary, carried out in that country between 1995 and 1997.[1] I begin by demonstrating why and in what form a production-based waste management struck roots in the first few years of Hungarian state socialism. Then I point out the pathologies of this waste politics, showing how these were thought to be remedied in reform socialism in the 1970s and 1980s. Following that, I demonstrate how after 1989 this revised production-based waste politics gave way to a politics of distribution in which the only legitimate subject of public discourse is the treatment, dumping and incineration of wastes. Finally, I evaluate the present waste practices in Hungary and offer some lessons for industrial ecology and the project of ecological modernisation.

**Waste as Use-Value in State Socialism**

It has long been argued that centrally planned economies were systemically wasteful. This is well documented by their records of material and energy intensities and their waste/GDP quotas, which were significantly higher than – often multiples of – equivalalent Western indicators [*Castoriadis, 1978; Filtzer, 1986, 1992; Gille, 1999; Goldman, 1972; Gomulka and Rostowski, 1988; Juhász, 1981; Kornai, 1980; Manser, 1993; Moroney, 1990; Nove, 1980; Reiniger, 1991; Simai, 1990; Szlávik, 1991; Ticktin, 1976, 1992*].[2] Despite this reputation, evidence is overwhelming that an elaborate system of waste registration, collection, distribution and reuse – and, to a lesser extent, reduction – already had taken root during the Stalinist years of state socialism. The conclusion of a piece of Hungarian propaganda material in 1951, according to which 'there's no such thing as waste', aptly captures the mentality underlying these efforts.

Accepting the coexistence of documented wastefulness with this progressive attitude to waste might be difficult at first. However, if we understand the economic and political tasks facing the new socialist state after 1948 and the conditions under which these tasks had to be accomplished, the emergent waste policies and waste politics appear as a logical solution. Some of these tasks were specific to state socialism. As the socialist state was establishing itself, it had to extend its redistributive power over the entire economy. With a rapid and aggressive wave of nationalisation, it expropriated private producers and hurried them and their former labour force into large state-owned or co-operative units where employees had to assume a new work discipline suitable for the forced speed of industrialisation. The Hungarian Workers' Party established a hierarchical bureaucracy that managed the economy through plans that were sanctioned as laws every three, and later five, years. At the point of production these plans were broken down into input and output quotas, and new disciplinary mechanisms enforced their fulfilment. In this new system, each individual, each and every nail was accounted for and could be utilised only according to its function as prescribed by the plans. Production wastes were no exception.

Other tasks however emanated from conditions of scarcity. The socialist state not only had to construct a detailed structure of power, but also needed a certain amount of legitimacy. It had to be able to create wealth and spur growth that was tangible to the working population even as the country was suffering severe post-war scarcities. Hungary had lost territories with raw material sources, its productive capacities were in ruins or transported off as compensation to the winners of the Second World War, and it suffered a Western embargo of metals exports to Hungary. Economising thus required drastic and comprehensive measures.

Planners designed and established several institutions dealing with waste. In 1950, the Central Planning Office founded the Bureau of Material Conservation, whose tasks included, among other things, devising ways to reuse wastes in industry. While this bureau was soon closed, another waste institution, the By-Product and Waste Reuse Company (MÉH), starting its operation simultaneously with the first five-year plan in 1951, turned into a permanent feature of Hungarian state socialism. MÉH was obliged to buy the wastes of enterprises and sell them to others according to plan. The circle of wastes collected was very wide, including metal, paper, textiles, batteries, leather, seeds and nuts, as well as animal bones and human hair.

Between 1950 and 1959, 34 central regulations on the collection, storage, delivery and price of waste materials were issued. The No. 102.700/1951 decree of the head of the Central Planning Office, the first waste law, obliged companies generating certain types of wastes to collect them and offer them for transfer to other firms in accordance with quotas. The 'waste-collecting organs' (ministries and councils) had to submit a monthly report to the Bureau of Material Conservation, and could 'use and trade waste materials under regulation only in accordance with the order of the Central Planning Office and the competent material management office'. Subsequent laws prescribed which wastes were to be delivered to which company, how to calculate the price of wastes, what to do with hitherto unregulated waste materials, and how much material reward could be given to those who collected wastes beyond the planned amount.

Other areas of administration were also scrutinised and modified to make them more adaptable for waste reuse. Scholarly journals published numerous treatises on the economics of waste reuse and on waste statistics and accounting [*Bóka, 1953; Csupor, 1954; Jávor, 1953; Lelkes, 1950; Pál, 1953; Valkó, 1951*].

Material conservation and waste recuperation were, however, not left only up to institutions, administrators and laws. Waste also became a key issue around which the public was mobilised. Numerous campaigns, organised either by MÉH or various party organs, aimed at collecting wastes and/or reusing them in factories, in agricultural co-operatives, in schools, and in districts of cities and villages. Metal-collecting weeks were organised; brigades dedicating themselves to waste reduction and reuse, waste-collecting stewards, and youth and female troops mushroomed and busied themselves, mostly after regular working hours.

The culmination of these campaigns was the Gazda movement: Géza Gazda, a metallurgical worker in the most prominent metallurgical works of the country, invented a new way to reuse scrap steel. Instead of melting down steel scraps, he designed a method and machinery to re-roll them into large sheets. The purpose was not only to save material and energy, but also to ease the burden on the overworked furnaces. Given the importance of metal waste in a country that was hastily trying to build up its heavy industry while deficient in iron ore amidst Western embargo on some metals, and given that the elimination of one or more phases in the labour process was an already preferred method of speed-up, this 'trick' immediately caught the eyes of party officials. In 1951 they ordered Gazda to 'initiate' a movement that would spread this new approach to waste reuse to other branches of industry.

This new approach, unlike recycling, gavc priority to reusing waste materials in their original quality and function without chemical or substantial mechanical transformation.

> We can also expect serious results from the innovation of Béla Märcz, a worker of the Artificial Leather Manufacturing Company. The essence of the innovation is that while so far the rejects … could be made recyclable only through grinding, now even the smaller pieces are sown back into sheets, and this way there seems to be a saving of 19,000 forints per year [*Leather Workers' Union, 1951: 4*].

As a result, while the party might have seen a speed-up trick in Gazda's movement, most workers came to associate it with the collection of wastes and their use in manufacturing goods and parts previously made from new materials. In a majority of cases, recycled goods did not fall under the plan or even the profile of a given firm. A metallurgical factory, for example, started using leather wastes for making small bags for bicycles. While such deviations from the plan were generally not advised, they were encouraged when wastes could replace raw materials imported from the West, and when they had the effect of lowering enterprises' raw material requirements, as in the following case:

> On the Day of Ideas this month, 21 out of the 32 submitted proposals targeted waste reuse. We recuperated so much wrought iron, otherwise difficult to obtain, that we withdrew our orders for the fourth quarter of this year and the first quarter of next year [*Iron Workers' Union, 1951: 29*].

Reducing wastes was often mentioned but received much less emphasis than reusing wastes in propaganda materials and in practice. Source reduction efforts concentrated almost entirely on redesigning the cutting pattern on sheets of metal, leather, and textile or changing their size so as to minimise the amount of material dropped during tailoring.

> Most of the wastes are produced during tailoring. In the Ganz Shipyard they cut covers for switchboards out of steel sheets. Before, when they needed such a cover, the sheet was cut in half in the warehouse into two 1,000 x 1,000 mm pieces. Why? There was a misconception that the remainder was 'easier to store'. Since then they realised that exactly three covers could be cut out from a whole sheet, because the size of a cover was just a third that of a sheet. This way then there is no waste from cutting [*Iron Workers' Union, 1951: 23*].

According to the party, the Gazda movement was among the most popular, if not the most popular, of all production-helping campaigns; indeed, it generated the greatest amount of propaganda materials and media attention. Perhaps this was due to the commonality of the emergent culture spread by central conservation programmes and the residual culture of the new industrial labour force, which grew up having experienced long periods of deprivation that fostered a thrifty attitude and ingenuity in economising. The popularity of this movement could also be attributed to the fact that it increased workers' leverage over managers and control over the labour process even as it facilitated the party-state's surveillance over the use of materials.

Claims about the movement, such as 'there are thousands of ways and possibilities to reuse wastes' and 'here ingenuity and creativity are given a free reign', indicate not just this expected 'from-below' initiative but also the perceived lack of limits to waste reuse. Clearly, the early socialist waste discourse treated waste as a useful material – what's more, a multi-purpose, pliable material – and as something not to be displaced from the sphere of production but rather to be reintegrated into it again aand again. This is not the Douglasian notion of waste as 'matter out of place', but rather a concept of waste as a material to be meticulously registered, collected, redistributed and re-used.

**Pathologies of Early Socialist Waste Policies**

The Gazda movement and other economic and administrative measures had two unintended consequences. First, the reuse of waste materials required additional raw materials, energy and labour, which, along with most products, were all in short supply. As a result, collected wastes were often left rotting and rusting on factory yards. This practice became clear to planners, who by 1954 passed a decree ordering that '[t]hose wastes that are usable within the Gazda movement without melting, breaking, etc. can be stored only in the amount that is sufficient for their processing for 90 days' (2.500-21/1954 decree of the president of the Central Planning Office).

However, reuse, even when it materialised, did not prevent wastes from becoming trash. These early recycled goods were in many cases not needed, and even chronic general shortages could not increase their appeal. As a contemporary brochure said, 'It may happen that they produce a good for which there is no demand, when in fact the same waste material could be used for making more important and more sought-after consumption goods' [*Leather Workers' Union, 1951: 12*]. As a result, by 1959, unsalable recycled goods piled up in such quantities that the National Office of Materials and Prices found it necessary to introduce discount 'Gazda movement prices'.

Not only did the movement and waste policies fail to reuse wastes effectively, but they also strengthened the tendencies of the centrally planned economy towards wasteful production. The approach that wastes are always reusable in an infinite number of ways led to a counter-productive attitude, which Gazda reproachfully caricatured this way: 'It does not matter if I produce rejects, since the spoiled product after some modification can still be reused. That's what the Gazda movement is for ...' [*Iron Workers' Union, 1951: 8*].

Furthermore, the waste quotas created an added incentive for workers and managers to produce with even higher waste ratios. This problem was admitted by central organs:

> In 1956 those ministries, and the companies subordinated to them, that generate non-ferrous metals stopped getting plan quotas. To wit, experience showed that it was impossible to prepare technically grounded and correct delivery plan quotas for the deliverers [of

> waste]. It happened that certain enterprises received quotas that were unrealistically high, which motivated them in the wrong direction, while other firms received loose delivery quotas, which they could fulfill much too easily [*Országos Tervhivatal, 1960: 1*].

'Motivation in the wrong direction' – that is, the *l'art pour l'art* production of wastes – reinforced the tendencies of the centrally planned economy towards wasteful production and a Sisyphean exhaustion of natural resources and the destruction of nature – resource use that produced no or very few benefits in return [*Gille, 1997*].

Yet, the negative environmental impacts of early socialist recycling efforts were not restricted to resource squandering. Another unintended consequence was manifested over the long run. Because waste was seen as a useful material, the state saw its role in assigning use-values to it and in encouraging production units to reuse wastes. Wastes were not to be thrown out but to be kept in production and circulated in the economy as long as possible. If technical conditions for reuse of certain wastes were missing this was viewed as a temporary problem that rapid technological development and the genius of socialist workers would solve shortly. Consequently, not only was the idea of useless wastes precluded, but so was the idea that wastes might be harmful. This limited perspective was rooted in the forced adaptation of a metallurgical concept of waste. Metal wastes, in general, could indeed be reused over and over again, and even if they were left to rust they did not constitute a significant source of air or water pollution. To deal with chemical wastes, however, whose reuse or recycling was technically more thorny, and whose accumulation in storage *was* an environmental threat, would have required a more flexible attitude. Problems created by such wastes called for more rigorous focus on source reduction and facilities for safe waste treatment and dumping. Source reduction was not prioritised because the metallurgical waste models suggested endless reusability. Requests for disposal were denied legitimacy because that would have meant renouncing a secondary raw material.

An example will demonstrate the consequences of this attitude. In 1968, the Budapest Chemical Works started producing a herbicide intermediary, tetrachlorobenzene, the production of which involved a high 45 per cent waste ratio. The company immediately started looking into reuse possibilities, each and every one of which, however, failed to bring a permanent solution. Despite this, in various upper management meetings reporting on talks with superordinate offices or scientific institutes on the waste problem, it was repeatedly stated that the primary goal was reuse. In 1972, the management decided to transfer its waste drums to a newly opened branch plant in the countryside, as they argued, purely for purposes of temporary storage: 'The by-products of tetrachlorobenzene must be stored in drums in our Hidas plant until they are reused' [*Budapesti Vegyimüvek, 1972*]. The last words of a 1975 report on negotiations held with the National Committee on Nuclear Energy and the central epidemiological authority on the possibilities of safe dumping were: 'The primary goal continues to be the reuse of wastes. The possibility mentioned above [dumping] must be maintained only for wastes that cannot be sold' [*Budapesti Vegyimüvek, 1975*]. And in 1978, even as management agreed on a dump site, it was emphasised that 'we must continue searching out reuse possibilities, and a contest with this objective must be announced. All proposals that would pay the half of the unit transportation costs from Budapest to Hidas to the reuser may be accepted' [*Budapesti Vegyimüvek, 1978*].

By the 1960s, non-reusable wastes piled up everywhere in the country in such quantities that factories were forced to find illegal ways to get rid of them. Local government officials now started complaining about illegal dumps appearing both next to dumps managed by councils (local governments) and on the territory of filled dumps already put back into cultivation. In 1967, the Executive Committee of the Council of Budapest was forced to admit that

> the question of [the disposal of] industrial wastes is unsettled in administrative respects. Neither the OÉSZ [National Construction Code], nor the Budapest Code of Urban Planning contains regulation concerning the neutralisation of waste and garbage of industrial origins [*Fövárosi Tanács Végrehajtó Bizottsága, 1967: 3*].

## Monetisation and Environmentalisation of Waste in Reform Socialism

These unintended consequences were increasingly acknowledged, as indicated by the introduction of the discount Gazda movement prices and revocation of waste quotas by the end of the 1950s. The concept of waste as useful material did not disappear, however. In fact, most waste-related institutions remained intact and waste collecting campaigns became permanent rituals of social life. Rather, the reforms led to a different understanding of the usefulness of wastes.

From the mid-1970s, the utility of waste was seen not in use-value but rather in monetary terms: waste was a cost of production. The merit of waste reduction and waste reuse, therefore, resided in decreasing such costs. Consider the differences in emphasis in the following two documents. The first one is from 1951.

> ... [O]ur party has now been keeping the cause of material savings on the agenda for a longer time, but especially since the congress. The leaders of our party have already emphasised many times that this is *not simply to be talked about as one tool of decreasing production costs, but primarily as the basic condition for fulfilling the plans* in many industries [*Ember, 1952: 27*, emphasis added].

The second is from the 1981 Waste and Secondary Raw Material Management Programme, which was implemented along with other rationalisation programmes in the early 1980s to combat the explosion of fuel and raw material prices on the world market.

> The more rational management of material resources received a priority among the economic policy goals of the sixth five-year plan. The program announced chronologically as the second among the austerity programmes wished to achieve an increased substitution of primary raw materials with wastes, and the exploration and mobilisation of economic reserves hidden in the recycling of secondary raw materials. *It treated waste utilisation as an activity reducing expenditures by which the per unit material costs could be reduced* in relation to all costs of production [*KSH, 1988: 10*, emphasis added].

The real novelty of this new discourse, besides monetising the concept of waste, is its acknowledgement of the environmental implications of waste problems. Its progressive elements reside in its retention of the earlier production-centred approach to waste problems, the recognition that waste problems create ecological ones, and the understanding that waste reduction can improve the state of the environment.

In Hungary in the early 1970s, the preventative effect of waste reuse was primarily seen in a slower exhaustion of natural resources, but after 1976 the benefits of waste recuperation were increasingly sought through the prevention of pollution by wastes. In line with this goal, the 1980s witnessed an effort to integrate waste reuse into economic considerations, as evidenced in many new policy tools. Among these were price increases for wastes, which made the collection and resale of wastes more worthwhile, and the new opportunity for firms to calculate prices for recycled products as if they had used new raw materials for them – which was supposed to encourage the use of wastes. The reuse of wastes now became the object of cost-benefit analyses; funds were established to motivate firms to apply waste-conscious technologies. The state financed at least two-thirds of the costs of reuse facilities. Firms were exempted from taxes on such investments. And 25 per cent of the interest on loans borrowed for such purposes were remitted. In addition to firms, individuals involved in waste collection (teachers co-ordinating children's waste collection, or concierges) were also entitled to special monetary rewards.

Parallel to this change in the concept of waste, the relationship between the enterprises and the state also went through a transformation. Instead of centrally calculated waste quotas, enterprises were now free to decide which wastes they wanted to reuse, sell, treat, or dump, and they were meant to be motivated by the above-mentioned financial incentives. As a corollary, the agents of industrial ecology changed as well. While in the early state socialist period, it was workers who were encouraged to redesign products and technologies and to invent waste reuse possibilities, now such tasks were transferred to professionals with economic and engineering expertise. Rather than consisting of small but ingenious adjustments to the existing product design and labour process, waste reduction and reuse now became subject to cost-benefit analysis and considered possible only after implementing high-tech innovations and large-scale modernisations. Investments included purchase of Western technology to reuse plastic wastes, facilities for reusing tyres, for recovering metals from dross, a muddy by-product, and the modernisation of a straw cellulose plant, just to mention a few examples [*Remetei, 1986*].

While the 1981 Waste and Secondary Raw Material Management Programme did not achieve its stated goal of substantially increasing the portion of secondary raw materials among industrial inputs [*KSH, 1988*], it turned out to be quite successful in finding uses for potentially dangerous wastes. Before the implementation of the programme, 17–18 per cent of hazardous wastes were reused or recycled [*Society for Scientific Education, 1980*], while in 1982 the figure was 21 per cent and in 1986, 29 per cent [*Árvai, 1990*].[3]

In sum, while the socialist waste discourse of the 1950s made sure that wastes were integrated into the economy (with quotas, distributive regulations, etc.), the 1980s' discourse, shaped by a monetary and environmental concept of waste, made sure that prevention relied increasingly on economic incentives as well as on formerly exclusive administrative measures. The Gazda movement and early socialist 'heroic' waste policies may or may not have earned the label, 'industrial ecology'. However, the waste policies of the 1980s, in which environmental concerns became central and a growing emphasis was placed on economic rather than administrative tools, could easily be endorsed by proponents of industrial ecology. Hungary, therefore, was in an excellent position to continue 'catching up with the West', at least in this one aspect, after 1989. Was this opportunity seized upon? This is the question the next section addresses.

**Post-Socialist Regression to the West's Past**

1989 was a year of hope for the people of Eastern Europe: not just a hope for a freer society without shortages, but also for a cleaner environment. There have been two discourses circulating about how this improvement can be achieved simultaneously with the more central processes of the transition, privatisation and democratisation. The initially strongest expectation has been that marketisation and privatisation will automatically improve the state of the natural environment. Reform economists in the 1970s were already arguing that harder budget constraints and more independence for enterprises in investment decisions – in sum, a free market – would force them to use natural resources more sparingly and with greater care.[4] This is the assumption that informs this transition discourse, which, adopting Dryzek's [*1997*] terminology, might be called the 'economic rationalism discourse'.

Many activists and experts, however, hoped that Eastern Europe, enjoying the Veblenesque advantage of late-comers, could draw lessons from the mistakes of Western capitalism and build an economic system in which environmental concerns were integrated from the beginning instead of being saved for later as add-on features. Taking this course requires a more active role for the state, at least a more direct relationship between industry and government, than the economic rationalism discourse sees as necessary. These expectations were later shared by decision-makers, some Western agencies and aid foundations. For example, the 1995 White Paper, addressing the conditions for the accession of former socialist countries to the European Union, declared that

> environmental policy and the internal market are mutually supportive. The EU Treaty aims at sustainable growth and high levels of environmental protection and that environmental requirements be integrated into the definition and implementation of other policies. An integrated approach to allow a more sustainable path of social and economic development is not only vital for the environment itself, but also for the long-term success of the internal market (European Union [*1995*], quoted in Klarer and Francis [*1997: 39*].

Let us call this discourse, after Hajer [*1995*] and Dryzek [*1997*], 'ecological modernisation'. Based on Mol's theory of ecological modernisation, such an integration requires a renewed and more horizontal relationship between industry and the state, more cooperation between civil society and the state as well as between civil society and industry, and a new approach to policies, akin to the principles of industrial ecology. To maintain the original focus of my paper, I will concentrate on this last aspect, asking whether an approach consistent with industrial ecology indeed is evident today in Hungary.

To the extent that the integration of economic with environmental policies have become an officially accepted and advocated approach to improving the state of the environment in Eastern Europe, the ecological modernisation, or more concretely, industrial ecology approach seems to have won. Its principles appear in various western written pieces of advice, such as the Environmental Action Programme for Central and Eastern Europe presented at the Environment Ministers' Conference in Lucerne in 1993, and in a series of reports by the Swiss Federal Department of Foreign Affairs prepared for the Hungarian Ministry of Environmental Protection and Regional Development in the first half of the 1990s.

While, based on these declarations, ecological modernisation, and with it industrial ecology, might even appear as yet another blueprint for Eastern Europe's successful transition to capitalism, nowhere has it been acknowledged that late socialist environmental policies were already heading in the direction of this more 'progressive', more prevention-centred environmental discourse. In fact, even the more moderate recommendation of a group of US and Hungarian experts that the 'assessment of the effectiveness and deficiencies of institutions established by the previous regime for the purpose of environmental protection, is an inevitable precondition of designing new institutions' [*Bochniarz and Kerekes, 1994: 13*] has fallen on deaf ears. As a result of this blind *tabula rasa* approach, the Hungarian path towards preventative policies was abandoned, the hopes for an environmentally sensitive transition to markets and democracy were broken, and, consequently, present developments point more in the direction of the economic rationalism discourse and of remedial, rather than preventative, environmental policies.

How did this reversal come about? Hungarian reform economists were blaming environmental degradation on the absence of enterprise freedom already in the 1970s, and indeed they had a strong impact on the environmental policies of the 1980s that operated with a monetised concept of waste and economic incentives. A radically reduced role for the state remained the primary point on their agenda after 1989, and supranational agencies, such as the World Bank and the IMF were, no doubt, more than supportive of this goal. From their perspective, every environmental policy tool or institution amounted to no more than a fetter on the operation of the market, and thus nothing seemed worthy of saving from state socialist waste policies either. Indeed, waste institutions, research projects, and policy processes were terminated either intentionally or as an inevitable result of the dismantling of the state.

This process of disintegration manifested itself through privatisations, elimination of state funding for certain environmental purposes, legislation, or as the case may be, lack or repeal of legislation, and the selective reorganisation of state environmental authorities.

*Privatisation*

Privatisation had multiple effects on waste practices. The privatisation of the almost 40-year-old state waste collection company, MÉH, for one, imposed radical limits on the scope of its activity, that is, the range of waste materials and circle of suppliers. As for other companies, their privatisation often followed a pattern that relieved new owners from old as well as new environmental liabilities. Initially, the Privatisation Act did not even address the question of environmental liabilities: until 1992, the privatisation agency provided the new owner with a guarantee that it would clean up whatever environmental damage was later discovered. In one case the amount of this guarantee was equal to the sale price of the firm, reducing the privatisation income of the state to zero [*Állami Vagyonügynökség, 1993*]. In addition, the amount and the time limit of this guarantee was usually based not on an independent expert assessment but rather on a bargain between seller and buyer. It was only in 1992 that a law (§35 of the LIV. Law) prescribed the obligation of to-be-privatised firms to prepare a plan for repairing environmental damages done in the past. This, however, was not effective with partial privatisations. In these, usually the most profitable and cleanest plants or departments were privatised, while those with deficits and environmental problems were left in the hands of the state, thus leaving the clean-up costs to those units that could least afford to pay them [*Állami Vagyonügynökség, 1993*].

*Elimination of State Funding*

The liberal transition paradigm looked at every state expenditure with suspicion. Environmental expenditures were no exception. Under the conditions of economic crisis and with the elimination of loans, the principal idea of industrial ecology that waste problems should be tackled in production through better product design and environmentally friendlier technologies could not be enforced. The government programme for waste reduction and reuse initiated in 1981 was abandoned, and the motivational system of credits and subsidies was subsequently eliminated under the pretext of normativity [*Bakonyi, 1991; Dworák, 1991*].[5] Some experts pointed out a missed opportunity in 1989 to implement new economic tools.

> With the introduction of the new tax system, there would have been a possibility, besides the previous direct subsidies, to establish a system of subsidies and regulation based on tax preferences (and naturally,

> extra taxes) similarly to all developed countries where such a system was successfully applied. The circle of tax preferences encouraging environmental ... and waste management activities, however, was finally drawn so narrowly and in such an unfortunate manner that in the case of waste reuse, for example, it functions as an actual counter-incentive. Just as with taxes, there was a failure to apply other economic regulative tools (loans, interests, etc.) to such ends as well [*Takáts, 1990*].

Such a failure compelled some others to call for increased state funding for environmental purposes [*Kindler, 1994*].

*Legislation*

The legislative area is perhaps the most telling of the direction of changes. While the number of laws enacted by the Hungarian Parliament in the year of 1992 was so high that it entered the Guinness Book of Records, a new comprehensive environmental act failed to be put forward by the Environmental Ministry [*Bochniarz and Kerekes, 1994*]. According to activists and others involved in the drafting, this was a conscious delaying tactic of the industrial lobby.[6] The lobby's interests lay in a privatisation process that would not be held back by ecological considerations, and in institutional arrangements in which environmental issues would be merely add-on features.[7] 'The Parliamentary Committee on Environmental Protection became increasingly frustrated with the impasse and finally decided to commission an independent draft,' which even the Minister supported [*Sajó, 1994: 38*]. Even so, the Comprehensive Act was not passed until 1995.

Even as comprehensive legislation was held up, older regulations were withdrawn. Most importantly, the decades-old system of deposits on packaging was eliminated. Riding on the wave of suddenly-open markets and formerly suppressed consumer desires, Western corporations such as Coca-Cola have brought in goods with non-recyclable packaging, and whose collection is no longer organised, and on which there is no longer a deposit. This is not an exclusively Hungarian phenomenon. As Kaderják says:

> With respect to deposit-refund systems, the common experience of the CEECs [Central and Eastern European countries] is rather that during the period of transition formerly well-performing systems have been partly destroyed (for example, deposit-refund systems on bottles or car batteries) [*Kaderják, 1997: 169*].

More in line with a production-centred approach are the environmental product charges on fuel, tires, refrigerators, refrigerants, packaging materials, and batteries (Act LVI of 1995) introduced in 1996.[8] The amount of the charge is reduced by half for those products that are environmentally friendly. The revenue from these go to and were expected to make up as much as half of the Central Environmental Protection Fund (CEPF) budget in 1996. This portion is earmarked for investments that abate pollution as well as for waste recuperation and recycling. Unfortunately, the draft new Waste Management Law, planned to be implemented at the beginning of 1998, says that this system of product charges is to be ended after the new recycling facilities are established. Greens are vehemently opposing this clause [*Bödecs, 1997*].

The policies of the 1980s had established a system of monitoring that made evaluation of a firm's waste activities easy and transparent by obliging firms to prepare material flow charts and material balances.[9] The obligation to prepare material flow charts was cancelled in 1992. Even the circle of firms obliged to prepare material balances, which is a device much easier to manipulate, was severely limited, and exemptions from this obligation became easy to acquire [*Romhányi, 1995;* Takáts, Interview, 15 July 1996, Budapest]. Such flow charts would be excellent tools for implementing the principles of industrial ecology because they reveal both material inefficiencies and the locations in a given labor process where hazardous substances are generated. A designer of the original law sees them as a mechanism of enforcing voluntary environmental quality standards, similar to those of the International Standards Organization [*Romhányi, 1995*]. However, such tools are eliminated under the excuses of production secrets and the sanctity of private property. Even the ministerial surveys of waste generation have come under the attack of new private firms invoking the same excuses.

In addition, while waste policies under most of state socialism concentrated on non-hazardous, and especially metal, wastes, at the present this group of wastes is subject to no regulation whatsoever, indicating yet another step backwards.

### *Reorganisation of Environmental Authorities*

The restructuring and reduction of various offices or departments in the state apparatus was inevitable during the radical transformations begun in 1989. However, the pattern of departments slated for the most drastic reductions suggests an administrative shift towards remediation and the state's withdrawal from waste issues. The Waste Department of the Ministry of Environmental Protection and Regional Development went through a series of reorganisations and reductions whose final balance indicates that it was the hardest hit among departments of the Environmental Ministry. A high-ranking official there suggested that the emaciation of this department under the pretexts of rationalisation and the so-called 'integrated approach to environmental protection' was encouraged by the industrial lobby.

> This is because there are a lot of problems with industrial wastes, whose solution requires a lot of money. The interest of various groups is to gain a share of the waste [treatment-disposal-incineration] market. But they only want the money from it without following the regulations. For permits, they have been going to the boss [rather than going through the required channels]. [Some] waste professionals in the ministry stood in the way of these intentions, so they had to be gotten rid of (Anonymous official, 1997).

This could have very practical consequences: a reduced staff is ill-equipped to enforce waste laws and is weak in the face of massive requests from domestic and foreign investors in waste facilities. Without enforcement, the explosion of the waste treatment sector acts as an explicit disincentive to reduce waste output and to recycle, ultimately leading to the marginalisation of the principles of industrial ecology. The consequences of this marginalisation are discussed in the following section.

## Catch-up or Clean-up: The Schizophrenia of Post-Socialist Waste Politics

History seems to repeat itself: the new discourse on waste once again produces the unintended consequence of wasting wastes. Under state socialism, the rigid use-value mentality and waste quotas encouraged the increased production of wastes. In addition, these wastes were reused without considering actual needs, and eventually turned into trash. Now, once again waste is piling up. Without potential reuse in sight, it has an equal chance of metamorphosing into garbage. While the waste reuse institutions have suffered a fatal blow, the so-called environmental services sector, comprised of businesses dealing with waste treatment, disposal and incineration, has experienced a growth spurt. A 1994 survey found about 400 companies that provided technical services (particularly engineering and design for air protection) and environmental products (especially products for air protection and laboratory services) (Dzuray [*1995*], summarised in Lehoczki and Balogh [*1997: 163*]).

While the growth of the environmental services sector is not a negative development, two circumstances do overshadow this increase. One is that a considerable number of these are so-called phantom firms, or at least non-licensed agents that do not follow waste dumping regulations, and when caught claim limited liability. The amount of wastes passed on to such entities, according to one estimate, may claim as much as a third of the generated hazardous wastes [*Lehoczki and Balogh, 1997: 144*]. The other shadow cast on this new ebullient sector is the end-of-pipe nature of the technologies they offer. Those environmental investments that require substantial initial capital and have slow turnover[10] will not materialise amidst shortage of capital, eliminated state subsidies, and strong pressures to invest in technologies with fast turnover [*Székely, 1991*].

> At the present, the state foregoes the reuse and the neutralisation of those wastes which might be more profitable but have slow turnover. These materials, by increasing the amount of hazardous wastes, cause damage, pollute the environment, and later will lead to serious problems in the whole economy [*Dworák, 1991: 26*].

Statistical data from 1992 indicate that waste generation continues to exceed waste treatment.[11] Between 1989 and 1992, the portion of annually generated hazardous wastes accumulated rather than treated increased from 17 per cent to 42 per cent.[12] That such practices are creating economic as well as ecological time bombs is becoming increasingly visible. Although no actual survey exists, according to experts, the number of illegal dumps has soared since 1989. Once again, albeit for different reasons, industrial waste is piling up and is being wasted.

Under these circumstances, policy-makers feel a tremendous pressure to find a quick treatment for the symptoms, that is, to facilitate dumping and incineration, instead of nursing back to life the preventative policy tools encouraging waste reduction and waste reuse. Rather than 'catching up' with the West's preventative mentality, Hungarians can call themselves lucky if they are able to catch up with their own proliferation of untreated and unutilised wastes.

The problem of the accumulation of wastes is exacerbated by the fact that total waste generation, especially of hazardous wastes, is on the increase. Despite predictions that waste generation will abate with the re-institution of market mechanisms, and after the initial drop in waste output due to economic decline, recovering industries are now quickly regaining their waste-generating capacities [*Lehoczki and Balogh, 1997*]. A statistical analysis by Kaderják and Csermely [*1997*], for example, concludes that sectoral changes in Hungary reduce the environmental damage of paradigmatic socialist industries, primarily metallurgy. Other industries exhibiting strong growth potential, such as the oil and plastics industries, however, are and will be responsible for a major growth in emissions. With respect to hazardous wastes more concretely, these authors similarly find that 'oil, pharmaceutical, sweets and the meat industries will become *growing* "producers" of hazardous wastes' [*Kaderják and Csermely, 1997: 27*, emphasis added]. Based on present emissions and the hazardous waste production of various industries, these predictions warn against waiting for the allegedly automatic benefits of structural shifts in the economy, and call for achieving waste reduction goals through technological changes, such as those advocated by industrial ecology.

Cleaning up old and new wastes, however, can easily torpedo efforts to implement preventative policies. By flooding Hungary with newly available, even if not universally affordable, waste treatment technologies, the already decimated incentives to reduce and reuse waste will lose all effect. Even as experts and policymakers are busy dreaming up plans for 'win-win' policy approaches and future visions consistent with ecological modernisation, reality is taking a different turn. Western investors, increasingly unable to sell their incinerator technologies on the saturated and environmentally more conscious Western markets, seized on the excellent opportunity to expand to Eastern Europe.

There is indeed little legal recourse against Western imports of waste facilities or wastes. The clause of 'best available technology' has been missing from laws, and while the 102/1995 waste law forbids the import of wastes for treatment, it allows it for reuse. Reuse, however, even according to the new draft on waste management, includes energy recovery from waste incineration [*Bödecs, 1997*].

Western incinerator firms are lured to the region by other factors as well: the desperation of local governments over the piling up of wastes; Western funds, especially soft loans, available for environmental purposes in postsocialist Europe; a relative scarcity of incinerator facilities in the region, thus the promise of at least initially monopolistic prices; loose and poorly enforced environmental standards for incinerators; the possibility of burning wastes imported from the West; and the imminence of the legal harmonisation between the EU and East-Central Europe, which requires an ample supply of waste treatment facilities. As an American banker investing in the export of waste treatment facilities answered to questions about whether East Europeans can pay for such technologies, 'If they want to become part of the greater European community, I don't see they have much of a choice' [*Schwartz, Koehl and Breslau, 1990*].[13]

A high-ranking official in the Hungarian environmental ministry, who wished to remain anonymous, described the resulting stampede by waste treatment businesses:

> Westerners think we don't have environmental impact assessment requirements. Then they find out through their tiny consultant firms what [regulations] we have, and they are shocked that they don't receive the permits and that the ministry is not negotiating with them. They think they can just bring whatever they want wherever they want. They think they can bring us fake pearls! (Anon., 1997).

And what a stampede it has been in the entire region! Between 1988 and 1996, an estimated 18 million tons of annual incinerator capacity was planned for Russia, the Baltics, Hungary, Poland, the Czech Republic and Slovakia, with about 93 per cent of these capacities offered for export by Western countries.[14] Put another way, there were about 187 facilities proposed for the region, ten in Hungary.[15] Without a comparable interest in waste recovery facilities, this pressure will most likely result in a return to remediation.

Even a quick glance at news media representation of waste problems reveals this conceptual shift. During state socialism, articles about discovered and implemented waste reuse possibilities and even muckraking reports about 'wasted materials' abounded. Now, all one can find is articles suggesting that waste is something inherently negative and harmful, and thus to be got rid of. 'Waste trade: rusty markets', 'Asbestos waste scandal', 'The best solution is to shoot the garbage up to the Moon' are some of the characteristic headlines. One never reads about political struggles over waste-generating investments, or waste reuse – the politics of waste production has been pushed out of public discourse by a politics of waste distribution (waste treatment, dumping, and incineration).

The sanctity of private property has transferred to the production of waste. Since waste is increasingly produced in private plants, public or state scrutiny over what is produced, what is made out of wastes, and what is done with the wastes generated is considered a violation of basic property rights. Citizens may have a say in where to dump or incinerate, and how the economic benefits of such facilities should be distributed, but they may never question 'purely economic decisions' about waste-generating investments, and about what should be done with by-products. Those few who try to expand the terms of the debate to issues of waste generation find themselves instantaneously charged with 'infringing on the principle of free market' [*Bödecs, 1996: ix*], being backward [*Bánhidy, 1996; Bödecs, 1996*], and as being 'religious fanatics' who base their arguments on emotions rather than on 'expert objectivity' [*Bánhidy, 1996*].

The present situation is the outcome of a misinterpretation of the scope and nature of state socialist waste politics. The common description in the literature of the state's disinterest and weakness in implementing and enforcing environmental policies has given the impression that environmental remediation will have to start from a blank slate. The progressive and promising previous initiatives by state authorities and professional research institutions have been erased and forgotten. From the perspective of liberal ideology and international monetarist interests, the state's efforts to encourage waste reuse and closely monitor the generation and movement of hazardous wastes constituted just another harmful intervention in the economy.

## Conclusion

In ancient Hungarian burial ceremonies, the deceased was buried together with his or her horse. The horse, even if healthy, was killed to accompany its master to the grave. A similar burial ritual is being conducted in postsocialist Hungary. With the party, centralised planning, and state ownership is buried a practical mentality, a material culture, that valued thriftiness and utilising waste. Gone are the many fledgling, not flawless, but nevertheless progressive and promising, policy measures, institutions, and laws aimed at reducing and reusing waste. The wastefulness of the present is perhaps best exemplified by the prodigality with which old ways of achieving worthy goals are discarded, irrespective of their potential for further use.

During Hungary's socialist era, from 1948 to 1989, policies treated waste as something useful and dealt with waste problems in production. In the 1950s, policies concentrated on waste reuse, encouraged with administrative means such as waste quotas and centrally organised waste redistribution. In the 1980s, waste was seen as a cost of production, and thus waste reduction and reuse were seen as steps to increase efficiency. Policy tools emphasised the financial motivation of waste producers and included credits, subsidies, and price manipulation. Professionals with economic and technical expertise were encouraged to participate in achieving these goals. It was in the 1980s that Hungarian waste policies first addressed waste problems not merely as an issue of material conservation but also as one of environmental pollution. The integration of economic goals with environmental ones, the emphasis on prevention, and the reliance on economic and technical expertise and technological innovation to achieve these goals are highly consistent with the objectives of ecological modernisation and industrial ecology.

These efforts suggest three lessons for Western and postsocialist Hungarian applications of industrial ecology and ecological modernisation. First, exaggerated expectations of reuse and recycling proved counterproductive in both analysed periods of state socialism. Increasing the proportion of reused and recycled wastes is of course a worthy goal, but, as the state socialist example proved, there are always waste materials for which reuse or recycling is technologically impractical, or there is no need. Many chemical and hazardous wastes fall into this category. In such cases, source reduction in absolute rather than relative numbers should be the goal. The promise of recycleability that tends to be the focus of projects hailed as successful examples of Western industrial ecology renders waste generation acceptable and may lead efforts away from source reduction, as it did in the Hungarian case.

Second, the Hungarian version of what today might be termed industrial ecology demonstrated that unless the underlying economic, organisational, and ideological-cultural causes of wasteful production are addressed, waste policies will fail. In the 1950s, plan fulfilment, especially that of taut plans,[16] remained the primary goal, even of material conservation campaigns themselves, leaving the relationship between taut planning and wasteful production unacknowledged. The economic logic that propelled producers to produce wastefully in order to fulfil taut plans eventually overwhelmed the material and waste conservation efforts, which therefore failed. Although planning became looser in the seventies, general shortages still prevailed, however, continuing to act against waste-conscious production.

As centrally planned economies tended to prioritise production and growth over efficiency, the reuse of wastes often proved economically disadvantageous and turned wastes into garbage. In the 1980s, efficiency became the accepted slogan for waste policies, but if the market dictated it, inefficient, wasteful, and environmentally unsafe production was continued – as in the case of the Budapest Chemical Works. This company did rather well in central material conservation and efficiency – increasing campaigns, as indicated by the awards it received. Yet with regards to its tetrachlorobenzene waste, all it could do was to minimally decrease the ratio of the by-product through modernisation in the second half of the 1970s, and to increase the sale price of the product in order to cover the costs of disposal. The state's need for hard currency, as well as the general political objectives of economic growth, overruled concerns over the waste product.

Bracketing the economic, organisational, and ideological and cultural causes of waste generation may already characterise Western industrial ecology. This inevitably leads not to the integration of environmental prevention with economic goals but rather to its subordination to them. Profitability and growth have never ceased to be the primary goals. On the micro level, for example, waste and energy reduction proposals usually have to meet certain requirements for return on investment [Nelson, 1994], and evaluations focusing on the macro level have already pointed out a certain narrow-mindedness in western industrial ecology.

> It is lopsided to imagine radical transformations of industrial products and means of production, while leaving change in human activities and institutions unexamined, to scrutinise carefully the capacity of the environment to respond to change, but to ignore the potential for human beings to transform their patterns of consumption. Technological change continually yields greater efficiency, but it is

> doubtful whether such improvements alone can make up for the growing scale and scope of the consumption of materials and energy [*Andrews et al., 1994: 475*].

While ecological modernisation theory calls for and documents the diluting of the profit motive, it has yet to address concerns about the counterproductive effects growth is bound to have on its proposed and welcome structural and discursive changes.

Third, in Hungary, in the Stalinist period, the lack of public control over production, or more precisely, over the generation and reuse of wastes, led to the misuse of wastes, that is, to the manufacturing of superfluous goods from wastes. Indeed Kaderják [*1997*], the only scholar who so far recognised the similarities between late socialist environmental policies and the industrial ecology approach, claims that there is a key difference between the two: the lack of a rule of law under state socialism which is a necessary legal environment for the implementation of Western industrial ecology. While the value of democracy cannot be overestimated, we must remain alert to the restricted use of this term. Kaderják, along with many others, is only concerned about the rule of law as a guarantee of the sanctity of private property, of normativity and of calculability for business. The rule of law, conversely, is never mentioned in the context of citizens' rights to participate in decision-making about technologies in the production, collection, and reuse of wastes and about clean-up choices. In my interviews, but in the news media as well, professionals tend to equate lay citizens' protests against certain facilities as 'too much democracy' or as 'the rule of the stupid over the clever'.

Proponents of Western industrial ecology have little to say about the role of civil society. This silence could direct presently still-promising transformations in industry towards what Hajer calls the technocratic version of ecological modernisation. Environmental improvements are expected solely from industry's voluntary measures that, in turn, should stay consistent with the traditional goals of profitability. Such goals, as we know, have never tolerated public scrutiny. Pearce [*1990*] warns about this when dissecting the notion of 'responsible corporations' – a moral pillar of industrial ecology – and recommends extending that term as well as the concept of 'political citizenship' this way:

> Large corporations claim that they are 'political citizens', socially conscious, useful members of society … , that they can be trusted to be self-regulating … Any meaningful concept of political citizenship must include a commitment to the continued viability of the community, which means a commitment to bring (net) benefits to it now and to help secure its future … As members of a democratic

> political community corporations have a right to participate in the political process and must therefore also recognize that others also have this right, even when this may involve a challenge to some corporate interests [*Pearce, 1990: 415, 425*].

As Hajer [*1995*] argues, ecological modernisation also may develop in a direction in which civil society is denied a say in implementing environmentally sound changes in production and the improvements in the structure of environmental protection. He calls for development of a 'reflexive' ecological modernisation in which workers, residents, and concerned citizens play an active role.

If a democratic version of ecological modernisation is indeed our objective, we must never stop asking the question of whose interests are served by changes consistent with industrial ecology or ecological modernisation, however progressive those may seem. In Hungary during the 1950s, the state's effort to establish and continually extend control over resources and to discipline workers was one of the key reasons for establishing the elaborate system of waste registration, collection, and redistribution. During the 1980s, proponents of the efficiency model of waste included economic reformers and enterprise managers eager to loosen the reins of the state. As waste policies were subordinated to such interests, they fell short of goals of material conservation and environmental prevention.

Today in Hungary, the commitment of industry to the citizens and communities that Pearce calls for continues to be absent because the so-called 'transition to market and democracy' has been understood primarily as democracy for the market. Even approaches consistent with industrial ecology and ecological modernisation, which require a certain amount of cooperation between state and industry, are viewed with suspicion. As Bochniarz and Kerekes put it:

> there is often very strong opposition against the incorporation of new environmental and economic tools, as a means of State interventions into a newly designed economic structure. A demagogic argument about returning to the communist means of State regulation, is often used in connection with this [*Bochniarz and Kerekes, 1994: 20*].

As long as Western industrial ecology is fraught with the same silences as the state socialist versions of industrial ecology was, the goal of catching up with the West's environmental discourse will be aiming too low for Hungary. Rather, a 'third road' might be the most practical and most desirable. This road should be built on a well-informed and democratic selection from the elements that worked in state socialism's production-centred approach to environmental protection, from the technical and economic tools of western industrial ecology, and on the above-mentioned extended notion of 'political citizenship'. Ecological modernisation theory must remain 'reflexive', and sensitive to global inequalities that could allow environmentally sustainable development in the West while post-socialist Hungary is being hijacked off the same road. Perhaps it is still not too late.

## NOTES

1. I discussed some of the findings of this research in Gille [*1999*].
2. These indicators support only one definition of wastefulness, more akin to economic notions of efficiency: waste output per unit of product. Wastefulness as a measure of wasteful consumption, illustrated by waste/capita indexes, is never even mentioned in these accounts, perhaps because if one were to interpret wastefulness this way, some of the most developed countries would be found actually exceeding most socialist countries [*Gille, 1999*].
3. These data do not include red mud. By comparison, OECD countries on the average recovered ten per cent of their industrial hazardous wastes in 1990. It is true, however, that Hungary's definition of hazardous wastes then was more inclusive than the OECD's today.
4. For detailed discussions and evaluations of the reform economists' approach to environmental problems, see DeBardeleben [*1985*] and Gille [*1997*].
5. Although in one year (1991), 20 per cent of the costs of environmental investments (not necessarily waste-related ones) could be retained [*Lehoczki and Balogh, 1997*].
6. According to a high-ranking official in the Environmental Ministry, the same obstruction can be observed now in the process of drafting Hungary's comprehensive waste management law.
7. The stakes involved in privatising and restructuring environmental protection are analysed by Andrews [*1993*], using the examples of the Czech Republic and Slovakia.
8. Product charges on transportation fuel had been in effect since 1992 [*Lehoczki and Balogh, 1997*].
9. A former ministry official told me that this tool was a Hungarian 'invention' and that Germany is only now implementing a similar tool.
10. Such as waste sorting and waste storage capacities as well as technologies for the reuse of certain wastes.
11. Laws that try to reduce the amount of time wastes spend in the plants and temporary storage are helpless against this trend. Enforcement and state-financed sites operating as public utility companies might help temporarily, but the ultimate goal would be to encourage source reduction.
12. My calculations, based on Lehoczki and Balogh [*1997: 144*].
13. The market for environmental services in Eastern Europe and the former Soviet Union was expected to double in the first half of the 1990s [*Trumbull, 1994: 8*].
14. My calculation from Greenpeace data [*Gluszynski and Kruszewska, 1996*].
15. In addition, in Hungary the only modern waste incinerator built before 1989 was sold to a French company. The country's first modern hazardous waste dump met a similar fate.
16. The term of taut planning has been used in the economic literature to signal the practice prevalent in state socialism in which enterprises were allocated only the exact volume of inputs that planners thought sufficient for the fulfilment of plans.

## REFERENCES

Állami Vagyonügynökség (State Privatisation Agency) (1993), 'A környezetvédelmi követelmények érvényesüléséröl az állami vállalatok privatizációja során' (On the Fulfilment of Environmental Requirements during the Privatisation of State Enterprises), *Környezet és Fejlödés*, Vol.5, No.4, pp.31–4.

Allenby, B.R. and D.J. Richards (eds.) (1994), *The Greening of Industrial Eco-Systems*, Washington, DC: National Academy Press.

Andrews, Clinton, Frans Berkhout and Valerie Thomas (1994), 'The Industrial Ecology Agenda', in Socolow, Andrews, Berkhout, and Thomas [*1994: 469–77*].

Andrews, Richard N.L. (1993), 'Environmental Policy in the Czech and Slovak Republic', in A. Vári and P. Tamás (eds.), *Environment and Democratic Transition: Policy and Politics in Central and Eastern Europe* (Boston: Kluwer Academic Publishers), pp.5–48.

Árvai, József (1990), *Hulladékgazdálkodás* (Waste Management), Budapest: Budapesti Müszaki Egyetem, Mérnöktovábbképzö Intézet.

Bakonyi, Árpád (1991), 'Hulladékok és másodnyersanyagok hasznosítása a gazdaságban' (The utilisation of wastes and secondary raw materials in the economy), *Anyaggazdálkodás és Raktárgazdálkodás*, Vol.XIX, No.1, pp.9–11.

Bánhidy, János (1996), 'Letter to Kukabúvár', *Kukabúvár*, Vol.2, No.2, p.15.

Bochniarz, Zbigniew and Sándor Kerekes (1994), 'Deficiencies in the Existing System of Environmental Protection in Hungary', in Z. Bochniarz, R. Bolan, S. Kerekes, and J. Kindler (eds.), *Designing Institutions for Sustainable Development in Hungary: Agenda for the Future*, (Budapest: Környezettudományi Központ), pp.11–22.

Bochniarz, Z., Bolan, R., Kerekes, S. and J. Kindler (eds.) (1994), *Designing Institutions for Sustainable Development in Hungary: Agenda for the Future*, Budapest: Környezettudományi Központ.

Bödecs, Barnabás (1996), 'szeméthegyen innen, termékdíjon túl' (Before Waste Piles, Over Product Charges), *Kukabúvár Melléklet önkormányzatok számára* (Supplement for local governments), Vol.2, No.2, pp.vii–ix.

Bödecs, Barnabás (1997), 'A Hulladek Munkaszovetseg javaslatai a hulladéktörvény szakmai koncepciójához' (The Proposals of the Waste Work Association on the Concept of the Waste Management Law), *Kukabúvár*, Vol.3, No.3 [Online.] http:\www.zpok.hu/kukabuvar/.

Bóka, István (1953), 'A Gazda mozgalom szervezési kérdései' (The Organisational Questions of the Gazda Movement), *Többtermelés*, Vol.7, No.3, pp.24–7.

Budapesti Vegyimüvek (Budapest Chemical Works) (1972), *Emlékeztetö 1972. augusztus 7. operativ vállalatvezetöi értekezlet* (Minutes, Operative Meeting of Management, 7 Aug. 1972.)

Budapesti Vegyimüvek (Budapest Chemical Works) (1975), *Emlékeztetö 1975. július 14. operativ vállalatvezetöi értekezlet* (Minutes, Operative Meeting of Management, 14 July 1975).

Budapesti Vegyimüvek (Budapest Chemical Works) (1978), *Emlékeztetö 1978. január 23. operativ vállalatvezetöi értekezlet* (Minutes, Operative Meeting of Management, 23 Jan. 1978).

Castoriadis, Cornelius (1978-79), 'The Social Regime in Russia', *Telos*, No.38, pp.32–47.

Csupor, Lajos (1954), 'Anyagtakarékossági mozgalmak és feladataink' (Movements of Material Conservation and Our Tasks), *Többtermelés* Vol.8, No.10, pp.22–5.

DeBardeleben, Joan (1985), *The Environment and Marxism-Leninism: The Soviet and East German Experience*, Boulder, CO: Westview Press.

Dömötör, Ákos (1980), *A MÉH nyersanyaghasznositó tröszt története (1950–1980)* (The History of the MÉH Trust [1950–1980]), Budapest, unpublished ms.

Dryzek, John S. (1997), *The Politics of the Earth: Environmental Discourses*, Oxford: Oxford University Press.

Dworák, József (1991), 'Magyarország hulladékgazdálkodása' (Hungary's Waste Management), *Anyaggazdálkodás és Raktárgazdálkodás*, Vol.XIX, No.2, pp.20–26.

Ember, György (1952), *Gazda-mozgalom a vasiparban* (The Gazda Movement in the Iron Industry), Budapest: Népszava, Szakszervezetek Országos Tanácsa Lap- és Könyvkiadó Vállalata.

Filtzer, Donald (1986), *Soviet Workers and Stalinist Industrialization: The Formation of Modern Soviet Production Relations*, London: Pluto Press.

Filtzer, Donald (1992), *Soviet Workers and De-Stalinization: The Consolidation of the Modern System of Soviet Production Relations 1953–1964*, Cambridge: Cambridge University Press.

Fövárosi Tanács Végrehajtó Bizottsága (Executive Committee of the Council of Budapest) (1967), *Jegyzökönyv az 1967. November 22-i ülésröl* (Minutes from the meeting on 22 Nov. 1967).

Gille, Zsuzsa (1997), 'Two Pairs of Women's Boots for a Hectare of Land: Nature and the Construction of the Environmental Problem in State Socialism', *Capitalism, Nature, Socialism*, Vol.8, No.4, pp.1–21.

Gille, Zsuzsa (1999), 'Conceptions of Waste and the Production of Wastelands: Hungary since 1948', in W. Goldfrank, D. Goodman and A. Szasz (eds.), *Environmental Issues and World-System Analysis*, Westport, CT: Greenwood Press.

Gluszynski, Pawel and Iza Kruszewska (1996), *Western Pyromania Moves East: A Case Study in Hazardous Technology Transfer*, Greenpeace, URL: http://www.rec.hu/poland/wpa/pyro-toc.htm.

Goldman, Marshall I. (1972), *The Spoils of Progress: Environmental Pollution in the Soviet Union*, Cambridge: Massachusetts Institute of Technology.

Gomulka , Stanislaw and Jacek Rostowski (1988), 'An International Comparison of Material Intensity', *Journal of Comparative Economics*, No.12, pp.475–501.

Hajer, Maarten (1995), *The Politics of Environmental Discourse: Ecological Modernization and the Policy Process*, Oxford: Clarendon Press.

Iron Workers' Union (1951), *A Gazda-mozgalom a vasiparban* (The Gazda Movement in the Iron Industry), Budapest: Népszava.

Jávor, Andor (1953), 'Az 'Egyéni Megtakarítási Számla-mozgalom' fejlödése' (The Development of the Individual Savings Account-Movement), *Többtermelés*, Vol.7, No.4, pp.28–32.

Jócsik, Lajos (1977), *Egy ország a csillagon* (A Country on the Star), Budapest: Szépirodalmi Könyvkiadó.

Juhász, Ádám (1981), 'Az anyagtakarékosság lehetöségei és feladatai az iparban – Tézisekk' (The Potentials and Tasks of Material Savings – Theses), in *Az információ 1981. évi helyzete Az anyaggazdálkodás helyzete és fejlesztésének irányai* (The Situation of Information in 1981: The Situation and Directions of Development of Material Savings) (The abbreviated version of talks given at the ninth itinerary congress, Szolnok, Hungary).

Kaderják, Péter (1997), 'Economics for Environmental Policy in the Central Eastern European Transformation: How Are the Context and Textbook Prescriptions Related?', pp.157–76.

Káderják, Péter and Ágnes Csermely (1997), 'Direct Impacts of Industrial Restructuring on Air Pollutant and Hazardous Waste Emissions in Hungary', in P. Kaderják and J. Powell (eds.), *Economics for Environmental Policy in Transition Economies: An Analysis of the Hungarian Experience*, Cheltenham: Edward Elgar, pp.15–38.

Kindler, József (1994), 'Evaluation of Economic, Social, and Political Preconditions for a Successful Implementation of the Institutional Reform', in Bochniarz, Bolan, Kerekes and Kindler [*1994: 119–51*].

Klarer, Jürg and Patrick Francis (1997), 'Regional Overview', in Klarer and Moldan [*1997: 1–66*].

Klarer, J. and B. Moldan (eds.), (1997), *The Environmental Challenge for Central European Economies in Transition*, Chicester: John Wiley.

Klimisch, Richard L. (1994), 'Designing the Modern Automobile for Recycling', in Allenby and Richards [*1994: 165–70*].

Kornai, János (1980), *Economics of Shortage*, Amsterdam: North-Holland.

KSH (Központi Statisztikai Hivatal – Central Bureau of Statistics) (1988), *Központi fejlesztési programok. A melléktermék- és hulladékhasznosítási program 1987. évi eredményei* (Central Development Programs. The By-product and Waste Reuse Program), Budapest: Központi Statisztikai Hivatal.

Leather Workers' Union (1951), *A Gazda-mozgalom elmélyítésével segítsük elö ötéves tervünk sikerét* (Let's Foment the Success of our Five-year Plan by Deepening the Gazda Movement), Budapest: Egyetemi Nyomda.

Lehoczki, Zsuzsa and Zsuzsanna Balogh (1997), 'Hungary', in Klarer and Moldan [*1997: 131–92*].

Lelkes, Gábor (1950), 'Az anyagtakarékosságról' (About Material Conservation), *Többtermelés*, Vol.4, No.10, pp.15–19.

Mándi, Péter (1951), 'A pártszervezetek feladatai az anyagtakarékosság terén' (The Tasks of the Party Organs on the Field of Saving Materials), *Pártépítés*, Vol.4, p.15.

Manser, Roger (1993), *The Squandered Dividend: The Free Market and the Environment in Eastern Europe*, London: Earthscan.

Mol, Arthur P. J. (1995), *The Refinement of Production: Ecological Modernization Theory and the Chemical Industry*, Utrecht: Van Arkel.

Moroney, John. R. (1990), 'Energy Consumption, Capital and Real Output: A Comparison of Market and Planned Economies', *Journal of Comparative Economics*, Vol.14, No.2, pp.199–220.

Nelson, Kenneth (1994), 'Finding and Implementing Projects that Reduce Waste', in Socolow, Andrews, Berkhout and Thomas [*1994: 71–382*].

Nove, Alec (1980), *The Soviet Economic System*, London: George Allen & Unwin.

Országos Tervhivatal (Central Planning Office) (1960), *Tájékoztató [a KGST-nek] a Magyar Népköztársaságban az ócska színesfém és hulladékgyüjtés megszervezésének megjavításával és a másodlagos színesfémkohászattal kapcsolatos kérdésekröl* (Report [to the COMECON] on the issues of organising and improving the collection of scrap non-ferrous metals and of the metallurgy of scrap non-ferrous metals in the Hungarian People's Republic), 19 Sept. 1960. XIX-A-16-u. Box 32/a. Topic 5. Document Number 2-00188/1960.

Pál, Károly (1953), 'Takarékoskodjunk az anyaggal' (Let's Economise Material Use), *Többtermelés*, Vol.7, No.4, pp.18–22.

Paton, Bruce (1994), 'Design for Environment: A Management Perspective', in Socolow, Andrews, Berkhout and Thomas [*1994: 349–57*].

Pearce, Frank (1990), '"Responsible Corporations" and Regulatory Agencies', *Political Quarterly*, Vol.61, No.4, pp.415–30.

Reiniger, Róbert (1991), 'Veszélyes hulladékok' (Hazardous Wastes), *Anyaggazdálkodás Raktárgazdálkodás*, Vol.XIX, No.6, pp.1–7 and Vol.XIX, No.7, pp.14–19.

Remetei, Ferencné (1986), 'A hulladék és másodnyersanyagok programjának eredményei és tapasztalatai 1981–1985' (The Results and Experience of the Waste and Secondary Raw Materials Programme 1981–1985), *Ipari és Építöipari Statisztikai Értesítö*, Vol.1986, Dec., pp.465–81.

Romhányi, Gábor (1995), Interview, Budapest.

Sajó, András (1994), 'Legal Aspects of Environmental Protection in Hungary: Some Experiences of a Draftsman', in Bochniarz, Bolan, Kerekes and Kindler [*1994: 35–44*].

Sarlós, Mihály (1952), *A vashulladék begyüjtés kérdései* (Issues of Iron Waste Collection) (Report for internal use of the National Planning Bureau). XIX-A-16a OT box 446, main group number 25, no file number, Új Magyar Központi Levéltár.

Schwartz, John, Koehl, Carla and Karen Breslau (1990), 'Cleaning Up by Cleaning Up', *Newsweek*, 11 June, pp.40–41.

Sekutowski, Janine C. (1994), 'Greening the Telephone: A Case Study', in Allenby and Richards [*1994: 171–7*].

Simai, Mihály (1990), 'Környezetbarát fejlödésünk' (Our Environment-Friendly Development), *Valóság* Vol.1990, No.9, pp.1–10.

Society for Scientific Education (1980), *Környezetvédelmi elöadói segédanyag* (Textbook for Environmental Protection), Budapest: Tudományos Ismeretterjesztö Társulat.

Socolow, Robert (1994), 'Six Perspectives from Industrial Ecology', in Socolow, Andrews, Berkhout and Thomas [*1994: 3-16*].

Socolow, Robert, Andrews, C., Berkhout, F. and V. Thomas (eds.) (1994), *Industrial Ecology and Global Change*, Cambridge: Cambridge University Press.

Spaargaren, Gert and Arthur P.J. Mol (1992), 'Sociology, Environment, and Modernity: Ecological Modernisation as a Theory of Social Change', *Society and Natural Resources*, No.5, pp.323–44.

Székely, Attila (1991), 'Az ipari hulladékgazdálkodás helyzete és problémái' (The State and Problems of Industrial Waste Management), *Anyaggazdálkodás és Raktárgazdálkodás*, Vol.XIX, No.12, pp.22–7.

Szlávik, János (1991), 'Piacosítható-e a környezetvédelem?' (Is Environmental Protection Marketisable?), *Valóság*, Vol.34, No.4, pp.20–27.

Takáts, Attila (1990), 'A hulladékok káros hatása elleni védelem jogi, müszaki és gazdasági szabályozása' (The Legal, Technical and Economic Regulation of Protection Against the Harmful Effects of Wastes), in J. Árvai (ed.), *Hulladékgazdálkodás* (Waste Management), Budapest: Budapesti Müszaki Egyetem, Mérnöktovábbképzö Intézet, pp.140–45.

Ticktin, Hillel (1976), 'The Contradictions of Soviet Society and Professor Bettelheim', *Critique*, Vol.6, Spring, pp.17–44.

Ticktin, Hillel (1992), *Origins of the Crisis in the USSR: Essays on the Political Economy of a Disintegrating System*, Armonk, NY: M.E. Sharpe.

Trumbull, Mark (1994), 'Global Cleanup Spots Find Revenue-Hungry US Firms at the Door', *The Christian Science Monitor*, Vol.86, 24 March, p.8.

Valkó, Márton (1951), 'Gazda-mozgalmi és újító hetek a Rákosi Mátyás Müvekben' (Gazda Movement and Innovation Weeks of in the Mátyás Rákosi Works), *Többtermelés*, Vol.5, No.12., pp.23–4.

White, Robert M. (1994), 'Preface', in Allenby and Richards [*1994: v–vi*].

*Developing Countries*

# Contradictions of Ecological Modernisation: Pulp and Paper Manufacturing in South-east Asia

DAVID A. SONNENFELD

A growing literature suggests we have entered an era of paradigmatic institutional change permitting a broad environmental transformation of society, including fundamental restructuring of manufacturing. To date, ecological modernisation theory has primarily examined advanced industrial societies. Few studies have taken into account its applicability in newly industrialising countries (NICs). This study does so, drawing on the author's research on South-east Asian pulp and paper industries in the late 1980s and early 1990s. Those industries improved efficiencies, reduced waste, and progressed towards clean production. At the same time, they failed to meet an important criterion of ecological modernisation, the dematerialisation of production. Rather, 'supermaterialisation' of pulp and paper production in South-east Asia may be helping facilitate dematerialisation in the North. The paper concludes that dynamics of ecological modernisation may be present in large-scale, export-oriented modern sectors of NIC economies, while being more problematic in small and medium-sized enterprises (SMEs). Such dynamics present serious challenges for ecological modernisation theory.

## Introduction

'Ecological modernisation' is a relatively new concept, coined in the early 1980s by German sociologist Joseph Huber [*Spaargaren and Mol, 1992;*

An earlier version of this contribution was presented at the annual meeting of the American Sociological Association, San Francisco, California, August 1998. The author wishes to thank Arthur Mol, Rajah Rasiah, Paul Gellert, Lucy Jarosz, and reviewers for *Environmental Politics* for helpful comments on earlier drafts. Field research upon which this study is based was supported by the Australian-American Educational Foundation (Fulbright Commission); University of California Institute on Global Conflict and Cooperation; Switzer Foundation; Australian National University; and University of California, Santa Cruz. The Center for Environmental Technology and Management (CENTEMA), Van Lang University, Vietnam; Chulalongkorn University Social Research Institute (CUSRI), Thailand; Washington State University; and University of California, Berkeley, provided conducive environments for completion of this paper. The views presented herein are the author's alone.

*Mol, 1995; Hajer, 1995*]. In simple form, it might be thought of as industrial restructuring with a green twist. Poetically, Huber suggests that 'The dirty and ugly industrial caterpillar transforms into an ecological butterfly' (Huber [*1985*] cited in Spaargaren and Mol [*1992: 334*]). Hajer [*1995*] defines ecological modernisation theory as 'the discourse that recognises the structural character of the environmental problematique but ... assumes that existing institutions can internalise the care for the environment'.

As a theory of social change, ecological modernisation suggests we have entered a new industrial revolution, one of radical restructuring of production, consumption, state practices, and political discourses along ecological lines [*Mol, 1995; Hajer, 1995;* also Mol and Spaargaren, in this volume]. As a 'normative theory' or 'political program' [*Spaargaren and Mol, 1992; Mol, 1995*], ecological modernisation advocates resolving environmental problems through 'harmonizing ecology and economy' [*Simonis, 1989*], and through 'superindustrialisation' rather than de-industrialisation [*Spaargaren and Mol, 1992*].

To date, ecological modernisation theory has been considered applicable primarily for advanced industrial countries, due to prerequisites for green industrial restructuring, 'e.g. the existence of a welfare state, advanced technological development ... a state regulated market economy ... and ... widespread environmental consciousness' [*Mol, 1995: 54*]. Ecological modernisation theory may also be 'increasingly relevant for Newly Industrialising Countries' [*ibid.*: 55]. The first empirical studies addressing this question are only now beginning to show results [cf. *Frijns et al., 1997; Mol and Frijns, 1999; Hengel, 1998*].

Building on a study of Indonesian, Malaysian and Thai pulp and paper firms' adoption of environmental technologies in the early 1990s,[1] this study aims to contribute to a discussion of the applicability of ecological modernisation theory to newly industrialising countries (NICs). Concerns are raised for scholarship on ecological modernisation especially in regard to dematerialisation, North–South equity, and the role of small- and medium-sized enterprises (SMEs).

## Operationalisation

A first step in examining the applicability of ecological modernisation theory to NICs is formally to define and operationalise the concept of ecological modernisation. Various authors have made such an effort. Early scholars focused on actual improvements in environmental performance. Paulus [*1986*], cited in Simonis [*1989: 347*], for example, suggests that 'Ecological modernization focuses on prevention, on innovation and structural change towards ecologically sustainable development ... It relies on clean technology, recycling and renewable resources ... .'

Jänicke *et al.* [*1989: 100*] suggest an important premise of ecological modernisation: 'that a reduction in the resource input of production will lead to a reduction in the amount of emissions and waste and also the costs of production'. Hajer [*1995: 25–6*] suggests three basic concepts emerging out of early work on ecological modernisation:

- 'Make environmental degradation calculable' [especially monetarily];
- 'Environmental protection is ... a "positive-sum game"'; and
- 'Economic growth and the resolution of ecological problems can, in principle, be reconciled'.

From these and other works, it might be suggested that ecological modernisation has three immediate and two ultimate technological/material objectives: in the short term, *waste reduction and elimination*, *resource recovery and reuse*, and *dematerialisation*;[2] in the long term, *resource conservation* and *clean production*.[3]

Later scholars have focused on mechanisms and broader social dynamics of ecological modernisation. Mol [*1995: 39*], for example, suggests that ecological modernisation's material objectives obtain both through '*economising the ecology*', using vehicles such as monetary valuation of natural resources, levying of environmental taxes, and establishment of market incentives; and '*ecologizing the economy*', via re-engineering production, improving industrial co-processing ('industrial ecology'), and boosting superindustrialisation.[4]

Mol goes on [*1995: 58*; also Mol and Spaargaren in this volume] to examine institutional and social dimensions of environmental transformation, arguing that in ecological modernisation: the state shifts from top-down regulatory intervention to negotiation with industry; non-governmental organisations play a key role, including through direct interaction with industry; political and economic globalisation are supportive; and 'counterproductivity' ('small is beautiful', co-ops, etc.) policy approaches are eclipsed.

Together, such technological/material objectives, mechanisms and institutional dimensions of ecological modernisation provide a useful starting point for evaluating the applicability of the theory to particular cases, including NICs. This analysis addresses two questions: to what extent has pulp and paper manufacturing in Indonesia, Malaysia and Thailand been 'ecologically modernised' in recent years? How have these changes taken place?

## Environmental Reform

One of the world's ten largest industries, pulp and paper manufacturing has received world-wide attention for environmental pollution for decades. In the 1970s, much public and regulatory attention focused on air quality, including the industry's 'rotten egg' (sulphur dioxide [SO2]) smell. More recently, the industry has received attention for use of chlorine in pulping and bleaching, especially of wood. In the mid-1980s, chlorine used in paper production was positively identified as associated with the production of dioxin, a highly toxic chemical. Regulatory attention shifted to the presence of this chemical in wastewater streams and to water quality more generally. Beginning in the late 1980s, Greenpeace launched an international campaign for the elimination of chlorine in pulp and paper manufacturing.

In South-east Asia, prior to the late 1960s, most pulp and paper manufacturing was very small scale, producing printing and writing paper, newsprint, cigarette paper, and packaging materials, largely from agricultural wastes (bagasse, a residue of sugar cane processing; rice stalks) and recycled paper (especially old corrugated containers [OCC], often imported from East Asia and even North America) [*Sonnenfeld, 1998c*]. Pulp and paper manufacturing expanded substantially in South-east Asia beginning in the late 1960s, an integral part of early industrial development in the region. Pollution became a major concern only with large-scale expansion of pulp manufacturing in South-east Asia beginning in the mid-1980s.

Protests in South-east Asia over pulp industry expansion and pollution in the late 1980s and early 1990s [*Sonnenfeld, 1998a, 1998b*] had a major impact however, leading to the establishment of new environment/technology regimes influencing adoption of newer, cleaner process technologies in the pulp and paper industry throughout the region [*Sonnenfeld, 1996*]. I focus here especially on adoption of advanced elementally chlorine-free (ECF) pulping and bleaching technologies. Such developments may be analysed in terms of the actual technological/material improvements, and the social dynamics forcing those improvements in each country.

### *Technological Improvements*

Many environmental reforms in process and end-of-pipe technology were adopted by Indonesian, Malaysian and Thai pulp and paper firms, c. 1987–96, especially in the large-scale, export-oriented segments of the industry. During this period, new mills were built and brought on-line in South-East Asia utilising some of the most advanced technologies in the world, and some older mills were modified. The most acute problems remain with the oldest, smallest facilities, some government owned.

Perhaps most dramatic was the construction of six new bleached kraft pulp mills using elementally chlorine-free (ECF) pulping and bleaching technologies in Indonesia and Thailand from 1992–96 (see Table 1). These mills operate more efficiently and with less pollution per unit of output than mills in many advanced countries [*Sonnenfeld, 1998b*]. During this period, no bleached kraft pulp mills were built without ECF technologies in these countries and Malaysia.

During the same period, companies improved environmental performance at older pulp mills in South-East Asia as well, through changing raw materials, upgrading pre-processing, modifying process technology, and improving waste treatment. Highlights include:

- Older Indonesian pulp mills added pre-treatment processes, reduced use of elemental chlorine, and improved wastewater treatment.
- When it was built in the late 1980s, Malaysia's only bleached kraft mill incorporated world-class technology including oxygen-activated sludge treatment. The mill progressively decreased use of elemental chlorine through the early 1990s.
- Thailand's first large-scale export pulp mill added oxygen delignification and upgraded wastewater treatment facilities. The country's largest pulp and paper group installed advanced wastewater treatment systems, and experimented with enzymes and bacterial pre-treatment for reducing elemental chlorine use.

The worst environmental problems exist among the smallest and oldest pulp and paper mills in the region, some government-owned. Efficiencies as well as pollution would dictate the closing of many of these operations. However, employment issues in high-unemployment economies have proved even more acute [cf. *Hanafi, 1994*]. Even in this troubled segment, national and international research and assistance programmes worked to make environmental improvements.

How did the new mills come to incorporate the latest technologies? What motivated some of the older mills to modify their process and end-of-pipe technologies?

*Social Dynamics*

Technological/material improvements in the manufacturing of pulp and paper in South-east Asia were the result of multiple social dynamics. At the local and country levels, community and environmental groups brought attention to companies' social and environmental practices; government agencies encouraged and sometimes forced companies to adopt cleaner production technologies; public and private sector research engineers

TABLE 1

CORRELATES OF ENVIRONMENTAL IMPROVEMENTS AT BLEACHED PULP MILLS IN INDONESIA, MALAYSIA, AND THAILAND, c. 1987–96[5]

| *opns date* | *manufacturer* | *country* | *env. imp.* | *begin date* | *pre-proc.* | *upg. wwt* | *reduce* $Cl_2$ | *ECF +* | *kraft* | *high capac.* | *strong regs.* | *strong mvts.* | *Nordic conn.* | *minority-owned* | *export-orient.* |
|---|---|---|---|---|---|---|---|---|---|---|---|---|---|---|---|
| 1968 | Siam Pulp & Paper | Thailand | Y | ? | Y | ? | ? | N | N | N | N | N | N | N | N |
| 1981 | Phoenix I | Thailand | Y | 1989 | Y | Y | Y | N | Y | N | Y | Y | Y | Y | Y |
| 1984 | Indah Kiat - Pulp Mill #1 | Indonesia | Y | 1987 | Y | Y | Y | N | Y | Y | N | Y | N | Y | Y |
| 1988 | Sabah Forest Industries | Malaysia | Y | 1989 | N | N | Y | N | Y | N | N | N | Y | Y | Y |
| 1988 | Inti Indorayon Utama | Indonesia | Y | 1993 | ? | Y | - | Y | N | Y | N | Y | Y | Y | Y |
| 1991 | Indah Kiat - Pulp Mill #2 | Indonesia | Y | 1992 | Y | Y | Y | N | Y | Y | N | Y | Y | Y | Y |
| 1993 | Siam Cellulose | Thailand | Y | 1993 | Y | Y | ? | N | Y | N | N | N | Y | N | N |
| 1994 | Phoenix II | Thailand | – | – | – | – | – | Y | Y | N | Y | Y | Y | Y | Y |
| 1994 | Indah Kiat – Pulp Mill #8 | Indonesia | – | – | – | – | – | Y | Y | Y | N | Y | Y | Y | Y |
| 1994 | Wira Karya Sakti | Indonesia | – | – | – | – | – | Y | Y | Y | N | Y | Y | Y | Y |
| 1995 | Riau Andalan Pulp & Paper | Indonesia | – | – | – | – | – | Y | Y | Y | N | Y | Y | Y | Y |
| 1996 | Advance Agro | Thailand | – | – | – | – | – | Y | Y | Y | ? | Y | Y | Y | Y |
| 1997 | Kiani Kertas | Indonesia | – | – | – | – | – | Y | Y | Y | N | Y | Y | Y | Y |

KEY:

opns date = date pulp operations started
env. imp. = implemented environmental improvements
beg. date = date environmental improvements began
pre-proc. = added environmental pre-processing improvements
upg. wwt = upgraded wastewater treatment
reduce $Cl_2$ = reduced use of elemental chlorine
ECF + = pulp produced without use of elemental (or any) chlorine
kraft = uses kraft chemical pulping process
high capac. = greater than 100,000 admt*/year of bleached pulp
*admt = air-dried metric ton
strong regs. = strong governmental environmental regulations/enforcement
strong mvts. = strong community, environmental, or other social movements
Nordic conn. = relationship with Pöyry, Nordic firms, or Nordic aid agencies
Nordic aid = recipient of Finnish or Swedish foreign aid, credits, loans, etc.
minority owned = mill owned by ethnic/national minorities, or expatriates
export-oriented = mill designed for production/sales to global markets

*Source*: Sonnenfeld [*1996*].

creatively developed process modifications; and local environmental conditions came into play. These processes played out in both similar and dissimilar ways in each of the countries studied. In addition, global and regional dynamics contributed to the ecological modernisation of pulp production in South-east Asia.

*Indonesia:* As an important part of the country's export-oriented industrialisation strategy, Indonesia's pulp and paper industry grew rapidly in the late 1980s and early 1990s. New greenfield[6] mills were built on the 'outer' Indonesian islands of Sumatra and Borneo in locales with long traditions of subsistence agriculture. Pulpwood plantations, established by industrial timber estate concession holders as a precondition to government licensing of new mill projects, spread over hundreds of thousands of hectares, disrupted the lives of tens of thousands of people. Pulp mills piped their liquid wastes into rivers used for drinking, bathing, fishing, washing, irrigation, as well as by other industries. The adoption of environmental technologies in Indonesia's pulp industry was directly linked to popular protest against impacts of rural industrial development.

Minority residents in North Sumatra have been engaged in conflict for years with the Raja Garuda Mas (RGM) group, owners of PT Inti Indorayon Utama, Indonesia's first new export-oriented pulp mill, built with substantial government support. What began as an ethnically-charged local struggle over land and forest tenure became a conflict of national and international significance with the involvement of environmental groups and organisations, including the umbrella group, Indonesian Forum on the Environment (WALHI). A boiler explosion at Indorayon in November 1993, led to demonstrations in Medan and elsewhere, and a government statement that henceforth all new pulp mills built in Indonesia would have to be 'ECF or better'.[7]

Under the Suharto government,[8] while having limited success in getting RGM to improve operations at Indorayon, activists' campaigns against the company helped advance the adoption of green technology elsewhere in Indonesia, including RGM's next new pulp mill, PT Riau Andalan, built in the early 1990s in east central Sumatra. PT Riau Andalan was designed and built by the Jaakko Pöyry group, from Finland, and incorporated the latest pulping, bleaching and wastewater technologies.

The Sinar Mas group, Indonesia's leading pulp and paper producers,[9] took advantage of cyclical downturn in the international economy and global pulp industry in the early 1990s to purchase advanced new technology at a substantial discount. Its subsidiary, PT Indah Kiat Pulp and Paper Co. (IKPP), also in east-central Sumatra, had been a target of community protest over land, forest, and water resources. IKPP, the

Indonesia Legal Aid Society (YLBH), WALHI, and BAPEDAL signed an historic memorandum of understanding in 1992, wherein the company agreed to clean up its operations and to assist development in surrounding communities [*Sonnenfeld, 1996: Appendix E*]. IKPP utilised in-house engineering expertise, consulting engineering services, and international aid to improve process technologies at its older mill. At the same time, it also adopted new, cleaner technologies at its new mill, in part to help gain access to green export markets.

A continuing problem area in Indonesia relates to older, smaller pulp and paper mills, many located in densely populated areas on the island of Java, home to the great majority of the country's 200 million people. In East Java, for example, the provincial government, working together with various international agencies in attempting to clean up heavily used and polluted waterways, sought to shut down a number of smaller (5,000 tons per year), older, relatively inefficient, polluting pulp and paper mills. It was unable to do so, however, at least in the present, due to very high levels of unemployment in the area, and the impact such mill closures would have on the local economy [*Hanafi, 1994*].

Environmental activists in Indonesia had a co-operative working relationship with Emil Salim, Indonesia's first Minister of Population and Environment, and first head of the Bureau of Environmental Impact Management (BAPEDAL). Activists' relations with BAPEDAL cooled in the mid-1990s, as the agency's second chief, Sarwono Kusumatmadja, steered the agency towards a more voluntary-compliance based relationship with industry.[10] Activists had a positive relationship with the staff of the government-run Institute for Research and Development of the Cellulose Industry (IRDCI), which works primarily with state-owned and smaller, older mills.

International consulting engineer and technology supply firms played an important role in the adoption of environmental technology in Indonesia's pulp and paper industry. The Jaakko Pöyry group has advised almost every new pulp mill project in Indonesia. Technology suppliers received strong backing from their home governments, and benefited from close working relationship with consultants. Some suppliers became joint venture partners with Indonesian producers.

At least a half-dozen bilateral aid programmes also contributed to the adoption of environmental technologies in Indonesia's pulp and paper industry. Canadian and Australian agencies supported development of national environmental regulations and administration in Indonesia. Austraid (Australia) and other agencies contributed to PROKASIH, the Clean Rivers Project, under which the province of East Java developed perhaps the toughest local water environmental standards in Indonesia. The

United States Agency for International Development (USAID) supported a clean technology assistance programme, which included participation by Indonesia's pulp and paper industry. Swedish and Japanese aid agencies supported environmental research at IRDCI.[11]

*Malaysia*: Sabah Forest Industries (SFI) was Malaysia's sole greenfield pulp mill during the period studied. Located in north-west Borneo, SFI is one of the most studied pulp mills in South-East Asia, with extensive baseline and follow-up environmental impact assessments conducted in association with the United Nations Environment Programme's (UNEP) Network on Industrial Environmental Management (NIEM) project [cf. *Murtedza and Landner, 1993*]. Principal concerns articulated in those studies related to effects of the mill's effluent on fisheries in Brunei Bay. Even more consequential than the mill's discharge, however, was heavy siltation in the Bay due to soil erosion from upland logging [*ibid.*]. It is not clear from Murtedza and Lander's study how much of that 'upland logging' was by SFI or its contractors.

During its start-up and initial years of operation, SFI was kept under strict environmental oversight by Malaysia's national government. SFI was required to conduct annual environmental audits and improve environmental performance. The close attention paid to SFI by Malaysia's national government may be partially explained by the mill's ownership by the state of Sabah. At the time, Sabah was one of only two states not led by the United Malays National Organisation (UMNO), Malaysia's ruling party.[12] SFI was caught in tension between local and national authorities on matters such as the granting of tariff protection and building of infrastructure.[13]

SFI used in-house research facilities to improve environmental performance, including reducing use of elemental chlorine and improving waste water processing. Technology and engineering supply firms, university researchers and the NIEM supplemented SFI's in-house efforts. The Swedish International Development Agency (SIDA) funded and provided training in Sweden for SFI engineers and machine operators.

Although there were no integrated pulp and paper mills in West Malaysia at the time of my fieldwork, there were a number of small and medium-sized paper mills utilising recycled paper as their primary raw material. Malaysia's weak environmental regulations governing the pulp and paper industry are meant at least in part to protect this small but economically important sector. In a recent white paper, for example, university researchers and government officials describe as 'valid' complaints from operators of small mills that they could not meet even existing weak environmental standards due to 'financial difficulty in installing efficient wastewater treatment plants as their production scales are too small' [*Murtedza et al., 1995: 12*].

*Thailand*: The pulp and paper industry in Thailand, too, has been at the centre of longstanding controversy. Small farmers, supported by NGOs, academics, urban professionals and the media, protested establishment of pulp plantations, industrial pollution, and loss of rural livelihood in Thailand. Two firms in the north-east region of the country received the greatest attention; one was perceived as foreign-owned, the other as minority-owned.

An industrial accident involving one of those firms had an important catalytic effect in public and governmental scrutiny of the industry. A massive fish kill in a nearby river was blamed on a mill operated by the Phoenix Pulp and Paper Co., joint venture of the 'European Overseas Development Corporation' (EODC)[14] and Ballarpur Industries (India). With an important election drawing near, the government responded to public demands for action, taking the unprecedented action of shutting down the Phoenix mill until it could upgrade its wastewater treatment system.

Phoenix management air-lifted a wastewater treatment system from Finland, got it operational in 30 days, and resumed operations. The company proceeded to add a second pulp line, supported with an interest-free Finnish loan. The new pulp mill incorporates the latest pulping and bleaching technology from Nordic supply firms. The construction project included upgraded pre-processing for the original mill.

With north-east Thailand beset by an extended drought, government agencies further required Phoenix to phase out all discharges into the nearby river. In response, the company developed a programme to use treated wastewater to irrigate pulpwood plantations in the vicinity of the mill. Optimistically, the company hoped to 'turn necessity into a virtue', by marketing its products to green markets internationally.[15]

The government's shutdown of the Phoenix mill rippled through the Thai pulp and paper industry. The Federation of Thai Industries' Pulp and Paper Club became involved in the USAID-sponsored Industrial Environmental Management (IEM) programme within weeks of the shutdown. Club members actively exchanged information about environmental technology, conducted environmental audits of each others' manufacturing facilities, and visited environmental technology suppliers in the USA.

The Soon Hua Seng group, a Sino-Thai business which had previously run into trouble establishing eucalyptus plantations, created a new subsidiary, Advance Agro Ltd., to develop a pulp mill, also in north-east Thailand. It hired Presko, a Thai–Finnish public relations firm, actively to monitor popular concern about the pulp industry in Thailand; and Jaakko

Pöyry as designer and general contractor for new mill. The mill, which began operations in 1995, was designed with ECF pulping and bleaching processes and advanced wastewater treatment facilities.[16] In building a 'green' mill, Soon Hua Seng hoped to forestall getting into political trouble as it had before.

The Siam Pulp and Paper group, division of partially crown-owned Siam Cement, was less frequently the target of public and political pressure to make environmental improvments to its operations. Nevertheless, the group took strong environmental leadership in its operations and in the Thai Pulp and Paper Industry Association. The group upgraded wastewater treatment facilities at its largest pulp and paper production site, and experimented with using enzyme and bacterial pre-processing to reduce chemical use at its pulp mills.[17]

In contrast to the experience of the new, large-scale pulp producers, pollution by smaller, older mills often 'slip through cracks' of regulatory enforcement in Thailand. An official of the Pollution Control Department told me, for example, that smaller pulp and paper facilities were not a high enforcement priority for his office, because there were relatively few such producers, and he had to deal with many far more pressing problems.

* * *

In both Indonesia and Thailand, one pulp mill's troubled beginnings set the tone for the environmental reform of an entire industry. While having different immediate outcomes, the cases set in motion similar country-level dynamics. These can be portrayed graphically as in Figure 1. Essentially, these dynamics entail a combination of social movement (NGO), state regulation, and business factors driving the innovation process at the national level.

The core dynamics of adoption of environmental technologies as depicted in Figure 1 include the original, 'landmark' conflict; the establishment of new standards/levels of expectation for industry environmental performance; the encouragement of both firm and supplier innovation; and implementation/adoption of the new, cleaner production technologies. Key participants in these processes are local community groups, domestic and international business interests, non-governmental organisations, regulatory agencies, bi- and multi-lateral aid agencies, and 'green' consumers. Particular local environmental conditions (such as the poor character of Phoenix's site) also are a factor.

In all three countries, there were markedly different dynamics in the large-scale, often export-oriented pulp industry segment as compared to the segment of smaller, older producers. Although the former received great

FIGURE 1

DYNAMICS OF ADOPTION OF ENVIRONMENTAL TECHNOLOGIES IN PULP AND PAPER MANUFACTURING IN INDONESIA AND THAILAND, c. 1987–94[18]

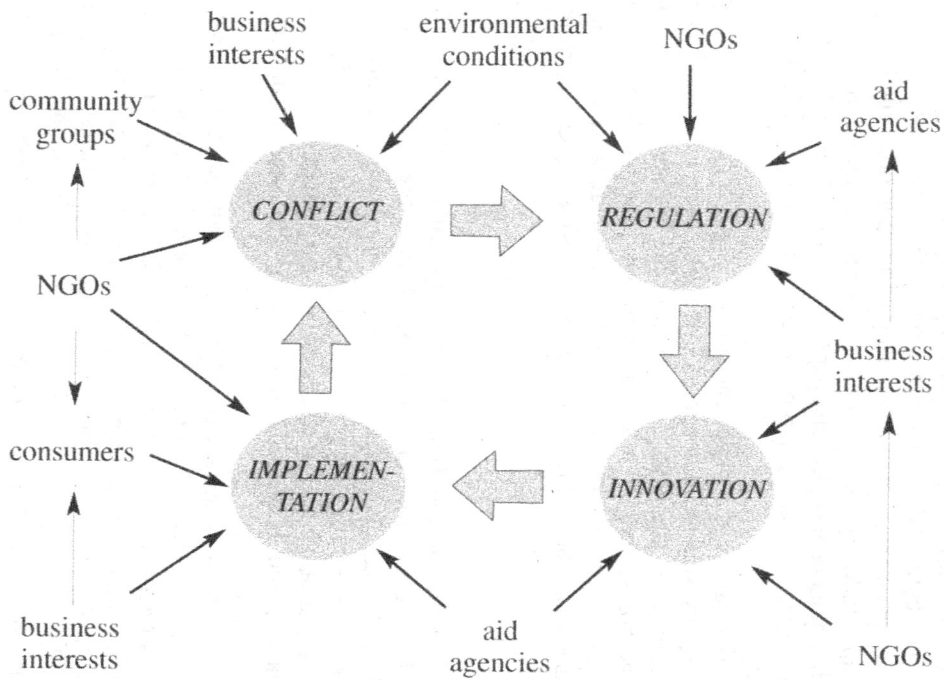


*Source*: Sonnenfeld [*1996*].

public and regulatory attention, the latter often 'slipped through the cracks' of both popular and regulatory oversight. Even when identified as causing serious pollution to public waterways, small and medium-sized pulp and paper mills typically were allowed to continue operating. In contrast, popular and government pressure was much greater on the larger, newer mills.

*Global and Regional Dynamics*

Global and regional institutions also contributed to environmental reform of (at least the large-scale, export oriented segment of) South-east Asian pulp manufacturing. Global market, finance, and technology considerations exerted influence on adoption of environmental technologies in the export-oriented, and financially dependent South-east Asian industries. International aid agencies and social movements played important roles as well.

On the one hand, various global factors motivated South-east Asian pulp producers to adopt the new, more environmental technologies. Environmental risk is factored in to the costs firms have to pay (in 'points' and interest) in world financial markets to issue corporate bonds. Maintaining access to global consumer markets also is important to export-oriented South-east Asian pulp producers. They pay attention to international eco-labelling efforts and aim to obtain eco-certification where possible. Phoenix, in Thailand, for example, was considering going after green markets in Japan and elsewhere with its non-wood (bamboo, kenaf), 'environmentally friendly' pulp.

On the other hand, global technology firms were highly motivated to sell their new technologies under favourable terms. When the new, ECF technologies were coming on the market in the early 1990s, Finland and Sweden were undergoing their worst economic (and political) crises in half a century; Europe and North America were in the midst of a major economic slump; and South-east Asia was one of the few regions of the world rapidly expanding. Technology supply firms and Nordic governments offered price discounts, trade credits, interest-free loans, and joint-venture partnerships to South-east Asian pulp producers to encourage them to adopt the new technologies.

Foreign consulting engineering firms played an active role in promoting adoption of environmental technology in Thailand's pulp and paper industry. Both new ECF mills in north-east Thailand were designed by Jaakko Pöyry and incorporate advanced, Nordic-sourced technology. H.A. Simons, a Canadian consulting engineering firm, was the major contractor for upgrading wastewater treatment facilities at the Siam Pulp and Paper group's mills.

International aid agencies, also, played an important role in the adoption of environmental technology in Thailand's pulp and paper industry. As noted above, the Finnish Agency for International Development (Finnaid) provided aid and trade credits for purchase of pulp manufacturing equipment, including at Phoenix Pulp and Paper in Thailand. USAID supported training in environmental management and waste minimisation in both Indonesia and Thailand.

Swedish assistance has been critical in carrying out the United Nations Environment Programme's Network on Industrial Environmental Management (NIEM). For more than seven years, the NIEM has conducted training, documentation, conferences, and workshops aimed at helping pulp industries in participating countries (China, India, Indonesia, Malaysia, the Philippines, Sri Lanka, Thailand, and Viet Nam) develop expertise and exchange information on environmental problems.[19]

International social movements played an indispensable role in the adoption of environmental technologies in South-east Asian pulp industries. The new, ECF technologies were developed in Sweden and Finland in response to European social movements' efforts to tighten environmental regulations and increase demand for green products. Transnational social movements helped disseminate environmental information to South-east Asian governmental, academic, and citizens' groups. Greenpeace, in particular, played a critical role in gathering information on the pulp industry and the environment, and making it available around the world. It consulted with pulp manufacturers, industry, governmental and university research institutes, regulatory agencies, and NGOs; and developed a good relationship with suppliers of environmental technology for the pulp industry, especially in Western Europe.

Expressed graphically, the global flow of environmental innovation in South-east Asian pulp manufacturing might look like Figure 2. While having important country-level characteristics due to unique cultural, political, economic, and environmental factors; there is also an essential global dimension of the diffusion of technologies, regulatory standards, and social movement influence in the pulp and paper industry. South-east Asia, as a late industrialising region, was affected by and able to take advantage of these dynamics [*Sonnenfeld, 1998b*].

Social movement action in North America and Europe galvanised development of the new, ECF technologies. New environment/technology regulatory regimes also diffused from North to South, both directly through North–South regulatory agency cooperation, and indirectly, through the establishment of global environmental norms and expectations, including among corporate and individual consumers. Australia is included in Figure 2 due to its geographical proximity to South-east Asia, and the direct

FIGURE 2

GLOBAL FLOW OF INNOVATION IN ENVIRONMENTAL PULPING AND BLEACHING TECHNOLOGIES, c. 1985–94[21]

| | | | |
|---|---|---|---|
| **N AMERICA** | Greenpeace USA; Greenpeace Canada | US & Canadian PPIs; US EPA | dioxin research; increasingly tight pulp mill regulation |
| **EUROPE** | Greenpeace Int'l | European PPI | Scandinavian co's develop new pulping & bleaching technologies |
| **AUSTRALIA** | Greenpeace Australia | Australian PPI; state & Common-wealth EPAs | scuttling of Wesley Vale project; NPMRP; local mill innovation |
| **SE ASIA** | community groups & NGOs | Indonesian & Thai PPIs; environmental ministries | new pulp mills built with ECF technology; what to do with old mills? |

*Source*: Sonnenfeld [*1996*].

involvements of Australian regulatory agencies, social movements, and industry consultants in South-east Asia. This characterisation is especially pertinent to the newest mills.

## Discussion

Based on this summary of research findings, to what extent can it be said that South-east Asian pulp and paper industries have 'ecologically modernised'? Let us return to the objectives, mechanisms, and institutional dimensions of ecological modernisation delineated above as a framework for addressing this question.

The first objective of ecological modernisation is *waste reduction and elimination*. South-east Asian pulp firms have made significant strides in reducing waste, at least in terms of the amount of waste per ton of product. The technologies utilised by the new pulp mills built and operating in South-east Asia are among the most efficient in the world. Modifications to reduce waste have been made to the previous generation of pulp mills as well. Where waste-reduction has least occurred is with the smaller, older pulp mills, some government-owned, which have been kept operating despite their inefficiencies and pollution. With regard to this criterion of ecological modernisation, South-east Asian pulp industries can be said to be well on the road to ecological modernisation, even while problems remain with small and medium-sized enterprises (SMEs).

The second objective of ecological modernisation is *resource recovery and reuse*. With regard to the pulp industry, several resources might be considered: water, chemicals, and fibrous raw materials. The new mills in South-east Asia show remarkable achievements in the reduction of the amount of water used per ton of pulp produced. Chemical recovery, as well, is highly advanced in the new mills. Historically, South-east Asian pulp industries were great *re-users* of fibre resources, utilising agricultural waste and wastepaper as raw materials. As the scale of production has increased, however, the industry has moved away from recycled inputs to greater reliance on virgin raw materials from native forests and tree plantations [*Sonnenfeld, 1998c*]. In sum, South-east Asian pulp manufacturers are making substantial progress in recovering/reusing water and chemicals; fibre is another story, discussed more below.

Perhaps the biggest Achilles heel of South-east Asian pulp producers with regards to ecological modernisation is the criterion of *dematerialisation*. In ecological modernisation, dematerialisation is achieved through the substitution of high-technology for raw material inputs, or alternatively, the substitution of recycled or recovered waste for virgin raw materials. As South-east Asia's pulp industries expand and modernise, less water and fewer chemicals are used and less waste is produced per ton of product. At the same time, producers have adopted wholesale the industrial model promoted by their Northern technology suppliers and funding sources, one based on large increases in both absolute and relative use of virgin raw materials.

Of greatest immediate concern are the proliferating greenfield pulp mills, especially in Indonesia, each of which must be 'fed' a tremendous volume of virgin fibre to maintain full production. Typically, native forests provide the first round of raw materials in the form of 'mixed tropical hardwood' (MTH) chips, while extensive fast-growing plantations have been established to address future raw material needs. Government policies which awarded very large forest concessions at little or no cost to well-connected companies and individuals only compounded and accelerated this dynamic.

*Resource conservation*, one of the long-term objectives of ecological modernisation, thus remains in the distant future of South-east Asian pulp industries. As with the shorter-term goals of waste reduction and resource recovery, South-east Asian pulp industries fare much better with regard to the second longer-term objective of *clean production*. Perhaps nowhere is the contradiction between these two objectives more evident than in South-east Asian pulp industries, where there is imaginably a future of clean production while resource conservation is nowhere in sight with regard to virgin fibrous materials.

Through what mechanisms have the industries' accomplishments been achieved? A number of *market factors* were significant in the adoption of environmental technologies in South-east Asia's pulp industries. South-east Asian producers made good use of downturns in the global economy together with the reconfiguration of European trade in the wake of the dissolution of the USSR to negotiate rock-bottom prices for the new technologies. International environmental standards and policies were a minor but nonetheless present factor. Faced with making large investments in fixed capital, South-east Asian producers did not want to be locked out of current or future markets requiring International Standards Organisation (ISO) certification, establishing environmentally preferred purchasing policies, or implementing eco-labelling.

*Science and technology* played a critical role both directly and indirectly. In South-east Asia, firm-based research and development laboratories, national industry research centres, and regional networking such as through the UNEP NIEM programme also contributed to the development, diffusion and modification of environmental pulp and paper industry technologies. South-east Asian producers also were able to take advantage of technological innovation in Finland and Sweden and other countries with major research and development efforts in pulping technology.

South-east Asian *states* played an interesting and important role in facilitating adoption of pulp industry environmental technologies. Though generally regarded as weak in their environmental regulatory efforts, in both Thailand and Indonesia, national environmental agencies were forced by popular pressure to intervene to halt production at pulp mills thought to be damaging the environment. They worked together with industry and NGOs, encouraging pulp producers to take preventative environmental measures so they could continue operations in a manner sensitive to community needs and less harmful to the natural environment. In Malaysia, national environmental officials pressed operators of Malaysia's only pulp mill at the time more than most businesses in the country.

Although regulatory standards were historically weak, government agencies, particularly in Indonesia, let industry know they would be setting progressively lower emission standards – giving industry added incentive to invest in cleaner technologies from the beginning. In Thailand, agencies established site-licensing and environmental impact assessment review procedures that strengthened more general and weak emission standards, and created a context for public input and oversight. In Indonesia and Malaysia, too, government agencies supplemented general environmental standards with site-specific licensing, audits, and environmental reviews. In this way, the pulp and paper industryy's big 'footprint' and high visibility made it an easy target for customised regulatory treatment.

The role of environmental and other *non-governmental organisations* (NGOs) was critical in the adoption of new environmental technologies in South-east Asian pulp production. Local community groups, often of ethnic and/or religious minorities, formed the 'first-line' of attack on the industry. National environmental, human rights, and alternative development NGOs took up the cause of local activists. Friendly journalists brought broad attention to the conflicts nationally and internationally. Together, local and national NGOs educated the public, forced regulators to act, negotiated with industry, and generally promoted an atmosphere in which producers were compelled to adopt new, environmental technologies.

Greenpeace International played a critical role in disseminating information about environmental risks associated with use of chlorine in pulp and paper manufacturing, spurring development of new technologies, and increasing regulatory scrutiny of the pulp industry. In South-east Asia, Greenpeace consulted with regulatory officials, NGOs, and even pulp companies interested in avoiding political hassles. Rather than be outside industrial and environmental planning, Greenpeace was very much 'inside' technology development (in Europe) and community oversight of industry (including in South-east Asia).

All of this took place in a context very much *interdependent with globalisation*. Technology, consumption, regulation, social movements all have global as well as local dimensions. None of this would have happened without the new technologies developed and available in the Nordic countries, green pulp and paper markets, converging global environmental regulatory standards, intergovernmental assistance programmes, and global interaction between social movements.

Lastly, no argument can be made that there has been a trend towards *de-industrialisation* in South-east Asia's pulp industry. There have been some calls for smaller-scale, appropriate technology in the global pulp and paper industry, for example the development of more mini-pulp mills to utilise agricultural waste rather than woody materials [*Marchak, 1995; Smith, 1997; Sonnenfeld, 1998c*]. However, even the late 1990s 'Asian flu' (global financial crisis) has not reversed development of additional, large-scale pulp manufacturing capacity, utilising cleaner production technologies in South-East Asia [cf. *APP 1998*].

Using criteria drawn from the literature, then, it can be argued that a partial ecological modernisation *has* taken place in South-east Asian pulp industries, with a major exception in regard to the dematerialisation of pulp and paper production, and continuing problems with small and medium-sized pulp and paper firms. What implications do these findings have for ecological modernisation theory and its applicability to newly industrialising countries?

## Conclusion

Clearly, ecological modernisation is on the agenda of newly industrialising countries. Cleaner production technologies are being used with beneficial effects relative to older technologies. Like their counterparts in advanced industrial countries, government environmental agencies in newly industrialising countries are moving toward collaborative relationships with producers. NGOs play a critical role both 'inside' and 'outside' environmental regulatory and management processes in South-east Asia as elsewhere.

Processes of ecological modernisation have specific characteristics in different types of economies and industrial sectors. In advanced economies, industrial restructuring involves modernising an ageing manufacturing base. With well-developed consumer markets and high wage and raw material costs, ecological modernisation in advanced economies combines improvements in environmental performance with productivity gains and manufacturing efficiencies. NICs, on the other hand, have the advantage of being able to use the latest, cleaner technologies from the onset of large-scale, modern manufacturing – 'leap-frogging', while benefiting from inexpensive raw materials and wages. Resource-extractive, export-oriented industries such as pulp and paper may be in a particularly strong position to benefit from such advantages. The marginal costs of adopting cleaner technology may thus be lower in the South, while the pull of international 'green' markets and standards may also be strong for export-oriented sectors.

The biggest problem for the applicability of ecological modernisation theory in NICs – and by extension for ecological modernisation theory more generally – may be in the area of dematerialisation. The cases examined here suggest that production is supermaterialising in the South, even if arguably dematerialising in the North. This is particularly troubling given the large areas of tropical rain forest being clearfelled to supply fibre and establish pulpwood plantations for this industry. This raises a critical question: Is ecological modernisation in advanced industrial societies dependent upon *increased* materialisation elsewhere? (Or more specifically, if global pulp and paper consumption continues to increase, where are those raw materials going to come from, if not native forests and timber plantations in the South?)[21]

A further concern is the applicability of ecological modernisation theory to small- and medium-sized enterprises, some of them government-owned. Such firms play an important role in NICs in providing employment and serving domestic markets. In South-east Asia's pulp and paper industries, many SMEs are older, use poorer technology, and are more polluting.[22] While it may make environmental sense to phase out some or many of such firms, doing so would have high social costs. Rather, governments and international agencies need to devise incentives to encourage technology firms to develop more ecological, smaller-scale production technologies.

As ecological modernisation theory further develops, it must take into account the entire globe – not only as marketplace for new ecological ideas and technologies, but also as locus of shifting and increasing material production. Also, the theory must be broadened to include small- and medium-sized, as well as super-industrial, technologies and enterprises.

## NOTES

1. Data were collected from 1992–96 through fieldwork in those three countries, Singapore, Australia and the USA, and from published and unpublished secondary sources.
2. The 'reduction in the resource input' discussed by Jänicke *et al.* [*1989*].
3. Production without (un-recycled) waste.
4. Substitution of high-technology for material inputs.
5. Elementally chlorine-free (ECF) mills are indicated by a grey background.
6. That is, 'being built on undeveloped land especially when unpolluted' [*Merriam-Webster, 1999*].

7. RGM transferred ownership of Indorayon and Riau Andalan to its Singapore-based subsidiary, Asia Pacific Resources Investment Holdings Limited (APRIL) in 1995, in part to facilitate raising funds via the international corporate bond market.
8. Conflicts in North Sumatra regarding the Indorayon mill continued after Suharto's resignation, with the new government intervening more strongly than its predecessor. See Reuters [*1998*].
9. The group now operates its pulp and paper companies through Asia Pulp and Paper Co., like its competitor, APRIL, a Singapore-based holding company.
10. The relationship between environmental activists and government officials, along with many other things, is being renegotiated in the post-Suharto era.
11. For further information on the Indonesian case, see Sonnenfeld [*1998a*].
12. Underlying the opposition-led state government were cultural, religious, and ethnic differences between Sabah residents and West Malaysia [*Vatikiotis, 1992; The Australian,* 1994].
13. UMNO recaptured the state government in 1995.
14. Primary shareholder of which is USA-born expatriate, George Davison.
15. Phoenix continues to have trouble. It was shut down most recently in July 1998, again for pollution of the Nam Phong [*The Nation, 1998*].
16. In 1998, the Soon Hua Seng group entered into a three-way strategic alliance in its Advance Agro subsidiary, taking on the Enso Group (Finland) and Oji Paper Co. (Japan) as major partners and board members [*Suwannakij, 1998*].
17. For further information on the Thailand case, see Sonnenfeld [*1998b*].
18. Stronger influence is indicated by a solid line; weaker influence with dashes.
19. Interview, Mr Mark Radka, Co-ordinator, Network on Industrial Environmental Management, United Nations Environment Program/Regional Office for the Asia-Pacific, Bangkok, 6 August 1994. This project was particularly useful in assisting smaller and medium-sized enterprises (SMEs), some government-owned, with more limited access to financial and technical resources than the newer, larger, export-oriented pulp firms.
20. The four columns represent geographical region, social movement organisations, government and industry actors, and historical developments with regard to the 'greening' of pulp and paper technology, respectively. The arrow indicates the direction of diffusion of environmental transformation of industry: both horizontally, from social movements to government and industry; and vertically, from North to South.
21. Arthur Mol suggests that one way of addressing this question is to look at dematerialisation in pulp and paper manufacturing as a function of the amount (or proportion) of virgin raw material in the ultimate products. Thus increased paper recycling and/or use of agricultural wastes or other raw materials could result in dematerialisation, even if absolute quantities of paper produced continued to increase.
22. Industry structures vary; for example, in electronics, smaller supply firms may be obligated by contractual relationships with their customers to follow international environmental management standards and practices.

## REFERENCES

Asia Pulp and Paper (APP) (1998), 'Summary of APP's 1998/99 Expansion Plans' [Online]. Available: <http://www.asiapulppaper.com/expansion1. htm>. Singapore [corporate information].

*Australian, The* [Sydney] (1994), 'Opposition Lonely Business in Malaysia', Jan., p.6.

Frijns, Jos, Paul Kirai, Joyce Malombe, and Bas van Vliet (1997), 'Pollution Control of Small-Scale Metal Industries of Nairobi', Department of Environmental Sociology, Wageningen Agricultural University, Wageningen, The Netherlands.

Hanafi Pratomo (1994), 'Impact of new government regulations on future existence of pulp mill', in *Proceedings, 48th Appita Annual General Conference, Melbourne, Australia, 1–6 May 1994*, Carlton, Australia: Australia-New Zealand Pulp and Paper Industry Technical Association.

Hajer, Maarten A. (1995), *The Politics of Environmental Discourse: Ecological Modernization and the Policy Process*, Oxford: Clarendon Press.

Hengel, Petra van der (1998), 'Textile Industry in Vietnam', M.Sc. thesis, Wageningen University, Environmental Sociology, The Netherlands, April.

Huber, Joseph (1985), *Die Regenbogengesellschaft. Ökologie and Sozialpolitik* [*The Rainbow Society: Ecology and Social Policy*], Frankfurt: Fisher.

Jänicke, Martin, Monch, Harald, Ranneberg, Thomas and Udo E. Simonis (1989), 'Structural Change and Environmental Impact', *Environmental Monitoring and Assessment*, Vol.12, No.2, pp.99–114.

Marchak, Patricia (1995), *Logging the Globe*, Montreal: McGill-Queens University Press.

Merriam-Webster, Inc. (1999), *WWWebster Dictionary* [Online]. Available: http://www.m-w.com/cgi-bin/dictionary. Accessed 14 Jan.

Mol, Arthur P.J. (1995), *The Refinement of Production: Ecological Modernization Theory and the Chemical Industry*, Utrecht: van Arkel.

Mol, Arthur P.J. and Jos Frijns (1999), 'Environmental Reforms in Industrial Vietnam', *Asia-Pacific Development Journal* (forthcoming).

Murtedza Mohamed and Lars Landner (1993), 'Sabah Forest Industries Sdn Bhd Pulp and Paper Mill: EIA Before Construction of Mill and Comparison with Actual Performance', in *Phase II Training Package 2, Volume II, Network for Industrial Environmental Management (NIEM)*. United Nations Environment Programme, Regional Office for Asia and the Pacific, Bangkok, March.

Murtedza Mohamed *et al.* (1995), 'Country Paper – Malaysia', paper presented at NIEM seminar on 'Regulatory Options for Fostering Improved Environmental Management in the Pulp and Paper Industry', Bangkok, 15–18 Nov.

*Nation, The* (1998), 'Phoenix Paper Mill To Be Closed' [Online]. Available: <http://203.146.51.4/nationnews/1998/199807/19980721/29181.html>. 21 July.

Paulus, Stephan (1986), 'Economic Concepts for Industry-Related Environmental Policies', in *Proceedings – Forum on Industry and Environmentt*, New Delhi: Friedrich Ebert Foundation.

Reuters News Service (1998), 'APRIL Says Indonesia Comments Surprise', [Online]. Available: <http://biz.yahoo.com/rf/981007/x.html>. 7 Oct.

Simonis, Udo E. (1989), 'Ecological Modernization of Industrial Society: Three Strategic Elements', *International Social Science Journal*, Vol.41, No.3 (Aug.), pp.347–61.

Smith, Maureen (1997), *The U.S. Paper Industry and Sustainable Production: An Argument for Restructuring*. Cambridge, MA: MIT Press.

Sonnenfeld, David A. (1996), 'Greening the Tiger? Social Movements' Influence on the Adoption of Environmental Technologies in the Pulp and Paper Industries of Australia, Indonesia, and Thailand', Ph.D. thesis, Sociology, University of California, Santa Cruz.

Sonnenfeld, David A. (1998a), 'Social Movements, Environment, and Technology in Indonesia's Pulp and Paper Industry', *Asia Pacific Viewpoint*, Vol.39, No.1 (April), pp.95–110.

Sonnenfeld, David A. (1998b), 'From Brown to Green? Late Industrialization, Social Conflict, and Adoption of Environmental Technologies in Thailand's Pulp and Paper Industry', *Organization and Environment*, Vol.11, No.1 (March), pp.59–87.

Sonnenfeld, David A. (1998c), 'Logging versus Recycling: Problems of the Industrial Ecology of Pulp Manufacturing in South-East Asia', *Greener Management International*, No.22 (Summer), pp.108–22.

Spaargaren, Gert and Arthur P.J. Mol (1992), 'Sociology, Environment, and Modernity: Ecological Modernization as a Theory of Social Change', *Society and Natural Resources*, Vol.5, No.4 (Oct.–Dec.), pp.323–44.

Suwannakij, Supunnabul (1998), 'Two foreign firms take 25.4% of AA', *The Nation* (Bangkok) [Online]. Available: <http://203.146.51.4/nationnews/ 1998/199809/19980929/32398.html>. 29 Sept.

Vatikiotis, Michael (1992), 'Federal Excess: Sabahans Feel Exploited by Central Government', *Far Eastern Economic Review*, 18 June.

# Ecological Modernisation Theory and Industrialising Economies: The Case of Viet Nam

JOS FRIJNS, PHUNG THUY PHUONG and ARTHUR P.J. MOL

As Ecological Modernisation Theory has been developed against the background of European industrialised societies, the value and applicability of Ecological Modernisation Theory for developing or industrialising economies is often questioned. This article explores this controversy by taking Viet Nam as a case study. The analysis of environmental restructuring of contemporary economic developments in Viet Nam focuses on three major issues: state-market relations, technology development, and environmental awareness. Regarding all three issues major deviations are found from an environmental reform trajectory as hypothesised by Ecological Modernisation Theory. The study thus concludes that Ecological Modernisation Theory is of limited value for analysing contemporary processes of and attempts at environmental reform in Viet Nam. However, recent developments in Viet Nam such as economic liberalisation, privatisation and internationalisation and the slowly starting democratisation of governance offer ecological modernisation-inspired opportunities to incorporate environmental considerations in the economic development of Viet Nam. If Ecological Modernisation Theory is to be used to outline a feasible path of environmental reform, it has to be refined, however, to fit the specific local conditions and institutional developments of industrialising countries.

## I. Introduction

Various scholars observe (see Mol and Sonnenfeld, this volume; *Spaargaren, 1997*) that Ecological Modernisation Theory has been developed against the background and in the context of Western European industrialised societies. To date, the socio-political, economic and cultural

The authors acknowledge the valuable comments of two anonymous referees, the participants to the 1998 ASA Round Table Discussion on Ecological Modernisation and especially the efforts of David Sonnenfeld in improving this contribution.

conditions of that geographical region have largely constituted its empirical foundation. Elsewhere, for example, we have identified essential institutional characteristics for ecological restructuring in Western Europe [*Mol, 1995; 2000*], including:

- a democratic and open political system;
- a legitimate and interventionist state with an advanced and differentiated socio-environmental infrastructure;
- widespread environmental consciousness and well organised environmental non-governmental organisations (NGOs) that have the resources to push for radical ecological reform;
- intermediate or business organisations that are able to represent producers in negotiations on a sectoral or regional basis;
- experience with and tradition in negotiated policy-making and regulatory negotiations;
- a detailed system of environmental monitoring that generates sufficient, reliable and public environmental data;
- a state-regulated market economy that dominates production and consumption processes, covering all the edges of society and strongly integrated in the global market; and
- advanced technological development in a highly industrialised society.

The admittedly Eurocentric basis of Ecological Modernisation Theory has been the focus of important recent criticism. Buttel [*2000*], for example, emphasises that political cultures, state structures and social institutions are not the same everywhere, and that in numerous countries the state exerts no regulatory power to push for more environmental practices. Others have put forward similar lines of criticism focusing on the relevance for developing countries [e.g. *Blowers, 1997; Blühdorn, 2000*]. This article begins to take up such challenges, exploring what perspectives Ecological Modernisation Theory opens up outside the small group of countries that have functioned as its regional base and empirical foundation. More particularly, this article analyses the usefulness of ecological modernisation concepts for evaluating and shaping pollution control efforts in newly industrialising economies, using contemporary Viet Nam as a case study. Elsewhere we have elaborated in detail on the core features of Ecological Modernisation Theory [cf. *Mol and Spaargaren, 1993; Mol, 1995; Spaargaren, 1997*] (see also Mol and Spaargaren; Mol and Sonnenfeld, this volume) and we will take this work as a starting point.

Earlier, we started this debate on the 'geographical scope' by expressing our ambivalences about the value of Ecological Modernisation Theory to *developing countries* and/or regions [*Mol, 1995*]. The institutional conditions in industrial societies differ in a number of vital aspects from those in developing societies, such as those in sub-Saharan Africa. The transfer of ecological modernisation ideas to the latter group of countries might be attended by the same risks as the transfer from industrialised to developing countries of technology without adapting it to the local social, institutional, ecological, and cultural context. Ecological Modernisation Theory may be more appropriate to some *transitional countries* in central and eastern Europe and *newly industrialising* or high-performance economies in, for instance, South-east Asia. In his study of environmental reform in Lithuania, Rinkevicius [*2000*; also this volume] has shown that the theory could bear fruit provided it is 'adapted' to or refined for the specific situation in what he calls a 'double-risk society'; that is, a society in transition facing both severe environmental and economic risks.

Something else is also at stake that might make Ecological Modernisation Theory relevant for non-European countries. In a globalising world, with modern industrialised countries still providing the dominant models of (economic) development, models of ecological reform that are believed to be inappropriate for non-OECD countries at first sight, might still be 'imposed' – for better or worse – upon these countries through a diversity of mechanisms. One could think of political mechanisms such as the international (sustainable) development programmes of various nation-states and international actors (such as World Bank and IMF), or the emphasis put on the transfer of environmental technology from industrialised countries to developing countries in the UNCED declarations, the Montreal Protocol and the elaboration on the Framework Convention on Climate Change. The so-called 'westernisation' of environmental reform is already in progress in developing countries. This is occurring through not only international environmental negotiations and assistance, and scientific-technological exchange of ideas and experiences, but also global firms operating in a world market (having learnt the Bhopal-lesson), and through international environmental NGOs such as Friends of the Earth International that stimulate their members to include the idea of 'environmental (utilisation) space' in alternative environmental reform models for their home countries all over the world.

This article takes Viet Nam as a case study for beginning to develop a model for ecological modernisation in newly industrialising countries. Viet Nam was selected not by accident or at random. Despite the 1998/99 crisis in their financial markets, South-east Asia is widely believed to become the leading economic region in the next century [cf. *Castells, 1997*] and continues to show patterns of amazing economic growth and industrialisation – at least in export sectors. Among those South-east Asian economies with high percentages of economic growth and rapid industrialisation, Viet Nam can be seen as a borderline case of a tiger economy. It is part of the new industrialising periphery that is trying to join the trail of such high-performance economies as Taiwan, South Korea, Singapore and Malaysia, while at the same time showing considerable differences from the institutional lay-out of North-west Europe.

In the next section we give an overview of industrial development and environmental pollution in contemporary Viet Nam. Then in section III, we examine the possibility of modernising pollution control strategies in Viet Nam, discussing three issues of importance to the environmental restructuring of institutions: the state–market relations, technology development, and environmental awareness. We conclude in section IV, by considering whether an environmental reform strategy based on the ecological modernisation model is of value to and applicable in Viet Nam.

## II. Industrial Pollution and Pollution Control in Viet Nam

This section describes the rapid industrialisation of Viet Nam, the accompanying environmental problems and the environmental policy developed to deal with them. The description of the current situation forms the foundation for the analysis of potential and actual environmental reforms from the point of view of ecological modernisation in section III.

### *Industrial Development and Pollution*

If we want to analyse the present and future environmental reform of Viet Nam's industrialisation, we should not only be aware of the typical characteristics of its contemporary economic and environmental development, but at the same time realise that Viet Nam is not 'just another developing country' (if such a classification exists at all). Viet Nam can be 'conceptualised' at the crossroads of two processes: the transformation from a command economy to a market-oriented growth model (in which it parallels to some extent other transitional economies), and the more traditional development from a less developed country to a new industrialising economy (in which it shares features with an increasing number of South-east Asian states). Only China seems to be at a more or less similar crossroad.

*Viet Nam: Between 'developmental state' and 'transitional economy'*: As in other new industrialising countries in South-east Asia, Viet Nam shows a rapid industrialisation. In that respect Viet Nam is a typical example of a South-east Asian tiger economy, although its economic development started later than and is definitely still lagging behind countries such as Taiwan, South Korea, Malaysia and Singapore. Notwithstanding the differences something special seems to be the case in the South-east Asian region. In his *magnus opus*, Castells [*1997: 206–309*] analyses the coming of a Pacific Era. Although he concludes that there is no institutional, cultural, political or even economic Pacific region, the South-east Asian developing economies share some similarities in economic development. The most significant of the commonalities is, according to Castells, the role of the state in the developmental process. The states in these South-east Asian countries had the political capacity and relative autonomy to push successfully their project of a strong state coupled with high levels of economic growth, low wages, limited internal democracy and the building of a national identity. The strong 'developmental state' seems a crucial factor in the economic success of the South-east Asian tiger economies, and Viet Nam forms no exception to that.

Viet Nam can be understood as a 'new industrialising periphery', having certain parallels with developments in other countries in South-east Asia. What is special about Viet Nam is the relation between a rapid industrialisation and a simultaneously changing economic system: the transition from a purely command economy to a market-oriented growth model. Sikor and O'Rourke [*1996*] provide a summary of the institutional transformations that structure current economic activities, following the economic reforms in Viet Nam since 1986 (better known as *Doi Moi*): the government has liberalised economic production and exchange; resource allocation has shifted towards market mechanisms for the purpose of increasing flexibility and efficiency; state enterprise reform, the 1993 Land Law, and tax reforms have transferred assets to, and strengthened the role of, the private sector; and international trade and investment have been liberalised through the 1987 Foreign Investment Law and more recent foreign trade reforms.

Regarding these institutional transformations, some similarities with other transitional economies can be expected, for one thing with respect to processes of environmental reform. In a number of Central and East European countries the transformation of command economies into more market-oriented growth models is paralleled with a reorganisation of existing traditional environmental policy systems into new ones that are believed to be better adapted to the changing situation. In most of these former command economies a detailed system of environmental institutions, laws and standards had been developed in close relation to the economic planning system. Although these environmental institutions sometimes looked impressive on paper, it has been widely acknowledged that their environmental effectiveness proved to be poor, owing to failing implementation, inappropriate price signals, lacking priority for environment, shortage of manpower and resources in environmental institutions, etc. [cf. *Ziegler, 1988*].[1] In contrast to these Central and East

European economies, an elaborated system of environmental policy-making was almost absent in Viet Nam before 1986 and had to be created as economic reform progressed. To some extent this has preserved Viet Nam from a difficult reform process of existing but inadequate and old-fashioned environmental institutions and legal systems. Without such additional complications, Viet Nam has at least a theoretical advantage to develop from scratch adequate environmental institutions that are directed at preventing and neutralising the devastating environmental consequences of a market-oriented industrialisation process. Before analysing Viet Nam's efforts thus far in that regard, we will introduce the economic and environmental outlook of Viet Nam's industrialisation.

*Industrial development in Viet Nam*: After the implementation of the economic reform programmes of the mid-1980s, Viet Nam at first went through a period of poor economic performance with high inflation and limited economic growth. Since 1990 however, Viet Nam has shown an average economic growth of about eight to nine per cent, and inflation had fallen below three per cent in 1996. That is impressive even by East Asian standards. Agriculture and forestry have traditionally been the major contributors to national income, providing 40 per cent of GDP, and the main sources of employment. Only about 13 per cent of the total employed labour force are in industry (1993). But, there is an increase in the economic importance of the industrial and service sectors relative to agriculture. Industry is rapidly expanding; it accounted for 25 per cent of GNP in 1996 and its share is expected to rise to 35 per cent by 2010 according to projections of the Ministry of Planning and Investment. The emphasis in industrial development had traditionally, before *Doi Moi*, been on heavy industry. Now, with the increasing importance of a market-economy system, the light industry with consumer products such as textiles and food processing is taking over dominance. A substantial part of the industrial development is taking place in newly established industrial zones in the Ho Chi Minh City and Ha Noi regions and Dong Nai Province.

Since 1993, industrial output has grown by nearly 13 per cent annually. The industrial sector consists of national and provincial state-owned enterprises and the non-state sector made up of co-operatives and private firms. The number of private firms in the formal industrial sector has grown, but most are small and individually operated. In contrast to what is often believed, state-owned enterprises (SOE) have proved to be quite dynamic since the early 1990s, showing rapid output growth while streamlining their workforce [*Irvin, 1996*]. This is not to say that there is no inefficiency or rent-seeking behaviour in SOE. Viet Nam operates only cautiously on the path of privatisation, and rather successfully modernises the SOE sector by acquiring new technology through state-owned joint ventures with foreign capital. Or as Jansen [*1997*] puts it, there has been a transition to the market, but only a limited transition to a private-sector economy. Apparently, foreign investors prefer joint ventures with SOEs to those with private firms, largely because of SOEs' privileged access to land and credit [*Irvin, 1996*]. Foreign Direct Investments (FDI) have grown steadily since the Foreign Investment Law of 1987 to some US$19,400 million in 1995, of which over half goes to the oil and gas sector and ten per cent to the industry sector [*UNDP, 1995*].[2] The oil and natural gas industries are dominated by foreign capital, as are food processing and light manufactures.

*Environmental problems related to industrial growth in Viet Nam*: *Doi Moi* has liberated market forces that have increasingly directed investments to sectors where Viet Nam enjoys a strong comparative advantage. These include both labour-intensive industries (such as textiles, clothing and footwear) and resource-intensive industries (such as fossil fuels, plantations, agricultural processing and building materials). Unfortunately, by their nature some of these sectors are not environmentally sound. For example, petroleum refining and petrochemicals can be associated with serious pollution problems. Agro- and food processing industry are major sources of organic pollution (water and waste), and textile dying, electroplating, leather tanning, and pulp and paper production can generate significant amounts of organic and hazardous waste.

The estimate is that in 1993 more than 3,000 major industrial enterprises in Viet Nam were discharging wastewater without any treatment. In the North, in Viet Tri Industrial Zone, for example, factories of pulp and paper, textile, food and chemical production discharge around 35 million m3/year of waste water containing 100 tonnes of sulphuric acid; 4,000 tonnes of chloric acid; 1,300 tonnes of sodium hydroxide; 300 tonnes of benzene; and 25 tonnes of pesticides [*Nguyen Cong Thanh, 1993*]. The same problem occurs in Bien Hoa I Industrial Zone in the South, where 30,420 m3/day of waste water containing 12,486 kg/day of BOD is discharged into Dong Nai River almost without any treatment [*Phung Thuy Phuong, 1994*]. As a result, in the major industrial centres, dissolved oxygen in surface water is always near zero, BOD and COD are high, and metals and other toxic substances are regularly found in high concentrations.

Urban air quality is also deteriorating, principally owing to the proliferation of motorised transportation coupled with the intensification of urbanisation and industrialisation. Ambient concentrations of pollutants such as CO, $NO_x$, $SO_x$, Pb, particulates, etc., exceed emission standards in many areas. Significant emissions of more than 50 toxic and hazardous gases have been recorded in Ha Noi, Ho Chi Minh City, Viet Tri and other industrial centres. In factories, workers are exposed to air pollution and associated health risk. Technical measures such as simple ventilation techniques and good housekeeping measures have often not yet been taken.

The majority of solid waste derives from domestic sources, only about 18 per cent are of industrial origin. The cities have difficulty providing adequate services for the collection and disposal of domestic waste. In Hanoi, less than half of domestic waste is collected, the remaining part is dumped in ponds, lakes, drainage canals and illegal dumping sites. About 400 tonnes/day of waste is dumped in the canals of Ho Chi Minh City [*Frijns and Truong Thi Kim Khanh, 1997*]. Hardly any data are available on industrial solid waste. There is no monitoring system that could provide detailed and reliable data on the amount, the contents and the origin and destination of different streams of industrial solid waste. Neither is there any form or system of hazardous waste management. Nguyen Phuc Quoc [*1999*] describes some current initiatives in Ho Chi Minh City and Dong Nai to set up an industrial solid waste management system, but until now these initiatives for collection, separation stations and treatment facilities have not surpassed the stage of desk studies, or – at best – pilot projects. Part of the industrial waste is disposed of together with domestic waste at disposal sites in urban areas, which are neither properly designed nor operated as sanitary landfills but simply as dumpsites. They generally suffer from poor siting, are unlined and have no leachate collection or treatment, thus putting surface- and groundwater resources at risk [*World Bank, 1995*].

In short, the rapid industrialisation has resulted in an increase of wastewater discharge, air pollution and energy consumption, and the release of industrial solid waste, causing serious environmental and health problems, especially in urban areas. A particularly difficult problem is the pollution of the numerous small-scale industries that are located in city neighbourhoods. Obviously, there is an urgent need for an effective environmental policy framework in Viet Nam.

### *Environmental Policy*

Less than ten years ago, in the heyday of the second international wave of environmental concern, Viet Nam was only on the eve of constructing a coherent institutional framework for environmental management and policy. The environmental measures were piecemeal and often set aside for the higher priorities of national economic development. Only in 1991 did the Council of Ministers approve the first National Plan on Environment and Sustainable Development 1991–2000, which aimed at developing a more coherent and integrated environmental policy [*SCS, 1991*]. It was supported by a framework Law on Environmental Protection (January 1994) and some new organisations, which together still make up the core of the institutional framework for environmental protection. We will now describe the institutional design, and then concentrate on the problems of the basically command-and-control approach of Vietnamese environmental policy.

*Organisational structure*: The Ministry of Science, Technology and Environment (MOSTE) is the central environmental organisation within the national government. Within MOSTE there exists a National Environmental Agency (NEA) established in 1993, whose exact tasks and responsibilities are still being elaborated.[3] MOSTE has regional departments in most provinces, namely, Departments of Science, Technology and Environment (DOSTE), usually staffed with no more than three to ten people working in environmental offices, depending on the priorities set within the province. The DOSTEs are linked on the technical level to MOSTE and on the political and administrative level to the provincial People's Committees.[4] Prior to the establishment of these provincial environmental authorities, the People's Committees of Ho Chi Minh City, Ha Noi and Hai Phong had set up local environmental offices that directly report to these local People's Committees. In Ho Chi Minh City, the Environmental Committee (ENCO) was responsible for environmental management at city level until 1998, when DOSTE of Ho Chi Minh City took over the care of environment. Within the districts of Ho Chi Minh City, the Environmental Bureau's take care of environmental management and implementation of ENCO/DOSTE guidelines at district level.[5]

Vietnamese environmental institutions, laws and environmental management practices are quite new, and suffering all the teething troubles that so often face new institutions and policy-making practices. The environmental institutions lack trained personnel and face financial constraints.[6] But more importantly, they lack the political power to put environmental concern on the agenda of the economic growth policy.

*Legislation*: The Law on Environmental Protection of 1994 provides the legal framework for MOSTE, NEA and DOSTE organisations. In general, it is an extensive act that sets obligations and duties.[7] It is a framework act that will have to be substantiated by numerous detailed guidelines and standards on the prevention and control of air, water and noise pollution, the production, transport and use of toxic substances, the siting and design of nuclear power plants, the transport and storage of wastes, etc. In some cases this has already been done but most legislative work on standards still lies ahead [cf. *UNDP, 1995*]. In 1995, MOSTE issued the national environmental standards of Viet Nam, known as TCVN1995. These standards involve air quality (ambient and emission), noise, and water (surface water, ground water and discharges). Yet these standards fall far short of the real needs and situation. The Law has opened prospects for economic instruments in that it provides for responsible parties to pay for environmental damage. Unfortunately, how levying fines will be determined and enforced has not yet been described. So far, the levels of pollution fines have been too low to have much effect. After 1994, MOSTE issued several sub-laws to form a legal corridor for state environmental agencies to implement their tasks. These include the decree on administrative punishments for violators of the Law on Environmental Protection, and decrees on immediate measures for the management of solid waste in industrial and urban areas.

The few sectoral laws dealing with environmental matters established already prior to this framework law were rather limited in scope, which gives Viet Nam the advantage of not having to cope with co-ordinating and harmonising different legal regimes in numerous sectoral laws. On the other hand, in the absence of the Law on Environmental Protection, several provinces and cities had promulgated environmental regulations over the years. Uncoordinated legal formulation and the lack of national guidelines or reference systems have led to conflicting environmental regulations.[8]

*Monitoring and enforcement*: A central precondition for effective environmental policy-making is extensive monitoring and enforcement. The monitoring of both environmental quality and emissions of pollutants is essential to set standards, to develop adequate strategies and measures, to control the behaviour of producers and polluters and to assess the effectiveness of certain policy instruments and programmes.

In Viet Nam, environmental monitoring activities are delegated to three institutions: CEETIA (Ha Noi Civil Engineering University) for Northern Viet Nam, EPC (Environmental Protection Centre) for Central and parts of Southern Viet Nam, and CEFINEA (Ho Chi Minh City Polytechnic University) for the Mekong Delta. But the budgets are limited and thus the monitoring stations are not well equipped, only a few locations are monitored and a limited amount of samples taken at each location. Industrial and other producers cannot be forced to monitor their activities and send adequate data to governmental agencies. MOSTE is obliged to produce periodic reports on environmental quality, but these fall short on adequate information and time series.

Although the Law on Environmental Protection sets clear tasks and obligations for environmental inspection and gives various instruments for criminal and administrative enforcement, in practice enforcement falls short for various reasons. The causes of limited enforcement in practice are related to legal discussions (for instance, on liability and civil enforcement)

[*UNDP, 1995*], but most important are the lack of manpower, the setting of other priorities within environmental agencies, the failing knowledge and experience on administrative enforcement within the governmental agencies responsible, and the lack of tradition on enforcement and control in environmental management.

There have been occasional upheavals in the control and inspection of industrial pollution. In the summer of 1997, for instance, over 5,000 major manufacturing plants were inspected countrywide. Of these, 46 per cent were fined a total of US$82,000, and 30 per cent ordered to stop production or relocate their establishments for severe violations of the environmental law [*Bich Ngoc, 1997*].[9] The continuity of these intensified inspections seems questionable.

*Problems of the command-and-control approach*: One must acknowledge that in a relatively short time period of less than ten years important initiatives have been undertaken to curb environmental degradation. In the tradition of a centrally planned economy but also of a developmental state, it was primarily the state that addressed environmental problems by setting up institutions and regulations for pollution control. Not unlike most second-generation industrialising economies in South-east Asia, this process started relatively late and has progressed slowly owing to dragging bureaucratic processes, a priority for industrial growth rather than environmental reform, and a lack of strong external pressure, for example from non-governmental organisations. The general picture is that of a still weak environmental management and policy-making system, which lacks proper monitoring and enforcement. According to the Deputy Minister of MOSTE [*Chu Tuan Nha, 1997*], the environmental policy structure is inadequate, the budget is far short of needs, and the capacity of state environmental agencies is extremely limited.

Environmental policy in Viet Nam has followed the traditional model of command-and-control which is characterised by laws, standards and regulations pertaining to emissions and products, as the main instruments, and a top-down implementation of legislation. Not unlike the policy models in western countries in the 1970s, this policy approach is a logical starting point but may be rather ineffective and bureaucratic, and fail to result in technological innovations. For the following reasons, a strict command-and-control approach does not work well in Viet Nam, no more than in the majority of developing countries. First, in a command-and-control approach non-compliance or violation of environmental regulations will result in fines or imprisonment. But in many developing societies, especially in Asia, for cultural reasons courts are a last resort, which means they are rarely used. Second, high-level regulation cannot be achieved under budgetary, manpower and administrative constraints. Third, a low degree of devolution of authority to representative local government reduces the capacity for monitoring and enforcement. In fact, neither MOSTE nor DOSTEs can effectively impose penalties on violators. Particularly difficult is the control of pollution caused by large state-owned enterprises. Local enforcement agencies may be reluctant to exercise their power on these enterprises. Fourth, fines are usually set too low to deter violators, they remain unchanged in nominal terms for years, and become eroded by inflation. The final and very damaging reason for a non-functioning command-and-control approach is the rent-seeking behaviour of enforcement officials.

So, while on paper the institutional layout has been improved and environmental laws and policy instruments have been developed, the implementation in practice falls short. This should not surprise us, since it seems the rule rather than the exception in most countries that rapidly develop their institutional framework of environmental management. The institutional framework and legal rules form an essential initial building stone for an effective system of industrial environmental management, and therefore to give it some priority is logical at a primary stage. To consolidate and mature this institutional structure, successes in implementation are essential. This means that most probably in the coming years more attention must be paid to implementation and enforcement processes.[10]

Although a strengthening and further maturation of the command-and-control approach remains essential, at the same time the difficulties it has entailed seem to call for possible complementary approaches of environmental management. In Western European countries, and theorised upon by the Ecological Modernisation Theory, alternative strategies for pollution control are being developed and implemented. Among the core features of these alternatives are more decentralised, participatory styles of governance, the activation of market instruments and dynamics, technological innovations by industry linked to more self-regulatory pollution control, and an increased involvement of environmental NGOs in environmental policy-making. The following sections examine the present state and possibilities of modernising pollution control strategies in Viet Nam along more or less similar lines.

## III. Possibilities to Modernise Pollution Control Strategies in Viet Nam

Few would disagree if we conclude from the former section that current strategies and efforts to redirect industrial development processes in Viet Nam in more environmentally sound directions are falling short. The question arises whether new, innovative approaches could be introduced and whether the Theory of Ecological Modernisation would be applicable. To assess the value of the Ecological Modernisation Theory for analysing the current transformations in pollution control as well as looking for future strategies to modernise environmental reform, we will concentrate on three features: the (changing) state–market relation, technology development, and environmental awareness.

*State Intervention and Pollution Control in Viet Nam*

Environmental policy in Viet Nam, which until now has relied heavily on a hierarchical command-and-control model, has been unable to deal with the pace of contemporary industrial growth. The Theory of Ecological Modernisation may provide a guideline – often labelled 'political modernisation' – for the analysis of opportunities for improved state intervention. This section will discuss the relevance of political modernisation to the state institutions of Viet Nam.

*Missing evidence of changing state intervention*: As pointed out above, Ecological Modernisation Theory does not so much deny the role of the state in radical environmental reforms, but emphasises its changing role in environmental management and policy in two ways. First, some tasks and responsibilities in environmental management shift from the state to the market. In other words, market dynamics and market actors play an increasing role in environmental reform. Second, the 'style' of state intervention in environmental policy shows relative changes from curative and reactive to more preventive; from closed policy making to a more participatory one; from centralised to decentralised; from hierarchical to more consensual, based on negotiations and intensive consultations. The emergence of market-based (economic instruments) and communicative approaches (voluntary agreement, mediation, negotiated rule-making, eco-labelling) illustrates this changing role of state and state policy.

If we analyse the development of state intervention in environmental protection in Viet Nam we come to the conclusion that very few of these changes have been observed or identified. We do not see much movement towards more consensual policy styles, more negotiated policy-making, a larger role of economic agents (such as suppliers and customers, credit organisations, insurance companies, branch associations) or economic mechanisms in environmental reform initiatives. The dominant development in Vietnamese environmental policy is rather the establishment or further refinement of a basically hierarchical and centralised command-and-control model, in which environmental laws and decrees, together with Environmental Impact Assessments, form the core. Until now, changes in environmental policy-making are characterised rather by the elaboration of this legalistic core into detailed environmental standards and individual environmental permits, than by a tendency towards economic instruments or communicative approaches [cf. *Mol and Frijns, 1998*]. From that observation, the value of Ecological Modernisation Theory for *analysing* contemporary state environmental reform in Viet Nam is limited. Contemporary developments and new initiatives in environmental policy-making in Viet Nam cannot be fully understood with the Ecological Modernisation Theory as it is framed now.

This leads us to the normative dimensions of Ecological Modernisation Theory: do the ideas and mechanisms grouped together under this umbrella provide us with useful strategies for future environmental reform in Viet Nam? And can we say anything about the possibilities and/or likeliness of introducing such innovations within the near future, especially given the contemporary economic and political institutions in Viet Nam? Several authors have pointed to the slow but steady changes in state reform strategies that have been observed recently in several new industrialising countries in the South-east Asian region. Their governments relied on hierarchical, closed environmental policy styles for years, but have begun to experiment with a range of alternative approaches, using economic and communicative instruments, to complement their regulatory efforts to deal with environmental problems [cf. *O'Connor, 1994; Rock, 1996*]. It goes without saying that these alternatives show clear differences with its parallels in European countries. Can Viet Nam learn from the experiences of Western countries and new industrialising economies in East Asia, taking into account the necessary adaptation to the specific local conditions? Obviously, a complete analysis of the advantages, disadvantages and opportunities of innovations in Vietnamese environmental policy-making is beyond the scope of this part of the paper. But a more limited and 'down to earth' analysis of the opportunities to extend the use of economic and communicative approaches in environmental policy-making, and a case study of industrial zones as gardens for practical experiments, might yet give us an idea of the value of applying ecological modernisation to environmental policy in Viet Nam.

*Market-based approaches*: Some types of economic instruments for pollution control have been used incidentally in Viet Nam, such as tax reduction for imports and for the installation of clean technology. Experiences with other kinds of economic instruments such as pollution charges,[11] product charges, user charges,[12] subsidies, deposit-refund systems,[13] and marketable permits are still lacking in Viet Nam [*Tran Thi Thanh Phuong, 1996*]. In addition, private economic actors such as insurance companies, credit institutions, branch and regional industrial interest organisations, producers and suppliers/customers are almost completely absent from environmental policy-making and reform. A considerable number of obstacles prevent the widespread application of economic approaches and instruments in Vietnamese environmental policy, to be divided into the more practical and the more fundamental/institutional constraints.

One practical constraint is the lack of knowledge of how to design and implement economic instruments and their impact on growth and income distribution. In addition, the use of economic instruments and approaches quite visibly raises the cost for the private sector and the wider public, whereas the costs of the command-and-control regulatory system are largely hidden from the public eye. For two reasons economic actors tend to stay aloof from environmental reform. For one thing they are completely preoccupied with economic activities and see no role for themselves in environmental reform in the near future. For another, they are hardly, if at all, organised in industrial associations and regional representational organisations. That takes us to the more institutional/fundamental constraints for such environmental policy innovations. The still considerable involvement of the state in economic developments and decision-making hinders the incorporation of economic dynamics and actors in environmental reform processes (although the opposite will not automatically happen).

Besides these obstacles, some openings for market-based approaches to environmental reform can be listed. In theory, some characteristics often attributed to economic instruments might benefit Viet Nam's environmental policy in the current situation (such as cost-effectiveness, reduced demand for human resources, easy enforcement, fewer opportunities for rent-seeking behaviour, revenue generation). As the Law on Environmental Protection seeks to apply the polluter pays principle, economic instruments have been considered an appropriate policy approach by the Vietnamese government. We share the opinion of Tran Thi Thanh Phuong [*1996*] and Zhang Lei [*1997*] that under the present conditions in Viet Nam as well as in China, the prospects of successfully applying economic instruments depend very much on the further (institutional) development of the market economy, the determination of the government, and the know-how about the design and application of economic instruments.

*Communicative approaches*: The stimulation and implementation of communicative models of environmental reform will encounter difficulties from two sides. First, Vietnamese companies are not very well informed and hardly know how to put forward and implement their own environmental goals, strategies and improvements. In general, they are not ready to move to more self-regulative approaches, which limits the chances of using voluntary agreements and the like. In addition the institutional structure of industrial production limits the opportunities for negotiating and implementing more communicative approaches. The absence of any organisations of industrial producers precludes a sectoral approach, often necessary to limit unequal competition and free riders. Voluntary instruments and strategies depend on green consumer demand or active public pressure, such as eco-labels or annual environmental reports. However, the demand for eco-products is poor, and there are no environmental NGOs that could effectively mobilise public pressure (see below). Only the introduction of more 'individual' instruments such as environmental management systems or ISO standards[14] is not blocked by the prevailing institutional structure of industry and economy.

Second, governmental and state-related barriers prevent a rapid and easy application of communicative innovation in environmental reforms. These constraints relate to the prevailing policy style and culture in Viet Nam, with a historic 'need' to be in control and a sense of authoritarian power. Indeed, the government continues to prefer a strong centralised state directing – among other matters – environmental policy. Not only the recent history of a central planning and command economy, but also the heritage of the traditional monarchy (which strongly influenced a culture of hierarchical behaviour) and the lack of experience in negotiations and joint policy-making, are unfavourable to a switch to more public and/or private participation in policy making. The characteristics of what Evans [*1996*] and others have labelled a developmental state combine a certain functional involvement of public organisations and economic actors with a strong state sector that tries to remain in control.

*Industrial zones as experimental areas for ecological modernisation*: In Viet Nam, industrial zones seem to be a good place to experiment with alternative governance models and state intervention strategies for environmental reform. The concentration of various industries in an area under the management of an investment company, that is, the parastatal Industrial Zone Management Board (IZMB), facilitates the application of alternative policy instruments and the shift of environmental governance tasks from public to more market-oriented private or para-statal organisations. Beside variation in types of industry, industrial zones are also characterised by diversity of ownership ranging from state-owned, private, joint venture, to foreign enterprises. The fact that in these zones the concentration of foreign companies – with better (access to) environmental knowledge, some experience with alternative governance models and a direct connection to the world market – is rather high, increases the chance of successful experiments with the normative strands of ecological modernisation. We will give some examples of (possible) innovations in Vietnamese environmental policy-making in these zones.

In industrial zones, where the cumulative effects of concentrating pollutants in small areas may cause severe problems, individual effluent standards or emission standards, without considering their collective effect, are less appropriate. In such cases tradable permits might be a better solution. User charges might be successfully applied in industrial zones, once common treatment facilities are established to take advantage of economies-of-scale in pollution control. In addition, lessons can be drawn from other new industrialising counties in tax-cum-subsidy schemes that combine carrot and stick [*O'Connor, 1994*]. As the IZMB of each industrial zone is already in charge of collecting land-rental fees from companies, their tasks can be extended to collecting pollution charges, allocating environmental funds, and co-ordinating the trade in emission permits. The administrative costs of the charge system (pollution charge, product charge, and user charge) will be low because they can be incorporated in the existing fee system of industrial zones. Especially in these industrial zones economic innovations for pollution control are promising, because as Tran Thi Thanh Phuong [*1996*] puts it: 'the transition to a market-based economy and the expansion of the private sector are good conditions for applying economic instruments in environmental management in Viet Nam'.

Communicative approaches and more specific voluntary agreements, at first sight do not seem very relevant to improve Vietnamese environmental policy. Nevertheless, experiments with such instruments and approaches could best start with foreign investment companies, or joint-venture companies, which in their home countries have had some experience in dealing with communicative approaches and instruments and have – or can mobilise – considerable environmental knowledge, which is essential to some form of 'self-regulation'. Industry associations, which play a bridging role in negotiations, consultations, consensus building and information exchange between government and individual companies in West European countries, hardly exist in Viet Nam. But IZMBs seem an appropriate Vietnamese alternative to fill the gap. These para-statal organisations have better access to, knowledge of and trust among individual enterprises than governmental agencies. While able to represent to some extent the economic interests of industrial zone enterprises towards the state, they do have sufficient distance to incorporate other than only short-term, narrowly-defined economic interests. Besides, industrial zones are a good setting for disseminating information and experiences on new regulation strategies, so that initial experiences with foreign investment companies can be replicated in other, similar companies in the same industrial zone. An environmental information centre should then be created for each industrial zone, so that companies, IZMBs and state environmental agencies can exchange experiences of improving environmental performances within and between industrial zones.

In these industrial zones IZMBs could even take over some of the tasks that currently come under the environmental responsibility of the provincial or local governmental agencies.[15] The provincial DOSTE-organisation (or in some cases the national MOSTE) should without any doubt remain in control of the final responsibilities, such as setting standards, issuing licenses and the final approval of EIA reports. But monitoring, environmental advice and information exchange, some enforcement tasks, collection of fees and fines, collection and treatment of solid waste, etc., could very well be delegated to these IZMBs or – in some cases – to private organisations. That would considerably lighten the tasks of the understaffed DOSTEs, which could then concentrate on both the 'black list' producers and the core government tasks. Such a rearrangement of tasks will not happen overnight, as it requires some institutional transformation and a reorganisation of state-market relations, both in the sphere of economic production and in the sphere of environmental regulation.

### *Technology Development and Pollution Control in Viet Nam*

Ecological modernisation emphasises the role of technology in the ecological transformation of production and consumption. Environmental technologies are often regarded as the means to keep economic development within ecological limits. Initially environmental technologies aimed at the treatment of waste streams, but increasingly the call is for process-integrated technologies that prevent waste streams. The concept of ecological modernisation highlights this shift from the so-called first-generation technologies, that is, end-of-pipe and clean-up technologies, to second-generation technologies, namely, cleaner production, product innovation and environmentally sound socio-technical systems. This section analyses the development and introduction of industrial environmental technology in Viet Nam with respect to that generation shift, and discusses the opportunities to eco-modernise technology.

Although this section deals with technology-induced transformations, one should keep in mind that these changes can develop successfully only in interdependence with socio-economic and political-institutional transformations. One trap into which Viet Nam could fall in its attempt to manage environmental problems would be to put the sole emphasis on the role of technology while neglecting the need for the simultaneous modernisation of the institutional setting in which the 'new' technologies are put to work. The consequence of such a development can be observed in

China. In numerous cases environmental technologies have been imported from developed countries, improving only parts of the industrial process, while the organisational reforms needed to support their functioning are not implemented [*Wang Ji and Ke Jin-Lian, 1992*]. The success of advanced technologies for pollution control depends not only on the technological development and diffusion itself, but also on the social context in which the technology is introduced, diffused and institutionalised.

*First- and second-generation technologies*: The development and implementation of environmental technology in Viet Nam is still in its infant phase, as even first-generation environmental technologies hardly exist there. Very little industrial pollution abatement can be found, except for some wastewater treatment facilities, which are operated marginally. Most new industries have not installed waste treatment facilities, let alone the older ones. Often, technical measures such as simple ventilation techniques to control indoor air pollution have not yet been taken by industries. Viet Nam has no hazardous waste management system at all. The absence of even first-generation environmental technologies is due to the total neglect of environmental concerns in industrial development, and the only recently emerging – but still weak – state and societal pressures on industry. The sheer absence of first-generation end-of-pipe technologies does not seem to be caused by limited knowledge of the development and creation of waste treatment facilities. In the region of Ho Chi Minh City, scientific knowledge of treatment technologies is reasonably developed and applied in, for example, wastewater treatment of textile industry and gas purification of the peanut-processing industry.

Not surprisingly, second-generation environmental technologies are even more underdeveloped, as is the knowledge base to develop and introduce them. On the whole, Vietnamese industry is equipped with old, often obsolete manufacturing technologies. These technologies do not ensure an efficient use of resources and have relatively high pollutant emissions. But, to survive in an increasingly competitive market, industries have to use more efficient technologies. To the extent that economic efficiency parallels environmental efficiency, cleaner production could be a promising approach for Viet Nam, just as it is gaining importance in the rapidly industrialising countries of Asia [cf. *Sakurai, 1995*]. Cleaner production is based on the 'pollution prevention pays principle'. Production processes that are more energy efficient, use fewer resources, and re-use waste materials not only reduce environmental impact but may also save costs. If not, then pollution prevention is in any case a more attractive approach for industries than treatment of waste by end-of-pipe measures that only add to the costs. As a first major step to more environmentally sound production processes, pollution prevention can be achieved relatively easily with only minor adjustments. Why, then, is it hardly applied in Viet Nam? First of all, it is not advocated and supported by Vietnamese government regulations. Regulatory emphasis is still on pollution abatement, for which end-of-pipe technologies are advocated, albeit without much success. Instead of command-and-control regulations that prescribe standard end-of-pipe technologies, cleaner production requires goal-setting and performance standards, and should be backed by governmental technical assistance and financial incentives. Secondly, although cleaner production and in general improved environmental efficiency might be economically advantageous (and necessary) in the long run, it still requires initial financial investments that many of the – especially state-owned – enterprises cannot afford. The many small-scale industries face similar financial, technical and attitudinal constraints in the introduction of cleaner production measures as we have seen in the textile sector in Ho Chi Minh City [*Frijns and Truong Thi Kim Khanh, 1997; Van Hengel, 1998*]. And thirdly, government officials and industrialists know little and are hardly aware of the concept of cleaner production, which is new for Viet Nam.[16] This could soon change, for a National Cleaner Production Centre with information dissemination as its prime objective was recently established in Ha Noi.[17] However, one should not forget that, even if cleaner technologies are introduced widely, overall pollution would most probably increase with industrial production growth.

*Technology leapfrog*: The fact that first-generation environmental technologies are still absent in Viet Nam does not mean that the second-generation technologies need to wait their turn. The historical development and introduction of environmental technology in Viet Nam might walk a different path.

Viet Nam is a typical example of a late industrialising country. This presents the opportunity to learn from earlier experiences elsewhere and to *leapfrog* to less costly and more effective industrial environmental technologies. In his study of the pulp industry of Thailand, Sonnenfeld [*1998*] concluded that this country's late industrialisation profited from the incorporation of advanced environmental technology. In Viet Nam, too, newer and less polluting production processes can be introduced, for one thing because of the high industrial growth rates and rapid turnover. This provides interesting opportunities to introduce cleaner production processes and control pollution while modernising the industrial sector. Privatisation and the shift to a market economy trigger changes in the internal organisation and attitude of companies, which might foster the implementation of cleaner production.

However, to leapfrog to more advanced environmental technologies requires openness to foreign capital and foreign technologies. Under *Doi Moi*, Viet Nam is opening its economy to foreign direct investment. The growing amount of foreign direct investment in industrial Viet Nam helps to modernise the industrial system, since in general new modern production processes move in. If these industrial investments are also modern in environmental terms some environmental reform in the production system might ensue. Notwithstanding their impact on the local environment, foreign industrial investments in general have a better environmental performance than the national industries of Viet Nam [*Wallace, 1996: 68*]. Consequently, international industries will not be among the environmental laggards in countries such as Viet Nam. That is confirmed by interviews with policy-makers and industrialists in Ho Chi Minh City, although it was pointed out that it was mainly true of industries from OECD countries, but not always of industries from new industrialised countries in the region.

Foreign direct investments may demonstrate environmental effects. Multinationals carry their home-countries' environmentally advanced practices with them when they establish production facilities in Viet Nam. They might even induce local suppliers of natural resources and semi-manufactured products to ecologise their products. The opening up of Viet Nam also results in economic and societal 'pressure' to adopt international industrial standards, including cleaner production practices and (voluntary) standards of environmental management. All these factors can contribute to the implementation of second-generation environmental technologies in Viet Nam in the near future. But, there is no reason to be too optimistic about the development of second-generation environmental technologies as a result of economic globalisation.[18] Studies have shown that even newly-established factories in Viet Nam, including those of joint ventures, import old equipment without proper pollution control facilities [*Ministry of Construction et al., 1995*]. As stated earlier, what counts is the change in the institutional framework in which possible technological innovations will operate.

*Supporting technology modernisation*: As a typical developmental state, rapidly industrialising Taiwan is an example of a country busily leapfrogging to more cost-effective environmental solutions. Its government has taken a pro-active role in reducing industrial pollution. The integration of environmental considerations in industrial policy strongly pushes environmental technology development in three principal ways [*Rock, 1996*]: by creating an indigenous environmental goods and services industry through a strategy of import-substitution; by heavily subsidising the industrial purchase of pollution control and abatement equipment; and by financing research into pollution prevention and providing industry with subsidised technical assistance in waste reduction.

If we now compare Taiwan with Viet Nam in terms of government support to industry in implementing environmental technology, we can only conclude that Viet Nam is doing very little. To be more precise, support for environmental technology has not been integrated at all in the government's industrial policy. In Viet Nam, an environmental goods and services sector, in particular for waste treatment facilities, is developing only slowly, without government support. No economic instruments are applied in developing environmental technologies. Only recently some experiments with tax reduction for the importation and installation of clean technology have been introduced.[19] Although economic instruments may effectively promote research, development and implementation of cleaner technologies, experiences in France, Germany and the Netherlands have revealed that both charges and subsidies lead primarily to the application of end-of-pipe technologies and not so much to the introduction of preventive technologies [*Cramer et al., 1990*]. Whether the same will happen in other countries nowadays can be questioned, but the assumption is warranted that the support for second-generation technologies requires additional measures, next to financial incentives. For example, in Korea in the early 1990s, the government applied a diversity of policy instruments to support clean technology programmes [*Chung, 1996*]. Among these were awards for clean technology development, an eco-labelling system, providing information and technical assistance, and shorter patent procedures for pollution prevention technology.

Although the Vietnamese government has an important role in supporting the development and implementation of modern second-generation environmental technologies,[20] it should be realised that the cleaner production approach is inherently connected with a movement away from command-and-control measures. Governments do not have the knowledge to prescribe innovations in the production process that could bring about major prevention of pollution. Governments should rather facilitate and provide incentives for industries to control (that is, prevent) pollution, without neglecting the monitoring and enforcement of final targets and goals. The initiatives to invest in the development of new, clean technologies have to come from industry itself, partly in its own interest and partly because of state and societal 'pressures'.[21]

To activate industries, it would be recommendable for the Vietnamese government to move towards a new relation with the private sector, which is rather difficult to establish, as we have stressed above. A change towards cleaner production requires support and 'pressures' from non-governmental organisations, such as trade associations, banks, labour unions, environmental organisations and consumers, and of course research institutes that develop technology and provide training. Indeed, technology development should be supported by changes in the Vietnamese economic and policy context. Unfortunately, until now, except perhaps for the research institutes, non-governmental organisations and industry have not done much to ecologise the economy by modernising the production technology. And by the same token, the government seems not (yet) willing to involve these actors in its economic and environmental policy making.

### *Social Protest and Awareness as Formative Factors in Environmental Reform*

In Western industrialised societies, environmental movements and concerned citizens have helped to reform destructive patterns of production and consumption from the late 1960s onward. Throughout the years, these actors have developed a variety of strategies and ideologies. Central to the ideas of ecological modernisation is the emphasis on the transformation of roles, strategies and ideologies of environmental non-governmental organisations and concerned citizens in processes of environmental reform, especially since the mid 1980s (see above).

Because Viet Nam is in some aspects a typical developing country, it should not surprise us that Viet Nam has throughout the years shown a pattern of development in environmental awareness and protest that diverges from the one found in Western countries. In developing countries – and also in Viet Nam – the organisation of environmental protest in NGOs took place later in time and is still limited. NGOs in developing countries are often hardly supported by (environmental) authorities and their resources in terms of money, manpower, knowledge and information are scarce compared to their Western equivalents. Environmental awareness is often less widespread among large segments of the population and priority is given to short-term economic goals in the narrow sense. And unlike in the developed countries, environmentally concerned actors in developing countries have understood from the beginning that environmental protection and economic development are two sides of the same coin.

On the other hand, being at the crossroads of an Asian 'developmental' state and a transitional economy, Viet Nam is atypical. So far, the transitional 'status' of Viet Nam has had mainly repercussions on the transformation of the economic system towards a more market-oriented growth model, but only to a minor extent affected political reforms. At the same time the characteristics of an Asian developmental state makes the Vietnamese state closely interwoven with societal organisations. In Viet Nam, societal organisations develop in strong interdependence with the state, which restricts the emergence of large-scale and widespread independent non-governmental organisations that have free access to information and the media. We can only analyse and understand the dynamics of non-governmental factors and actors that play a formative role in environmental reform by taking the typical and a-typical conditions in contemporary Viet Nam into account.

*Local protest*: In some interesting studies O'Rourke and others [*1997; Sikor and O'Rourke, 1996; Nghiem Ngoc Anh et al., 1995*] point at the dominance of what they have labelled 'community driven regulation' in Viet Nam. Environmental authorities often refrain from controlling, regulating and enforcing industrial polluters until they receive frequent and strong complaints from local communities about the severe deterioration of their living conditions. In some instances local community members – sometimes organised, often unorganised – have been quite successful in altering the destructive production patterns. In line with the now widespread idea that environmental reforms need not run counter to economic developments, such protests did not so much aim at closing production sites but rather at improvements within the boundaries of economic feasibility. The local protests proved especially successful when directed at transnational companies and backed by strong support from environmental NGOs in the home country of the multinational.[22]

In our own case studies we found similar patterns of 'community driven regulation' in the Ho Chi Minh City region [cf. *Le Van Khoa and Boot, 1998*]. But equally often those living in the vicinity of these production units – especially medium and small size industries – were economically or socially dependent on them, which hampered complaints addressed at the authorities [cf. *Van Hengel, 1998*]. At first glance these mechanisms look similar to what happened during the first wave of environmental concern in the Western industrialised countries. In the late 1960s and early 1970s it were in most cases also local (organised) concerned citizens who forced authorities to environmental action. But several differences have to be taken into account. Environmental authorities in present-day Viet Nam – no matter how poorly equipped and empowered – are usually fully aware of the environmental risks and threats. Their country boasts a system of environmental legislation that reflects some of the experiences and knowledge accumulated throughout the world during the past decades of environmental policy-making. And, finally, whilst in Western industrialised countries local environmental concern developed from the late 1960s onward into first local and a little later national environmental organisations, this has not yet happened in Viet Nam.

*Absence of domestic environmental NGOs*: Only quite recently have environmental organisations gained a foothold in Viet Nam. But Viet Nam lacks national environmental NGOs that should be the link between local protests and international environmental NGOs. As Eccleston and Potter [*1996*] conclude, there are hardly any domestic Vietnamese environmental NGOs. Almost all environmental NGOs that operate at the national level, or NGOs with a clear and specified environmental mission, are national branches of international organisations. The Viet Nam NGO Directory 1997/1998 [*Ruijs, 1997*] gives evidence of the large number of international NGOs in Viet Nam, several of which pay attention to environmental issues. ENDA, IUCN and WWF are probably among the best known. These national branches of international environmental NGOs show specific characteristics: they almost all focus on either nature protection or the raising of environmental awareness in schools and other institutions, and hardly (can) pay attention to campaigning, directing protests at industries, or adversarial attempts to influence governmental policy.[23] In the best tradition of ecological modernisation, most national branches of international NGOs collaborate with Vietnamese governmental agencies, multilateral UN organisations or sometimes private industries to trigger environmental improvements. ENDA, IUCN and WWF work together closely with the National Environmental Agency and several provincial environmental authorities in environmental programmes and projects. They neither oppose the governments (as was the case in the 1970s and early 1980s in developed countries) nor leave them aside as we often see in African countries and in some Asian countries such as Bangladesh, where NGOs take over traditional state tasks, for instance the provision of collective goods such as health, education and environmental quality. The background of a developmental state and a transitional society explains the position of international environmental NGOs.[24]

At the same time most traditional, national societal organisations that are closely linked to the state, such as those organising the youth, women or workers, seem hardly involved in environmental protection in residential areas or occupational health hazards at production sites. To a limited extent they organise environmental campaigns, but more often they participate in campaigns on environmental awareness and training, information dissemination and practical clean-ups organised and co-ordinated by local governments, such as 'Green Sundays' and 'Green Weeks' involving the Youth League, or 'Women and Home Sanitation' and 'Clean Houses, Nice Streets' campaigns involving the Women's Union of Ho Chi Minh City. These are rather incidental activities and do not touch the core of these organisations. As such the traditional mass organisations do not seem to fill the gap left by the absence of domestic one-issue environmental NGOs. The hypothesis of Eccleston and Potter [*1996*] that the existence of large mass organisations under communist rule in Viet Nam left little room for specific domestic NGOs, as the former effectively mobilised concerned citizens to participate in policy-making, explains the absence of large/mass domestic environmental movements, but not the non-emergence of smaller national professional organisations. Eccleston and Potter conclude that with the transition of the economy, the monopoly of the mass organisations in representing societal interest as well as their links with the state and the main political party may decline, leaving room for the emergence of more independent one-issue environmental organisations. Although Viet Nam is well on its way to economic transition, and several civil society groups are expanding their autonomy from the state [cf. *Tran Thi Lanh, 1994; Sidel, 1995*], up till now there is little evidence of the emergence of domestic environmental NGOs in Viet Nam. At the same time, in many of the surrounding developmental South-east Asian countries with more or less similar characteristics, environmental NGOs were established some years ago. Sonnenfeld's [*1996*] analysis of the potential role of NGOs in forcing technological innovations in the pulp and paper industries of Indonesia and Thailand shows how important such environmental NGOs are to industrial restructuring and environmental reform.

In Viet Nam, in the absence of strong environmental NGOs, it is primarily the state, the media and scientists that are raising environmental awareness. MOSTE, in collaboration with organisations of culture, information and fine arts, has repeatedly organised national contests on the environment with active public participation. The mass media have vigorously given more publicity to environmental matters. Viet Nam television and radio have separate programmes on science and education focused on environment, and a number of environment bulletins have been put into circulation. At schools and colleges, environmental issues have become part of the curriculum. It is the scientists and the results of their studies that have highlighted the significance and urgency of environmental issues for the country. At present they are the leaders in building the agenda of Viet Nam's environmental problems.[25]

*Inducing environmental reform*? In consideration of the (changing) role of environmental protests and awareness as formative factors of environmental reform and industrial restructuring in Viet Nam, we must conclude that the analytical ideas of ecological modernisation as developed for industrialised countries seem to be of limited value. The internal/domestic developments have their greatest momentum in local community protests, albeit often in a non-organised way and on ad-hoc basis. These community protests have never been very adversarial (in the positive sense) either in strategy or ideology; complaints and protests were incorporated successfully in the policy-making system through 'controlled' mass organisations and the political party, while at the same time they were able to trigger some environmental reforms at that local level. On the national level, the Vietnamese branches of international NGOs have contributed to redirecting the course of industrial development. If ecological modernisation as an analytical category proves to have any value for Viet Nam, it is through such global interaction patterns: the strategies and ideologies that have become dominant in the West are transplanted in Viet Nam by these international NGOs. At the same time, the co-operative strategies of international NGOs find fertile ground in a society that is characterised by a developmental state with close co-operation patterns between the state and societal actors. The question of the normative value of ecological modernisation ideas, that is the possibility and desirability of constructing an environmental reform strategy for Viet Nam along the lines of (Western) ecological modernisation models, is of a different kind and will be dealt with in the last section of this analysis.

## IV. Towards an Ecological Modernisation Approach in Viet Nam

The general conclusion from our analysis would be, as we will argue below, that ecological modernisation ideas cannot shed much light on the contemporary dynamics of environmental reform in Viet Nam. But that is not the end of the story. We will also ask whether it is useful to design future environmental reforms according to Ecological Modernisation Theory. Or, to put it differently, can we identify transformations in Viet Nam's institutional development that could make the country more 'receptive' to ecological modernisation ideas in the near future?

### *The Value of Ecological Modernisation Theory for Viet Nam*

In the first section of this study, we reviewed the development of Ecological Modernisation Theory in the context of Western industrialised societies, and listed several characteristics of environmental reform processes as described by that theory. In assessing the value of Ecological Modernisation Theory for present-day Viet Nam, we concentrated on the role of the state, technological innovations and the contributions of environmental NGOs. An analysis of contemporary processes of environmental reform in Viet Nam proved that the analytical concept of ecological modernisation has limited value for all three points. We conclude that Viet Nam's environmental policy system is incapable of detailed monitoring for lack of financial and human resources, and has failed to develop and implement environmental technologies. Independent environmental NGOs are scarce and environmental activities among economic agents almost absent. Environmental awareness among the public is neither high nor widespread enough to trigger social movements that could put sufficient pressure on governmental and industrial organisations for radical ecological reform. State and social organisations are tightly interwoven, which prevents the emergence of independent non-governmental organisations with free access to information and the mass media. The role of the state is very solid[26] and vital in the economic and environmental field, as is characteristic of most South-east Asian economies. Even though some environmental tasks have been formally institutionalised in the state system, environmental policy-making is still at an infant stage and mainly of a hierarchical policy style. More innovative strategies of environmental policy-making based on community participation, negotiated rule-making, close government-industry co-operation, and economic/market mechanisms, have not yet been established.

Notwithstanding these observations, some developments suggest the start of a modernisation of environmental transformation in Viet Nam. First, the polluter pays principle endorsed in the Law on Environmental Protection provides a legal framework for the introduction of market-based instruments, and a further diversification of environmental policy instruments may be expected. Second, the transition to a market-oriented economy and the slower path of privatisation may result in a change in environmental practices of industries along the lines of ecological modernisation. Third, the new policy of opening up to foreign direct investment can help some national industries to adopt advanced technologies and meet international industrial standards. Fourth, through competitiveness in the international market, international environmental standards will become a stronger challenge to the export-oriented national industries. Fifth, the growing number of branches of international NGOs may help redirect the course of industrial development and pave the way for national environmental NGOs. Sixth, Vietnamese scientists are making a significant contribution to the dissemination of environmental information, resulting to some extent in governmental actions. Seventh, the mass media have recently paid more attention to environmental problems, which helps to increase environmental awareness among the public and put some pressure on local polluters. And finally, community protests against industrial pollution have the potential to influence and redirect the environmental performance of industries, especially when these are given more 'elbow-room' by the prevailing political structure. In the concluding section, the likeliness of such developments will be put into the perspective of overall changes in contemporary Viet Nam.

### *Current Developments that Provide Opportunities for Ecological Modernisation*

Ecological Modernisation Theory was originally developed as a theory for describing and analysing how environmental considerations are becoming part and parcel of processes of societal and economic development. Although in developed countries this theory has proved its merit for the analysis of environmental reform processes, we have seen that for an industrialising economy like Viet Nam its descriptive and analytical connotation is less valuable. In our view, and quite obviously, this has to do with Viet Nam being a developing economy, having all the typical differences with the West-European countries, which after all are the birthplace of Ecological Modernisation Theory. As Viet Nam is in the process of industrialisation and economic transformation, and slowly moving towards a more democratic system, Ecological Modernisation Theory might gain relevance for Viet Nam, particularly as a prescriptive policy perspective. As we have already paid sufficient attention to the ongoing industrialisation, we will now concentrate on the other two developments: the transition of the economy through liberalisation, privatisation and internationalisation, and the steps towards more democratic governance.

Viet Nam is a transitional country moving towards a market economy. The country is opening up, the state is relinquishing its control of the market, and the industrial sector has started to privatise. Internationalising the economy offers opportunities for modernising environmental technology through foreign direct investments, as we have analysed and described, and international environmental treaties are to some extent triggering environmental improvements in Viet Nam.[27] We have also analysed opportunities for incorporating cleaner production as the industrial sector is being privatised, or commercialised. According to Hettige *et al.* [*1996*] the current wave of privatisation in South and South-east Asia makes pollution-intensive public enterprises less significant.[28] In addition, privatisation might separate the production from state administration, so that the government is no longer both polluter and controller of pollution. On the other hand, privatisation of state-owned enterprises may result in less state control in general and thus less control of pollution. In that regard Viet Nam has the advantage over other transitional economies that most state-owned enterprises are not being privatised but merely becoming more market-oriented and thus seeking more resource-efficient, less wasteful production.

If globalisation has had an impact on the economic transition of Viet Nam, it has not yet transformed the political system. Only slowly has Viet

Nam started to democratise its governance structure. Not only, as Rondinelli and Le Ngoc Hung [*1997*] conclude, is the current administrative structure still inconsistent with economic transformation, but local levels of administration have also limited authority and budgetary responsibilities, and thus find it hard to respond to local needs for (environmental) infrastructure services. The problems are worsened by the lack of formal channels other than the politically dominated People's Councils[29] for local desires to be reflected in central government decision-making. For a well functioning market economy, as well as for adequate environmental reforms, some decentralisation of the state administration system seems required and might be expected in Viet Nam. One step already taken towards decentralisation of environmental policy in Viet Nam is that local environmental authorities can to some extent design their own environmental regulations specific for their local area, provided they do not conflict with national regulation. We have described the advantages of more co-operative environmental strategies between government and industry to control pollution, and how economic and communicative policy instruments could operate more effectively at a decentralised level. Decentralisation would also provide opportunities for financing environmental policy. Local environmental administrative bodies that have the authority to enforce pollution control and use the revenues of pollution fees for their own administration and monitoring activities, are more likely to perform satisfactorily than a centralised system of environmental policy implementation. Since many local authorities lack skills and knowledge in environmental matters, capacity building is then a must. Finally, the Vietnamese state still holds a strong position in economic and societal development, a phenomenon it shares with other so-called developmental states [*Evans, 1996; Castells, 1997*]. In developmental states, public officials connect to citizens and private organisations, and that interwovenness has played a pivotal role in the successful transformation of the economies of East Asia. If this 'state–society embeddedness' is attended by further political democratisation and independence of civil organisations, environmental management and policy-making might very well profit.

In short, if the transition of the economy and democratisation of governance proceed, Ecological Modernisation Theory might become more applicable in designing mechanisms for environmental reform in Viet Nam. Nevertheless, Viet Nam and other developing economies will always mirror their specific local conditions and institutional layouts. This calls for caution in transferring Ecological Modernisation Theory to societies for which it was not originally developed. And of course the question remains whether developments that were set in motion after *Doi Moi* will indeed continue in the face of problems like the recent financial crisis. We might conclude that Viet Nam will (have to) define its own programme of environmental reform, appropriate to its specific time-and-space-bound situation. In doing so, it may find the Theory of Ecological Modernisation of definite value; lessons can be learnt from the fore-runners in environmental reforms, and globalisation processes will ensure that environmental deterioration and reforms become increasingly globally interdependent.

But an Ecological Modernisation Theory for Viet Nam will differ on some major points from the one that we currently find in the European literature. Institutional transformations similar to those identified for Western Europe by the Ecological Modernisation Theory could occur in Viet Nam but shaped according to its own identity: (1) an increased role of the Vietnamese academic community, not only in developing appropriate environmental technologies and strategies that are economically feasible with short-term returns and built upon already widespread recycling activities, but also in putting the environment on the political agenda; (2) an increased importance of economic dynamics in environmental reform, not so much through privatisation, but by a state controlled orientation to a market economy; (3) increased consultation with the private and decentral public sector by the central government in pollution control, but within the hierarchical policy style typical of a developmental state; (4) an increased role for civil society in environmental decision-making, not necessarily through the occurrence of new environmental NGOs but through community pressure and local initiatives and along the current lines of public-state interaction. These could be the prospects for ecological modernisation in Viet Nam, resembling environmental reform developments in Western Europe but having its own characteristics. The joint challenge to social scientists and environmental policy-makers is to design such geographic variations of ecological modernisation models.

## NOTES

1. On some environmental points, however, these European centrally planned economies performed better than their competitors on the other side of the Berlin wall (for example, public versus private transport, reuse and recycling of household waste).
2. Recent reviews on investments in 1998 showed that Ho Chi Minh City presents the highest investment in Viet Nam with 591 projects and total investment capital of US$9,084 million; the capital Hanoi, the second, with 289 projects and total capital of US$7,261 million; and Dong Nai Province, the third, with 196 projects and investment capital of US$4,093 million. Industry accounts for 61.9 per cent of the total projects and 46.2 per cent of the investment capital. In Ho Chi Minh City and Dong Nai Province, 40 per cent respectively 96 per cent of the investment capital is provided by foreign investors (*Viet Nam Investment Review*, 17 May 1998).

3. MOSTE's environmental responsibilities centre on the formulation of laws and regulations; international co-operation; co-ordination of intersectoral co-operation and standardisation; environmental monitoring; and co-ordination of environmental research and technology development. NEA has five technical divisions: EIA Review and Environmental Technologies Division, Environmental Monitoring Division, Environmental Pollution Control and Inspection Division, Environmental Information and Training Division, and Research and Technology Division. In 1995, MOSTE had a total of about 1000 staff, of which around 100 were part of NEA. Also other ministries, such as those of Planning and Investment, and of Education and Training, have formed departments for environmental protection.
4. The People's Committees of some provinces, such as Dong Nai, Hau Giang and Phu Khanh, have set their own regulations concerning air and water quality standards, effluent quality standards, requirements for EIA, duties of concerned parties in environmental protection, and outlines of pollution control procedures and technology [*Nguyen Cong Thanh, 1993*].
5. The tasks of ENCO and DOSTE of Ho Chi Minh City include environmental inspection and monitoring, EIA of (smaller) investments projects, programmes on minimisation of industrial pollution, environmental education, and resolving public complaints related to pollution. ENCO and DOSTE provide technical instruction to the District Environmental Bureau's and monitor their activities. The District Environmental Bureau's have their own legal entity, and draw resources from fines, among other sources. The Bureau's look after environmental complaints and are involved with public extension programmes.
6. The Ministry of Planning and Investment envisages an increase in environmental expenditure from about 0.3 per cent of GDP in 1995 to 0.5 per cent in 2000, the ultimate target being one per cent [*GOV, 1995*].
7. The Law on Environmental Protection also lays down the instructions for Environmental Impact Assessment procedures. All major investment projects (both foreign and domestic, both private and state) are required to make an EIA along narrowly defined criteria and guidelines and send that to the relevant authorities that use it as one of the bases to approve projects or authorize their implementation. Furthermore, the Law regulates the harmonisation of Viet Nam's environmental policy with international regulations in treaties signed by Viet Nam.
8. For example, in the absence of any serious co-ordination efforts, industries which have been refused permission to locate in Ho Chi Minh City for reasons of alleged water pollution, have obtained permission to locate in the province of Dong Nai, upstream Dong Nai river and thus polluting Ho Chi Minh City [*ADB, 1996*].
9. In Ho Chi Minh City, ENCO selected in 1995 a total of 43 industrial enterprises, judged to be the worst polluters. These enterprises were put on a 'Black List' and were given six months time in which to prepare a detailed plan for complying with relevant environmental standards. Industries failing to submit acceptable plans would be subject to possible closure or other penalties. By the end of the period, ENCO had received and approved environmental reports of 18 of these enterprises, and by the end of 1995, 4 factories have initiated their own pollution control programmes. The 'Black List' has been enlarged to 87 major industrial pollution sources [*ADB, 1996*].
10. It will also be necessary to promote environmental awareness in line agencies, as so far there has been inadequate incorporation of environmental considerations into the process of macro-planning, for example, the spatial development plans of urban centres [cf. *Frijns and Truong Thi Kim Khanh, 1997*] and the industrial development plans for light and heavy industry [cf. *Ho Chi Minh City Construction Department, 1996*].
11. In 1997, ENCO of Ho Chi Minh City proposed to impose a charge on facilities that lack pollution treatment measures. In Ha Noi, a 10 per cent surcharge is added to the water bill to help meet the costs of sewerage and pollution control. The actual revenues from this surcharge are small, but at least the principle has been established [*ADB, 1996*]. Recently, some research on the pollution charge scheme was done by the Department of Economics, Environmental Economic Unit of the Viet Nam National University, Ho Chi Minh City. Among which are the study on the feasibility of industrial effluent charge for food processing

industries and an ongoing study on effluent charge for textile industry [*Do Thi Huyen et al., 1997*].

12. User charges will be developed in the near future for common wastewater treatment systems and solid waste treatment facilities.
13. In fact, a deposit refund system has been practised for quite a long time. However, this system is only applied for beverage containers.
14. ISO 9000 was first introduced in Viet Nam in 1995, in the first workshop on the quality of Viet Nam's products held by the General Department of Quality Measurement and Standardisation in Ha Noi. Since then, 10 companies have received an ISO 9000 certificate, and about 100 companies are expected to get the certificate in the near future [*Nguyen Thi Sinh, 1998*].
15. In Bien Hoa, the IZMB Sonadezi has already taken up environmental management tasks.
16. The Economic Development Institute of the World Bank is currently executing an Industrial Pollution Prevention Program that consists of training, seminars, and working tours for Vietnamese officials.
17. *Vietnam Development News* Vol.9, No.3, p.17, 1997. UNIDO assists (through the Cleaner Production Programme and in the future through the establishment of a National Cleaner Production Centre) industries from the Black List in establishing cleaner production measures, as it was realised that forcing them to use end-of-pipe measures did not work. UNDP is also shifting their focus away from end-of-pipe to clean technology.
18. There are conflicting views on the environmentally soundness of multinationals operating in developing countries. In contrast to findings of Sonnenfeld [1998], a survey of Hettige et al [1996] of the pulp and paper sector in four Asian countries revealed that multinational branch plants do not perform better on cleaner production than their domestic private counterparts.
19. According to the ADB [*1996*], however, no special financial assistance should be provided to industries for technology upgrading, particularly when the major benefit is increased productivity, as this is inconsistent with current market reforms. On the other hand, currently, the Vietnamese government only provides loans for investments that produce strong short-term returns, and this makes it thus difficult for enterprises to finance pollution control equipment or longer-term investments in pollution prevention technologies [*Sikor and O'Rourke, 1996*].
20. For example, by providing technical assistance, economic incentives, training and environmental awareness programmes, and developing a system of environmental auditing.
21. In Thailand, the government initiated a cleaner production project that included training workshops, demonstration projects, audits and establishments of an information centre. Yet, the broader scope of cleaner production activities failed to become self-promoting within the enterprises. Regarding this, Bunyagidj and Greason [*1996*] recommend cleaner production promotion to be an integral part of an ISO 14001 certified environmental management system. Thai industries are interested in adopting the ISO 14001 standard on a voluntary basis to maintain competitiveness in export markets. The standard requires periodic auditing and reviews, and this might ensure continuous application of cleaner production principles.
22. O'Rourke [*1997*] describes the successful community complaints at the Nike production site in Vietnam, as they were backed by international and American protests on the poor production conditions for workers at Nike companies in developing countries.
23. ENDA seems to come most close to a real grassroot organisation, working basically in local communities together with local authorities and mass organisations – such as youth and women organisations – on environmental and other community development topics.
24. Sometimes a more formal reason is put forward: the fact that international NGOs are since 1997 officially acknowledged through a legal Decree that regulates their functioning, while this is still not the case for domestic NGOs, would be an important reason for the continuing absence of national NGOs. However, before this legal Decree had been published, international NGOs had appeared on the Vietnamese stage.
25. A group of environmental scientists has organised themselves into the Viet Nam Association of Nature and Environmental Protection. They provide consultancy for state environmental agencies, international co-operation in environmental technology transfer, training and

extension (they publish three magazines). Furthermore, there exists an Association of Water and Environment, and an Association of Urban Environment.

26. In fact, one can notice an institutionalisation of environmental concerns by the state, having not only state regulatory organisations, but also state-owned enterprises and popular organisations linked to the state like the Youth League, all which form part of the environmental 'organisational structure' in Viet Nam.
27. UNDP [*1995*] lists the international environmental treaties that have been signed by Viet Nam, among which are those on CFC reduction and phase out.
28. Publicly owned pulp and paper companies in that region appeared to undertake far less abatement compared to the private sector.
29. Members of the People's Councils are elected, and they nominate the members of the People's Committee that is the executive body. Complaints of people are directed to the People's Council, but complaints on environmental issues usually go to DOSTE.

## REFERENCES


ADB (1996), *Ho Chi Minh City Environmental Improvement Planning*, Final Report prepared by Parsons and JT-Envi, HCMC: ENCO.

Bich Ngoc (1997), 'Authorities Turn Up Heat on Dirty Factories', *Vietnam Investment Review*, 8–14 Sept., pp.12–13.

Blowers, A. (1997), 'Environmental Policy: Ecological Modernisation and the Risk Society?', *Urban Studies*, Vol.34, Nos.5–6, pp.845–71

Blühdorn, I. (2000), 'Ecological Modernisation and Post-Ecologist Politics', in Spaargaren, Mol and Buttel [*2000*].

Bunyagidj, C. and D. Greason (1996), 'Promoting Cleaner Production in Thailand: Integrating Cleaner Production into ISO 14001 Environmental Management Systems', *UNEP Industry and Environment*, Vol.19, No.3, pp.44–7.

Buttel, F. (2000), 'Classical Theory and Contemporary Environmental Sociology: Some Reflections on the Antecedents and Prospects for Reflexive Modernisation Theories in the Study of Environment and Society', in Spaargaren, Mol and Buttel [*2000*].

Castells, M. (1997), *End of Millennium* Vol.III of 'The Information Age: Economy, Society and Culture', Malden/Oxford: Blackwell.

Chu Tuan Nha (1997), 'Fight to Save VN Environment Hampered by Lack of Resources', *Vietnam News*, 2 Aug., p.15.

Chung, J.S. (1996), 'General Environmental Policies and Clean Technology in the Republic of Korea', *UNEP Industry and Environment*, Vol.19, No.1, pp.43–5.

Cramer, J., Schot, J., van den Akker, F. and G. Maas Geesteranus (1990), 'Stimulating Cleaner Technologies through Economic Instruments: Possibilities and Constraints', *UNEP Industry and Environment*, Vol.13, No.2, pp.46–53.

Do Thi Huyen, Le Ha Thanh, Nguyen Thi Lan, Phung Thuy Phuong, Nguyen Tran Quan and Ky Quang Vinh (1997), *The Feasibility of an Industrial Effluent Charge in Viet Nam. The Case of Food Processing Industry*, report to Economy and Environment Program for South East Asia, Ho Chi Minh City: Viet Nam Research Network on Environmental Economics.

Eccleston, B. and D. Potter (1996), 'Environmental NGOs and Different Political Contexts in Southeast Asia: Malaysia, Indonesia and Vietnam', in M.J.G. Parnwell and R.L. Bryant (eds.), *Environmental Change in Southeast Asia. People, Politics and Sustainable Development*, London: Routledge, pp.49–66.

Evans, P. (1996), 'Government Action, Social Capital and Development: Reviewing the Evidence on Synergy', *World Development*, Vol.24, No.6, pp.1119–32.

Frijns, J. and Truong Thi Kim Khanh (1997), *Environmental Assessment of the Tau Hu Canal Rehabilitation Pilot Project, Technical Advisory Report*, VIE/95/051 Project: Strengthening the Capacity for Urban Management and Planning in Ho Chi Minh City.

GOV (Government of Viet Nam) (1995), *Socialist Republic of Viet Nam, Socio-economic Development and Investment Requirements for the Five Years 1996-2000*, Government

Report to the Consultative Group Meeting, Paris.
Hettige, H., Huq, M., Pargal, S. and D. Wheeler (1996), 'Determinants of Pollution Abatement in Developing Countries: Evidence from South and Southeast Asia', *World Development*, Vol.24, No.12, pp.1891–904.
Ho Chi Minh City Construction Department (1996), *Scheme of Planning Concentrated Industrial Zones in Ho Chi Minh City*, Vol.1–3a.
Irvin, G. (1996), 'Emerging Issues in Viet Nam: Privatisation, Equality and Sustainable Growth', *European Journal of Development Research*, Vol.8, No.2, pp.178–99.
Jansen, K. (1997), *Economic Reform and Welfare in Viet Nam*, Working Paper No.260, The Hague: ISS.
Le Van Khoa, and S. Boot (1998), 'Industrial Environmental Management. The Case of the Tapioca Processing Industry in Vietnam', REFINE M.Sc. Thesis Series No.2, Wageningen: Wageningen University.
Ministry of Construction, ADB, UNCHS and UNDP (1995), *Urban Sector Strategy Study Report*, Hanoi.
Mol, A.P.J. (1995), *The Refinement of Production, Ecological Modernisation Theory and the Chemical Industry*, Utrecht: Van Arkel/International Books.
Mol, A.P.J. (2000), 'Globalisation and Changing Patterns of Industrial Pollution and Control', in S. Herculano (ed.), *Environmental Risk and the Quality of Life*, Rio de Janeiro (forthcoming).
Mol A.P.J. and J. Frijns (1998), 'Environmental Reforms in Industrial Viet Nam', *Asia-Pacific Development Journal* (forthcoming).
Mol, A.P.J. and G. Spaargaren (1993), 'Environment, Modernity and the Risk Society. The Apocalyptic Horizon of Environmental Reform', *International Sociology*, Vol.8, No.4, pp.431–59.
Nghiem Ngoc Anh, Hai, Vu Manh, Evans, P.B. *et al.* (1995), *Environmental and Industrial Renovation in Vietnam. A Report from Vinh Phu Province*, working paper No.4 of the Institute of International Studies, Berkeley, CA: University of California.
Nguyen Cong Thanh (1993), 'Viet Nam Environment Sector Study', prepared for Asian Development Bank, Bangkok.
Nguyen Phuc Quoc (1999), 'Industrial Solid Waste Prevention and Reduction', REFINE M.Sc. Thesis Series No.4, Wageningen: Wageningen University.
Nguyen Thi Sinh (1998), 'Tinh Kha Thi cua Viec Ap Dung He Thong Quan Ly moi Truong Theo Tieu Chuan ISO 14000 doi voi Doanh Nghiep Viet Nam' (The Feasibility of the Application of ISO 14000 to Viet Nam's Enterprises), B.Sc. Thesis, Ho Chi Minh City: Environmental Economic Unit of the Viet Nam National University.
O'Connor D. (1994), *Managing the Environment with Rapid Industrialisation. Lessons from the East Asian Experience*. Paris: OECD.
O'Rourke, D. (1997), 'Community-Driven Regulation. Towards an Improved Model of Environmental Regulation in Vietnam', paper prepared for the Conference on Sustainability, Degradation and Livelihood in Third World Urban Environments.
Phung Thuy Phuong (1994), 'Pollution Control Strategy for a Selected Industrial Estate in Viet Nam', M.Sc. thesis, Bangkok: AIT.
Rinkevicius L. (2000), 'The Ideology of Ecological Modernisation in "Double-Risk" Societies. A Case Study of Lithuanian Environmental Policy', in Spaargaren, Mol and Buttel [*2000*].
Rock, M.T. (1996), 'Towards More Sustainable Development: The Environment and Industrial Policy in Taiwan', *Development Policy Review*, Vol.14, pp.255–72.
Rondinelli, D.A. and Le Ngoc Hung (1997), 'Administrative Restructuring for Economic Transformation in Vietnam', *International Review of Administrative Sciences*, Vol.63, pp.509–28
Ruijs, O. (ed.) (1997), *Vietnam NGO Directory 1997/1998*, Hanoi: NGO Resource Centre.
Sakurai, K. (ed.) (1995), *Cleaner Production for Green Productivity: Asian Perspectives*, Tokyo: Asian Production Organisation.
SCS (State Committee of Sciences) (1991), *National Action Plan for Environment and Sustainable Development 1991–2000: Framework for Action*, Hanoi: SCS.
Sidel, M. (1995), 'The Emergence of a Non-profit Sector and Philantrophy in the Socialist Republic Vietnam', in T. Yamamoto (ed.), *Emerging Civil Society in the Asia Pacific*

*Community: Nongovernmental Underpinnings of the Emerging Asia Pacific Regional Community*, Singapore: Institute of South East Asian Studies/Tokyo: Japan Centre for International Exchange.

Sikor, T.O. and D. O'Rourke (1996), 'Economic and Environmental Dynamics of Reform in Vietnam', *Asian Survey*, Vol.36, No.6, pp.601–17.

Sonnenfeld, D.A. (1996), 'Greening the Tiger? Social Movements' Influence on Adaptation of Environmental Technologies in the Pulp and Paper Industries of Australia, Indonesia and Thailand', Ph.D. thesis, Santa Cruz: University of California.

Sonnenfeld, D.A. (1998), 'From Brown to Green? Late Industrialisation, Social Conflict, and Adaptation of Environmental Technologies in Thailand's Pulp Industry', *Organisation and Environment*, Vol.11, No.1, pp.59–87.

Spaargaren, G. (1997), 'The Ecological Modernisation of Production and Consumption. Essays in Environmental Sociology', Wageningen: Department of Environmental Sociology WAU (dissertation).

Spaargaren, G., Mol, A.P.J. and F. Buttel (eds.) (2000), *Environment and Global Modernity*, London: Sage.

Tran Thi Lanh (1994), 'The Role of Vietnamese NGOs in the Current Period', paper presented at the Vietnam Update 1994 Conference *Doi Moi*, the State and Civil Society. Australian National University, Canberra, 10–11 Nov.

Tran Thi Thanh Phuong (1996), 'Environmental Management and the Policy-Making Process in Viet Nam', paper presented in Seminar on Environment and Development in Viet Nam, Australian National University.

UNDP (1995), *Incorporating Environmental Considerations into Investment Decision-making in Vietnam*, Ha Noi: UNDP.

Van Hengel, P. (1998), 'Why Producing Waste if it is Thrown Away Anyway? Prevention of Waste and Emissions of Small and Medium Scale Textile Dying Enterprises in HoChiMinh City', REFINE M.Sc. Thesis Series No.3, Wageningen: Wageningen University.

Wallace, D. (1996), *Sustainable Industrialisation*, London: Royal Institute of International Affairs/Earthscan.

Wang Ji and Ke Jin-Lian (1992), 'Towards Cleaner Production in China', *Nature and Resources*, Vol.28, No.4, pp.11–16.

World Bank (1995), *Vietnam: Environmental Program and Policy Priorities for a Socialist Economy in Transition*, Report No.13200-VN, Vol.1 of 2, Washington, DC: World Bank.

Zhang Lei (1997), 'Challenges and Opportunities. A Study on Environmental Management of Township and Village Enterprises in China', M.Sc. thesis, Wageningen: Wageningen University, Department of Environmental Sociology.

Ziegler, C.E. (1988), *Environmental Policy in the USSR*, London: Frances Pinter.

# Notes on Contributors

**Maurie J. Cohen** is Visiting Assistant Professor in the Environmental Studies Program, Binghamton University (State University of New York), USA and Associate Fellow at the Oxford Centre for the Environment, Ethics and Society, Mansfield College, Oxford University, UK.

**Jos Frijns** is an environmental management expert at the Dutch consultancy firm, Grontmij, in the department of environmental management.

**Zsuzsa Gille** is Assistant Professor of Sociology, University of Illinois at Urbana-Champaign, USA.

**Pekka Jokinen** is Senior Lecturer in the Department of Regional Studies and Environmental Policy, University of Tampere, Finland.

**Arthur P.J. Mol** is Senior Lecturer in environmental sociology and heads the environmental policy group within the Department of Social Sciences, Wageningen University, The Netherlands.

**David N. Pellow** is in the Department of Sociology at the University of Colorado at Boulder, USA.

**Phung Thuy Phuong** is lecturer in environmental policy at the National University of Ho Chi Minh City and carrying out Ph.D. research on industrial transformation in Viet Nam at Wageningen University, The Netherlands.

**Leonardas Rinkevicius** is Associate Professor and Head of the Sociology Department at the Institue of Policy and Public Administration, within the Faculty of Social Sciences, Kaunas University of Technology, Lithuania.

**Allan Schnaiberg** is in the Department of Sociology at Northwestern University, Chicago, Illinois, USA.

**David A. Sonnenfeld** is Assistant Professor of Sociology, Washington State University; and Ciriacy-Wantrup Visiting Scholar, Institute of International Studies, University of California, Berkeley, USA.

**Gert Spaargaren** is an environmental sociologist working at Wageningen University in The Netherlands. He also holds a chair on the policy-dimension of environmental education at Tilburg University, The Netherlands.

**Bas van Vliet** is Associate Researcher in Environmental Sociology and preparing his Ph.D. thesis in the Department of Social Sciences, Wageningen University, The Netherlands.

**Adam S. Weinberg** is in the Department of Sociology and Anthropology at Colgate University, Hamilton, New York, USA.

# Index